THE RAPID GROWTH OF HUMAN POPULATIONS 1750 – 2000

Histories, Consequences, Issues Nation by Nation

William Stanton

Dedication

To the memory of Thomas Robert Malthus, 1766–1834, whose inspired warning, that excessively rapid population growth would cause terrible human poverty and distress, has been tragically ignored.

Acknowledgements

Roger Martin, Andrew Ferguson and James Duguid of the Optimum Population Trust, Hugh Ivimey-Cook and Jim Hanwell, geologist and geographer respectively, and my wife Angela, criticised and improved draft versions of this book. Colin Campbell, oil geologist, provided his latest analyses of oil and gas depletion. Jane Swinyard, librarian, set my initial researches on a sound footing. Peter Bacon, graphic designer, elaborated the graphs. Bill Hughes and his team at Multi-Science Publishing were undaunted by a text that rejects much conventional wisdom and political correctness. To all of them, I offer my best thanks.

Contents

The Evidence (235 graphs and briefings)

The Commentary (examining the Evidence)

Using This Book

Running through the book on the *upper parts* of the pages is new reference material: an assemblage of 235 graphs and tables showing, in detail, the recorded growth of population in each of the world's independent nations, as well as in 49 geographical regions which have, separately, useful and relevant records. Each graph is accompanied by a brief historical summary of political, cultural, ethnic and environmental change. The graphs appear in alphabetical order in four groups based on current population size, as indicated on the edge of each right-hand page.

Running through the book on the *lower parts* of the pages is the Commentary, which discusses how and why the evidence of the graphs amplifies or is at variance with conventional wisdom and political correctness, and goes on to outline scenarios of population development based on the new evidence.

Three novel concepts are introduced and are often referred to, usually as abbreviations:

WROG (Weak Restraints On Growth)
The global WROG period is the two and a half centuries since about 1750, during which populations have exploded (Section 4.5).

DC (Death Control)
DC population surges, characteristic of the WROG period, were made possible by science-based improvements in agriculture, industry and medicine (Section 4.3).

VCL (Violent Cutback Level)
A population cannot rise naturally above this level, due to civil strife, genocide, ethnic cleansing, etc. (Section 4.2).

Cross-referencing is made easy by naming target nations in capitals, e.g. RWANDA, and including section numbers at the top of the page.

Chapter 1

Introduction and Overview

In 1750, give or take a few years, *Homo sapiens* began its potentially disastrous population explosion. We had been on Earth for at least 150,000 years, and world population was about half a billion, when technologies born in the Industrial Revolution eased the severe restraints (hunger, disease and Darwinian conflict) on population growth. In the next 200 years world population surged to three billion. Now, only 50 years later, it is six billion, and the severe restraints are returning, in force. Lemmings and locusts are classic examples of animals whose numbers explode when conditions are favourable and crash when they run out of resources. Does a population crash, lemming style, await our species?

In the fossil record, new species appear, achieve importance, then decline and become extinct. During my geological career, searching for metal ores and trying to protect fresh water and hard limestone, I became consumed with curiosity about the prospects for our own species as it uses up, or fails to protect, the planet's resources. Curiosity led me to begin the investigation that is the subject of this book.

Born and raised in the lush countryside of southern England, I learned the fascination of Nature, the complex world of ants, eels and frogs, caterpillars and butterflies, hillsides, hollow trees, caves, snakes, glow-worms and mushrooms, sticklebacks and sundew – to name just a tiny fraction of all those unpretentious living things and their habitats that comprise, in the dull but useful modern jargon, *biodiversity*. Then as a field geologist I became familiar with more exotic wildlife, first in Angola's savannahs and rainforests, and then in the scrublands and rocky wastes of southern Portugal.

A great change in the English countryside occurred while I was geologising abroad. It was, in a word, *intensification*. When I returned to Somerset in 1970 I found that familiar fields, which before World War Two were lightly grazed flower-filled meadows, had become grass monocultures heavily dosed with agrochemicals. Streams and ponds, that during my childhood had been home to darting fish and healthy water plants, were now effectively dead, clogged with green and brown algae, polluted by nutrients draining off the land. The stone quarries of my local limestone hills had upped their extraction of crushed rock fifteen-fold, devastating the landscape. The reasons were obvious: farmers and quarrymen had to increase production to satisfy an exploding national demand for food to eat and stone to construct buildings and roads.

What caused the burgeoning demand? England had recovered from the privations of World War Two. People had money to spend. The newly prosperous citizens were building larger houses, buying more cars, and improving the roads and other infrastructure to suit their expansive lifestyles. British farmers, trying to feed the whole nation, were using the latest agricultural techniques. And England's population was increasing by 300,000 per year. Every day there were 820 more consumers, according to the Office of Population Censuses and Surveys (now the Office for National Statistics).

So the environment that I cherished was deteriorating because of increasing prosperity and

CHINA Area 9,537,000 km²

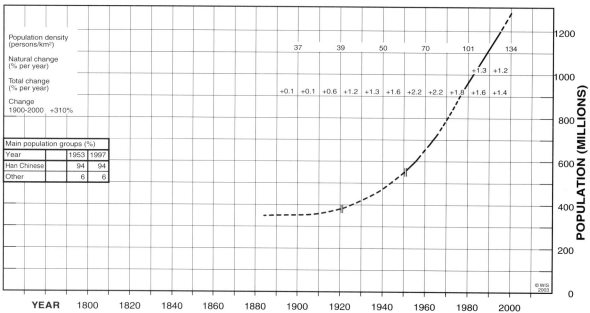

The transition from hunter-gathering to farming took place in north China about 8000 years ago; later than in Mesopotamia's fertile crescent but contemporaneously with central America. By 4000 years ago China had cities and complex societies. Thereafter dynasty followed dynasty, with chaotic intervals, recorded as written calendar history from about 1500 BC. In the 5th century BC China adopted Confucianism, a philosophy based on traditional morality. Art and literature flourished during the Han dynasty (202 BC to 220 AD). The Song dynasty from 960 AD was overthrown in 1279 by conquering Mongols, who were themselves expelled in 1368 at the start of the Ming dynasty which lasted until 1644.

European (Portuguese) navigators arrived in 1513, but trade with the West was restricted by the Ch'ing dynasty from 1757 until its defeat in the Opium Wars (1839–42 and 1856–60). Korea was captured by Japan in 1895 and Mongolia seceded in 1921. Japan conquered great swathes of eastern China in the 1930s, causing up to 35 million deaths, only to relinquish them in 1945. Tibet (which was independent before and after a period of Chinese control, 1720–1923) was annexed in 1951. The Chinese civil war between nationalists and communists was won in 1949 by Mao Zedong who established the communist nation that still endures, a major world power with a huge but largely poverty-stricken rural population engaged in agriculture, a flourishing urban industrial economy, and immense mineral and fossil fuel resources. Government control over the vast rural economy is sometimes weak. There are serious water shortages at times in the north. Ploughing and overgrazing of dry grassland is causing dust-storms and desertification of c. 4000 square kilometres every year.

China was the largest and most advanced country in the world 800 years ago, with a population of about 115 million, the maximum that its peasant farmers could feed (Ponting, 1993). Small improvements in agricultural productivity allowed a slow population increase, controlled by frequent famines and plagues, to about 350 million in the late 1800s. Thereafter, population growth became exponential as China began to adopt Western science and technology, initiating a DC surge. Population estimates varied widely before the 1953 census. In the late 1970s, when for several decades annual growth had been around 2%, China introduced the notorious 'one child per family' policy. Later it was partly relaxed, although it was supported by well-informed sections of the population who understood why it was desirable. Rapid growth continues, with 17% more boy babies than girls, by choice (in some rural areas, 80% of young children are boys). In 2030, it is predicted, the population will be about 1.6 billion. However, in 2000 about a million adults were HIV-positive, and the disease was expected to spread rapidly.

When flooding affects the most densely populated regions it is an immensely expensive anthropogenic disaster, as in 1998 when 5 million hectares of crops were destroyed, affecting 220 million people. Capital punishment is commonplace, with more than 1000 executions in 2000.

increasing population. Prosperity excited everyone. The creation and distribution of wealth is a science in its own right: economics. Population growth was seldom considered. Only busybodies raised that subject, it seemed; the number of babies a couple had was their business and nobody else's. After a while I realised that public discussion of population growth and its consequences was affected by a complex taboo (section 6.7).

Two hundred years ago, however, the sudden increase of population in nations affected by the Industrial Revolution had been a subject of great interest. In 1798 an English economist, the Reverend Thomas Malthus, wrote his controversial *Essay on the Principle of Population*, arguing that expanding populations would outgrow their food supplies, so controls would be necessary to prevent catastrophe. Malthus underestimated the food-producing potential of the new technologies. The perception that his doomsday prediction has not come

INDIA (excluding Kashmir) Area 4,110,000 km² to 1947, then 3,165,000 km²

Population density (persons/km²)		69	76	93	136	213	318
Natural change (% per year)						+2.1	+2.0
Total change (% per year)	+0.2 +0.6 +0.4 +0.7 +1.5	+2.2	+2.4	+2.7	+2.3	+2.1	
Change 1950-2000 +186%							

Main population groups (%)		
Year	1901	1997
Hindu	70	80
Muslim	21	11
Christian		4
Sikh		2
Buddhist	3	1
Other	6	2

India

Indian Empire

WIS 1999

POPULATION (MILLIONS): 0, 200, 400, 600, 800, 1000, 1200

YEAR 1800 1820 1840 1860 1880 1900 1920 1940 1960 1980 2000

> 180 million

When the first European explorer, Vasco da Gama, reached India in 1498 he found a country with at least 5000 years of history. Some of the world's first farmers migrated to the Indus valley from Mesopotamia around 3500 BC (but a submerged city perhaps 9500 years old was found off Gujarat in 2001). City states and kingdoms developed that practised Hinduism and were over-run successively by Aryans, Huns and Muslims. India's caste system, defining social status, developed as Aryan incomers from the north-west imposed their own hierarchies (priest / warrior / trader / artisan) on the indigenous peoples. Roma (gypsies) migrated from India to Europe between the 9th and 14th centuries. From 1206 a Muslim dynasty ruled northern India. It fell in 1398, with bloody massacres, to the Mongol invader Tamerlane, whose descendant Babur founded the Mogul or Mughal Empire, embracing almost all India, in the 16th century. Invading Persians destroyed Mogul cohesion in the 18th century, and subsequent wars between rival princedoms were exploited by British merchants who gained control of all India for the British Empire in 1858.

British India was relatively short-lived and was marked by growing Hindu-Muslim rivalry. Independence came in 1947, accompanied by partition and huge massacres as Muslims fled to Pakistan and Hindus to India. The death toll was about 2 million, out of 12 million migrants. Hindu extremism is a powerful and growing political force in India, intolerant of other religions. Hindu-Muslim fighting over the Ayodhya religious site killed thousands in 1992 and again in 2002.

Since partition, India's population has trebled. Governments have tried to popularise small families since the 1970s. There are fewer females than males (about 0.9 to 1), due to female infanticide and abortion. Several hundred thousand women, mostly lower caste, are sterilised each year, following official policies. India is predicted to overtake China as the world's most populous nation before 2050. Had British India not split up, its population would have surpassed China's in 1997.

Much of northwest India was tropical forest 2000 years ago, but woodland clearance for agriculture has turned 250,000 square kilometres of it into the Great Indian Desert. Some 70% of the population tills the land, so given the high population density it is not surprising that 6 billion tonnes of soil are being lost annually in monsoon rainstorms. Massive floods, rendering millions of people homeless, are occurring with increasing frequency in the north-east coastal states. They alternate with droughts. Pressure to occupy the forest reserves, habitat of the few remaining tigers, is intense. In 2000, c. 3000 tigers remained in India (compared to more than 100,000 in 1900), and about 100 are taken each year by poachers, who even steal them from zoos.

The introduction of new high-yielding strains of wheat and rice in the so-called Green Revolution succeeded in doubling the output of these cereals in the 1960s and 70s. Hunger was alleviated for a few decades, but millions of smallholders were forced to sell their land because they could not afford the extra agrochemicals and irrigation that were necessary. By 2000 crop yields were starting to decline. About 30% of the population lives in extreme poverty. Many rural workers have fled to the cities. Some 4 million persons were HIV-positive in 2000.

India fought a short border war with China in 1962. Jammu and Kashmir, the northern state claimed by both India and Pakistan, is divided by the 'Line of Control' along which a state of war has existed, off and on, since 1948, causing c. 20,000 deaths. There are 8 million inhabitants on the Indian side and 1.5 million on the Pakistani side. Separatist guerrillas of the Bodo and Naga tribes have been active in Assam, in the northeast, since the 1950s.

true after 200 years allows optimists to assert, when challenged by neo-Malthusians, that technology will always be able to outwit starvation. But many populations would indeed be starving in 2002, were it not for foreign aid.

It has been fashionable, since about 1950, to counter population doomsayers like Paul Ehrlich (*The Population Bomb*, 1968) by quoting a particular theory which holds that world population should soon stabilise, naturally and painlessly. As people become more

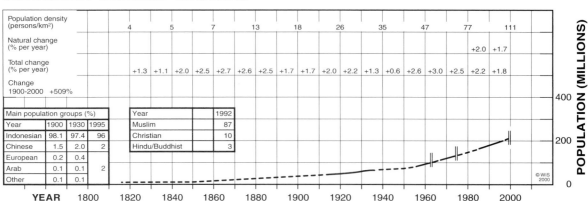

INDONESIA   Dutch East Indies   Area 1,919,000 km²

Population density (persons/km²)			4	5	7	13	18	26	35	47	77	111
Natural change (% per year)											+2.0	+1.7
Total change (% per year)			+1.3	+1.1	+2.0	+2.5	+2.7	+2.6	+2.5	+1.7	+1.7	+2.0	+2.2	+1.3	+0.6	+2.6	+3.0	+2.5	+2.2	+1.8
Change 1900-2000 +509%													

Main population groups (%)

Year	1900	1930	1995
Indonesian	98.1	97.4	96
Chinese	1.5	2.0	2
European	0.2	0.4	
Arab	0.1	0.1	2
Other	0.1	0.1	

Year	1992
Muslim	87
Christian	10
Hindu/Buddhist	3

POPULATION (MILLIONS) — 400, 200, 0

YEAR 1800 1820 1840 1860 1880 1900 1920 1940 1960 1980 2000

© WIS 2000

More than 3000 years ago, waves of migrants from Southeast Asia were moving east through the 13,600 islands of Indonesia, displacing or merging with its original Melanesian population and diverging to form some 300 ethnic groups. Hindu and Buddhist kingdoms and empires developed and were converted to Islam by Arab spice traders from the 13th century onwards. The first European visitors to the densely forested volcanic archipelago were Portuguese navigators in 1511, but Dutch traders and colonists cornered the spice trade in the 17th century and established the first coffee, sugar and tobacco plantations in the 18th. Chinese immigrants set up many small businesses. By the early 20th century the 'Dutch East Indies' comprised nearly the whole archipelago. Dutch rule, often harsh and repressive, continued until World War Two which saw Japanese conquest in 1942 and occupation until 1945. Indonesia declared independence in 1945, resisting Dutch attempts at recolonisation. Irian Jaya was added by UN decree in 1963. President Suharto's accession in 1967 involved the killing of up to a million 'communists'. East Timor was taken by force in 1975.

Indonesia's DC population surge at around 3% per year began in the 1950s. The central islands, especially Java, soon became so overcrowded that huge 'transmigrations' were organised, in which millions of peasants were sent to develop oil palm and other plantations in outlying islands like Sumatra and Irian Jaya. Rain forests and peatlands were burned, indigenous tribes were displaced and rare animals (e.g. the orang utan) endangered. Corruption and nepotism are rife, with legally protected forests logged, burned and replaced by plantations often owned by high officials. In 1997 and 1999 forest burning was so intensive that Malaysia and Singapore suffered serious air pollution.

Unlike most Muslim nations the Indonesian government has run campaigns promoting small families, with slight success. Indonesia is now the fourth most populous nation in the world. The economy, based on agriculture, logging, gas and oil production, manufacturing and tourism, prospered until the Asian recession of 1997, when it collapsed. The foreign debt reached $120 billion in 2001. There is poverty and violence in outlying areas as ethnic minorities like the Dayaks react against the loss of their traditional lands. Before fighting began in the Moluccas in 1999 the islands were celebrated examples of ethnic harmony, but now their Chinese and Christian communities are attacked by expeditionary forces of Muslim fundamentalists, intent on jihad (religious cleansing), with hundreds killed. In 1999 Christian East Timor seceded, claiming independence after 24 years of repressive occupation during which more than 100,000 Timorese were killed. In 2001 separatist revolts were ongoing in Aceh and Irian Jaya, and there were 1.2 million internal refugees. Censuses: occasional from 1920.

prosperous, the theory says, they will have fewer babies and growth will cease. I could see few signs of increasing prosperity in the developing world, and even in developed nations the populations were growing fast, so I was not convinced by this *demographic transition theory* (and I examine the harm it has done in section 5.2).

In the later 20th century, *political correctness* (*PC*) dominated Western attitudes (section 7.2). On the one hand PC stifles debate on population growth and its consequences; on the other it promotes practices that, the public is assured, will conserve Earth's resources. One of these is *sustainable development*, which 'enables the present generation to satisfy its needs while ensuring that future generations can satisfy theirs' (section 5.3). Leaving aside the glaring fact that the needs of a majority of the present generation, living in the developing world, are painfully unsatisfied, most of the proposed sustainable practices could only work if population growth was negligible.

Sustainability requires (for example) reduced emissions of greenhouse gases to combat global heating (section 6.1), but a reduction of, say, 10% in emissions per average person would be ineffective if the number of average persons increased by more than 10%. Nothing could be more obvious. Nevertheless, eco-warrior organisations such as Friends of the Earth and Greenpeace have been intent, since the 1980s, on saving the planet from global heating and climate change by campaigning for emission reductions alone. They have achieved no

UNITED STATES OF AMERICA Area 9,159,000 km² (land)

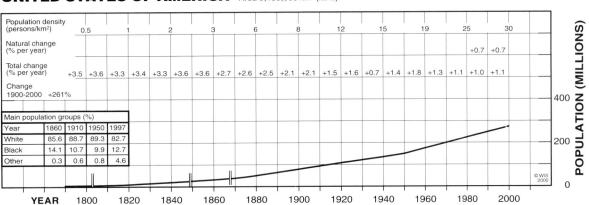

| Population density (persons/km²) | 0.5 | 1 | 2 | 3 | 6 | 8 | 12 | 15 | 19 | 25 | 30 |

| Natural change (% per year) | | | | | | | | | | +0.7 | +0.7 |

| Total change (% per year) | +3.5 | +3.6 | +3.3 | +3.4 | +3.3 | +3.6 | +3.6 | +2.7 | +2.6 | +2.5 | +2.1 | +2.1 | +1.5 | +1.6 | +0.7 | +1.4 | +1.8 | +1.3 | +1.1 | +1.0 | +1.1 |

Change 1900-2000 +261%

Main population groups (%)				
Year	1860	1910	1950	1997
White	85.6	88.7	89.3	82.7
Black	14.1	10.7	9.9	12.7
Other	0.3	0.6	0.8	4.6

POPULATION (MILLIONS)

© WIS 2000

YEAR 1800 1820 1840 1860 1880 1900 1920 1940 1960 1980 2000

∧ 180 million

North America was unpopulated by humans until late in the Ice Age, perhaps 40,000 years ago, when Mongoloid tribes migrated from Asia across a land bridge (formed by low sea levels) into Alaska, from where they dispersed over the whole continent. When the first Spanish explorers reached Florida in 1513 the Amerindian population of what is now the US numbered between 4 and 10 million, but it was vastly reduced in the 16th and 17th centuries by European diseases introduced by Spanish settlers in the south, French in the Mississippi valley and British on the eastern seaboard. The survivors were harassed and their lands were conquered and ethnically cleansed as the Europeans pushed inland. By the late 18th century the British eastern states controlled the eastern third of the country. Millions of African slaves were imported to work on cotton and tobacco plantations. After wresting independence from Britain in 1783 the USA purchased French territories west of the Mississippi in 1803, took Mexican lands in the southwest by force in 1848, and bought Alaska from Russia in 1867. The civil war between northern and southern states (1861–65), fought to preserve the Union, brought an end to slavery.

Mass immigration began in the 19th century and continued through the 20th, mainly from Europe at first but eventually from all over the world. Many immigrants were entrepreneurs seeking opportunities denied them in overcrowded and resource-poor Europe (like miners from Cornwall, England, who had exhausted their rich veins of copper and tin, or peasants fleeing centuries of repression in Ireland and elsewhere). The apparently inexhaustible resources of the New World allowed innovators full expression, with the result that American business grew to dominate world trade by the mid-20th century. The USA intervened decisively, if late, in the 1914–18 and 1939–45 world wars, gaining strength by the exhaustion of the other participants. The 'brain drain' from

other nations to the lavish facilities and salaries of US research establishments nourished American technical and commercial advantage, vital in the relentless struggle maintained by the USA during the second half of the 20th century against the other world superpower, the communist USSR, which eventually collapsed in 1991.

Even in 2000 the USA is sparsely populated by world standards. The low population density has allowed the designation of huge national parks and wildernesses and the production of immense food surpluses to sell abroad. Some vital resources, particularly oil, have been depleted by the habitual prolific consumption of a nation accustomed to being rich, but the USA now has all the power it needs to act, as in the Gulf in 1991, to safeguard its access to strategic resources.

A fast-growing population that already emits far more greenhouse gases than any other cannot complain if, due to climate change, its naturally hot summers and cold winters have become more extreme in recent years. Added to the carbon dioxide produced by traffic and industry is natural carbon dioxide from volcanic regions (Yellowstone National Park emits 44 million tonnes per year, equivalent to 10 coal-fired power stations). In 2001 the USA, with 4.5% of the world's population, was emitting 25% of the world's greenhouse gases, but President Bush rejected international calls to reduce emissions because the US economy would suffer. The increase of US population has been comprehensively documented by censuses every tenth year from 1790. Immigration still plays a major part; thus in 1994 immigrants comprised 52% of the 2.7 million total increase. About 1.3 million abortions are performed annually. Compared to other population groups, white non-Hispanic growth has slowed in recent decades. There were 7 million Muslims (2.6% of the population) in 2001.

significant reductions on the global scale, but global population, about which they are silent, has increased by 25% since 1985.

PC people who ignore population growth condemn themselves never to understand the causes of congestion and environmental deterioration (sections 4.8, 6.2, 8.9). They set themselves impossible humanitarian goals (section 6.5). If they run life-saving charities, they fight their battles against famine and disease in ways that can succeed only in the short term. This is because birth control, the taboo subject, is not part of their package (sections 8.3, 8.8).

An intriguing research opportunity beckoned. I felt sure that population growth must have been a major factor in the build-up to historic disasters like famines and wars, and in longer-term problems such as deforestation, wildlife loss, soil erosion and desertification, but conventional history books seldom devote much space to demographic change. Detailed historical analyses of population pressure and its effects, worldwide, were not in general

UNION OF SOVIET SOCIALIST REPUBLICS
Soviet Union   Area 21,250,000 km² to 1939, then 22,400,000 km²

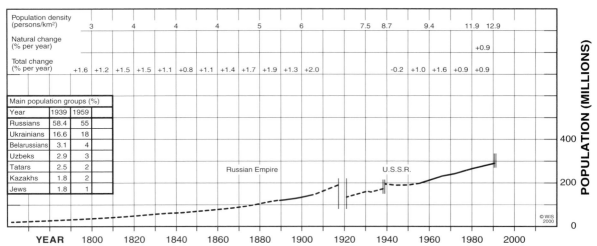

Population density (persons/km²)		3		4		4		4		5		6			7.5	8.7		9.4		11.9	12.9		
Natural change (% per year)																					+0.9			
Total change (% per year)	+1.6	+1.2	+1.5	+1.5	+1.1	+0.8	+1.1	+1.4	+1.7	+1.9	+1.3	+2.0			-0.2	+1.0	+1.6	+0.9	+0.9				

Main population groups (%)

Year	1939	1959
Russians	58.4	55
Ukrainians	16.6	18
Belarussians	3.1	4
Uzbeks	2.9	3
Tatars	2.5	2
Kazakhs	1.8	2
Jews	1.8	1

Russian Empire

U.S.S.R.

POPULATION (MILLIONS)

400

200

0

© WIS 2000

YEAR   1800   1820   1840   1860   1880   1900   1920   1940   1960   1980   2000

The Russian Empire (see RUSSIA) ruled by the Tsars of the Romanov dynasty was terminated in 1917 by the Bolshevik Revolution. The USSR emerged after 5 years of civil war having lost much territory in eastern Europe. From 1924 Joseph Stalin directed Soviet policy according to Communist theory, with massive industrialisation and collectivisation of agriculture; inefficiency and disorganisation on a vast scale led to famine and many millions of deaths. Opponents of the regime were ruthlessly purged. The German invasion of 1941, ending in 1945, caused enormous loss of life (the Soviets calculated that in 1959 the population was 42 million less than it would have been, had World War Two not occurred). The USSR retained much territory in eastern Europe after the war. It became a world superpower rivalling the USA in the "Cold War" with its immense military might and technical prowess, but the 5 Year Plans of the Communist system proved inferior to capitalism as regards economic and political competence. Communist populations perceived themselves as residents of police states lacking individual freedoms. In 1991 following liberal reforms by Mikhail Gorbachev individual republics began to secede and the USSR collapsed. Some but not all of the republics joined the new Commonwealth of Independent States (CIS), headed by Russia.

circulation (until Clive Ponting's masterly *Green History of the World*, 1991). Basic data seemed hard to find, but this was because I set about the research in an unmethodical fashion, to satisfy my own curiosity. I didn't discover a comprehensive work, *The Atlas of World Population History* (McEvedy & Jones, 1978) until quite late in my investigation. It addresses population change over millennia. My graphs and tables cover much shorter periods, in more detail.

I browsed through second-hand bookshops looking for old encyclopedias and world atlases which gave population figures for each nation at the time of printing. A few industrialised nations began regular population censuses around 1800. Census data for England, Ireland, Scotland and Wales are available for 1801 and every tenth year thereafter except 1941. Earlier population estimates for England appear in several histories (see References). In the short term (every year since 1981) the British charity *Population Concern* has distributed an annual "World Population Data Sheet", compiled by the *Population Reference Bureau* of Washington, USA, which gives a mass of up-to-date statistics on every nation.

In 1991 I joined the *Optimum Population Trust*, whose objectives were to calculate the optimum population (number of people) for any country or region compatible with a good quality of life for everyone. For the OPT I drew up graphs of population growth in the world (1993) and in England (1994) in their historical contexts (reproduced here, updated, as WORLD and ENGLAND LONG TERM graphs).

(*Note: in this book, when a nation is named in* CAPITALS, *the reader is referred to its graph and briefing for additional relevant data.*)

England's long-term population graph was a real eye-opener. It illustrates, unambiguously, the onset and development of the human population explosion. For 2000 years until the mid-18th century England's population occasionally reached carrying capacity, about 5 million (section 4.1), but it was usually held below that level by lawlessness, warfare or disease. Then, abruptly, England's carrying capacity began a huge expansion. Ample food

WORLD POPULATION GROWTH

Key events on the graph

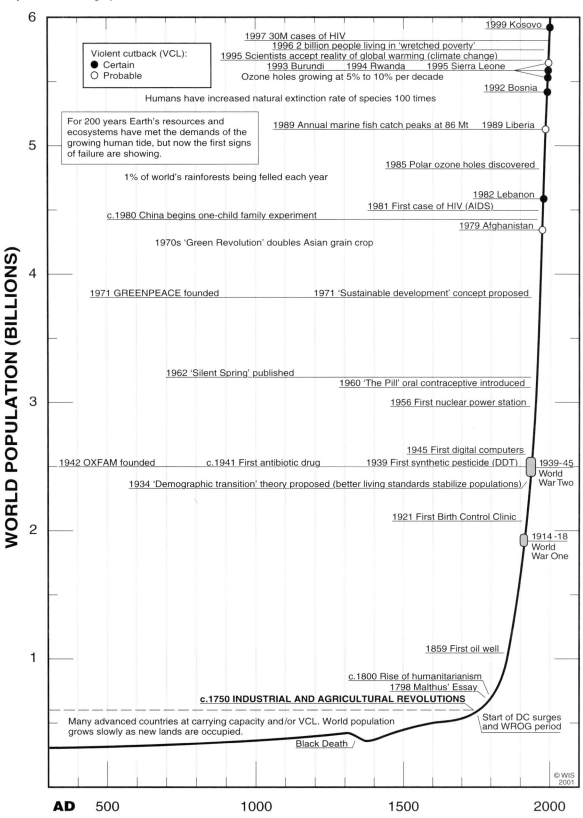

Violent cutback (VCL):
● Certain
○ Probable

1999 Kosovo

1997 30M cases of HIV
1996 2 billion people living in 'wretched poverty'
1995 Scientists accept reality of global warming (climate change)
1993 Burundi   1994 Rwanda   1995 Sierra Leone
Ozone holes growing at 5% to 10% per decade

1992 Bosnia

Humans have increased natural extinction rate of species 100 times

For 200 years Earth's resources and ecosystems have met the demands of the growing human tide, but now the first signs of failure are showing.

1989 Annual marine fish catch peaks at 86 Mt   1989 Liberia

1% of world's rainforests being felled each year

1985 Polar ozone holes discovered

1982 Lebanon
1981 First case of HIV (AIDS)
c.1980 China begins one-child family experiment
1979 Afghanistan

1970s 'Green Revolution' doubles Asian grain crop

1971 GREENPEACE founded   1971 'Sustainable development' concept proposed

1962 'Silent Spring' published
1960 'The Pill' oral contraceptive introduced
1956 First nuclear power station

1945 First digital computers
1942 OXFAM founded   c.1941 First antibiotic drug   1939 First synthetic pesticide (DDT)   1939-45 World War Two
1934 'Demographic transition' theory proposed (better living standards stabilize populations)

1921 First Birth Control Clinic

1914-18 World War One

1859 First oil well

c.1800 Rise of humanitarianism
1798 Malthus' Essay
c.1750 INDUSTRIAL AND AGRICULTURAL REVOLUTIONS

Start of DC surges and WROG period

Many advanced countries at carrying capacity and/or VCL. World population grows slowly as new lands are occupied.

Black Death

WORLD POPULATION (BILLINGS) — vertical axis: 1, 2, 3, 4, 5, 6

AD   500   1000   1500   2000

> 180 million

© WIS 2001

AFGHANISTAN Area 652,000 km²

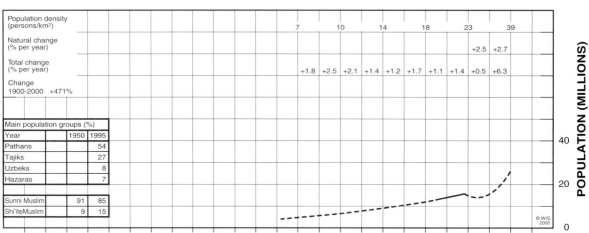

Population density (persons/km²)						7	10	14	18		23		39	
Natural change (% per year)											+2.5	+2.7		
Total change (% per year)					+1.8	+2.5	+2.1	+1.4	+1.2	+1.7	+1.1	+1.4	+0.5	+6.3
Change 1900-2000 +471%														

Main population groups (%)		
Year	1950	1995
Pathans		54
Tajiks		27
Uzbeks		8
Hazaras		7
Sunni Muslim	91	85
Shi'ite Muslim	9	15

YEAR   1800   1820   1840   1860   1880   1900   1920   1940   1960   1980   2000

POPULATION (MILLIONS) 40 20 0

© WIS 2000

Mountainous arid Afghanistan was part of the Persian Empire in the 6th century BC, after which it fell to invaders including Alexander the Great, Indians, Greeks, Huns and, in the 7th century AD, Arabs spreading Islam. Genghis Khan's Mongols overran the region in 1220. Their descendants are the Hazaras. Mongol control lasted nearly 3 centuries.

It is said that Afghans are only happy when they are fighting someone. They threw off Indian domination in 1747 and founded a kingdom under Ahmad Shah, who conquered parts of Persia and India. Revolts and civil wars disintegrated the kingdom, making it vulnerable to Russian and British expansionism in the 19th century. Between 1838 and 1880 Anglo-Indian forces twice invaded Afghanistan to set up puppet regimes, which failed to maintain their influence for long in face of Afghan militancy. Independence was restored in 1919 but tribal and dynastic revolts and coups delayed economic progress until Zahir Shah (1933–73) brought relative stability, marred by droughts and famines. A family coup deposed Zahir in 1973 and installed a republic, which was replaced in 1978 by a communist one. Muslim militias, the *mujahedin*, mostly Pathan (or Pashtun) tribesmen, attacked the communists, who persuaded Soviet forces to occupy the country in 1979.

The following decade saw intense warfare between the Soviet-backed communist regime and the mujahedin who used smuggled American weapons. Up to six million civilians fled to Iran and Pakistan. Soviet support faltered and Soviet troops withdrew in 1989, but civil war between the various ethnic groups and warlords intensified. Between 1994 and 2000 a militant Muslim fundamentalist movement, the Taliban 'students of religion' (including Arabs and Pakistanis), conquered 95% of the country and set up a strict Islamic regime. Kabul, the capital, was known in 1996 as the city of widows and 100,000 orphans. Many refugees returned, but others fled. In 1999 there were still 2 million in Pakistan and 1 million in Iran (a high birth rate is normal in refugee camps). In 1998 the Taliban government refused to facilitate humanitarian aid to earthquake victims (4000 dead) in the unsubdued north.

After two millennia of ethnic and religious conflict Afghanistan was a poverty-stricken nation ruled by a Muslim fundamentalist government that outlawed music, playing cards, television and the Internet, repressed women, and even forbade footballers to wear short trousers! The Taliban destroyed the world's tallest statue (53 metres) of Buddha, and ordered the destruction of all non-Islamic shrines. Late in 2001 Taliban rule was ended by an alliance of mujahedin warlords backed by US air power. The USA was reacting to the September 11 attack on New York organised by the terrorist Osama bin Laden, whom the Taliban were sheltering. Afghanistan's ethnically divided population appears to be far above carrying capacity and VCL, but it continues to increase, thanks to Western aid, foreign peacekeepers who try to restrain ethnic strife between warlords, and the income from drug sales (Afghanistan is the world's main grower of opium).

In 2000 a severe drought causing food shortages estimated at 2.3 million tonnes was alleviated by foreign aid.

and (later) better health care had become available, thanks to the Industrial and Agricultural revolutions. Malnutrition could now be relieved, so that its attendant diseases, such as pneumonia, dysentery and tuberculosis, were less commonly fatal. Death became more controllable, and the population began its explosion. When I found that the same phenomenon had affected every nation at some stage in its recent history I began to refer to it as the *DC* (*Death Control*) population surge (section 4.3).

DC population surges have been humankind's response to unfamiliar material plenty. They have characterised a global period of *Weak Restraints On Growth* (*WROG*), discussed in section 4.5. The start of a nation's DC surge was the start of its own WROG period. The world's WROG period has been an anomalous never-to-be-repeated 2.5 centuries during which humans have multiplied their numbers ten times by plundering Earth's finite resources, especially fossil fuels.

The Industrial Revolution unlocked vast storehouses of wealth, enough to ensure prosperity for everyone throughout the WROG period if world population had remained at

ALGERIA Area 2,382,000 km²

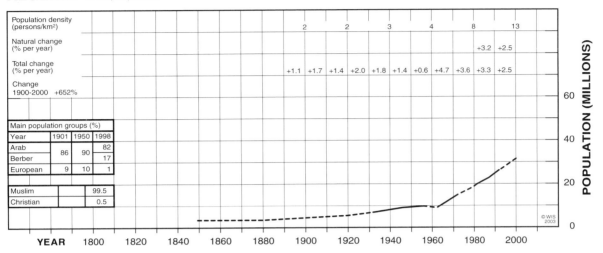

Population density (persons/km²)			2	2	3	4	8	13
Natural change (% per year)							+3.2	+2.5
Total change (% per year)	+1.1	+1.7	+1.4	+2.0	+1.8	+1.4	+0.6	+4.7	+3.6	+3.3	+2.5
Change 1900-2000 +652%								

Main population groups (%)

Year	1901	1950	1998
Arab	86	90	82
Berber			17
European	9	10	1

| Muslim | | | 99.5 |
| Christian | | | 0.5 |

POPULATION (MILLIONS)

60 40 20 0

© WIS 2003

YEAR 1800 1820 1840 1860 1880 1900 1920 1940 1960 1980 2000

Carthage conquered the Berber nomads of Algeria in the 9th century BC and was conquered in turn by Rome in 140 BC, As Numidia, it was an important Roman granary. Vandals and then Byzantines captured the coastal areas in the 5th and 6th centuries but the most significant invasion was by Arab armies, who converted the people to Islam, in the 7th century. From the 16th century, when Ottoman Turks conquered the region, the Barbary (Berber) Coast was feared for its pirates and slave traders. France occupied northern Algeria by force between 1830 and 1870 and suppressed a rebellion in 1871. European settlers, mainly French, arrived in large numbers, taking the best land and assuming economic and political power. From 1954 the native Algerians waged a bitter civil war against France, achieving independence at the cost of a million lives in 1962. Most of the settlers returned to Europe.

Independent Algeria adopted secular socialist principles and, thanks to the income from big oil and gas exports, became a prominent non-aligned African power. From 1992, however, Muslim fundamentalists have waged a savage campaign of terror against a succession of part-military, part-elected governments. By 2000 the fighting had caused about 100,000 civilian deaths, mostly in villages, and was impoverishing the nation.

In 2001, riots by the Berber ethnic minority against unemployment and poverty caused at least 60 deaths. The high birth rate (more than half the population is under the age of 20) has created a huge pool of unemployed young people, many of whom fight for change, or emigrate, mostly to France, in search of a better life.

18 – 180 million

its late 18th century level (less than one billion). But population growth has outpaced economic growth globally, so most world citizens are poverty-stricken (section 6.4, the *Micawberish Rule*). The well-off minority of humankind has discovered the self-satisfaction of living ostentatiously virtuous and unselfish politically correct lives, rarely achievable in the pre-WROG Darwinian world of strong restraints (section 6.7).

Most of us still live in our national WROG periods, during which Darwinian competition for survival between individuals, nations, races and religious groups has eased. Nations co-operate, for the common good, in international associations (section 6.3). Western societies have come to believe in human rights and the sanctity of human life (section 5.6), and that the quality of life can only improve (sections 8.1, 8.2). The future seems assured, allowing sentimentality to oust sense in many aspects of Western thinking (sections 8.4, 8.5, 8.6). But the WROG period is temporary. Strong restraints on growth have already returned, with chaos and misery, to several overcrowded multicultural nations (e.g. RWANDA).

For the first time I realised how wide and significant is the variation between the population densities of different nations, and how this affects national approaches to all manner of subjects. For example, a writer in *New Scientist* magazine (29 September 1988) complained that National Parks in England are less pristine than those in the USA. Hardly surprising, I replied, considering that little England, with (at that time) 356 people per square kilometre, was fourteen times as crowded as the vast USA, with only 26 people per sq. km. Then again, the English high-speed rail link from London to the Channel Tunnel has been difficult and controversial to build because it cuts through many densely populated areas, whereas in France (102 people per sq. km. in 1988) the link to Paris, snaking through sparsely populated farmland, was completed quickly with little fuss.

ARGENTINA   Area 2,780,000 km²

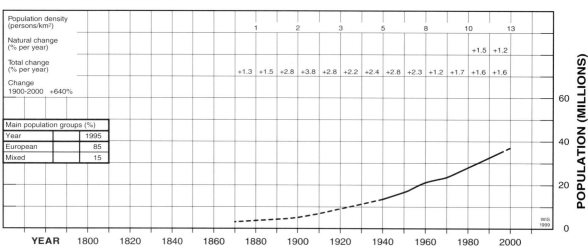

Population density (persons/km²)									1		2		3		5		8		10		13	
Natural change (% per year)																					+1.5	+1.2	
Total change (% per year)								+1.3	+1.5	+2.8	+3.8	+2.8	+2.2	+2.4	+2.8	+2.3	+1.2	+1.7	+1.6	+1.6		
Change 1900-2000 +640%																						

Main population groups (%)		
Year		1995
European		85
Mixed		15

POPULATION (MILLIONS)

YEAR   1800   1820   1840   1860   1880   1900   1920   1940   1960   1980   2000

WIS 1999

Spanish explorers reached Argentina in 1516. Spanish settlers soon overcame Amerindian resistance and Spain ruled the country until 1816 when independence was declared. Significant population growth began about 1890 with waves of European immigration. Argentina is the second largest country in South America and the fifth most sparsely populated. The somewhat fragile economy depends on agriculture, especially livestock raised on the vast steppe-like *pampas* that cover most of the country, and on mining (including oil and gas), manufacturing and fishing. A growing financial crisis, beginning in 2001, saw the country defaulting on its $140 billion foreign debts. In recent years the Antarctic ozone hole has extended so far north that the southern half of Argentina is subjected, every spring, to excessive ultra-violet radiation.

High population densities in regions liable to earthquakes, volcanic eruptions, floods, droughts or hurricanes can result in ordinary natural events becoming *anthropogenic disasters* (section 4.6).

The city librarian at Wells, Somerset, made my research much easier. She advised me that the Bristol Central Reference Library has a nearly complete set of *The Statesman's Yearbook*. This weighty tome, published every year since 1864, has pages of data on every nation in the world, updated year on year, including the latest population estimates and census results. The oldest issues refer to earlier population figures where known. *Whitaker's Almanack* gives similar but less comprehensive population data. Thereafter I spent many days in the Bristol library, with the result that the graphs on these pages are pretty nearly as accurate as the records permit.

As the 235 individual graphs neared completion, significant trends and groupings confronted me. Some defied conventional wisdom. For example, the *demographic transition theory* (section 5.2) has been largely confounded, as a national population stabiliser, by international migration. In reality, the only set of nations showing population stability or actual shrinkage consists of the economically troubled ex-Soviet Slav nations and the one-time Soviet satellites of eastern Europe.

Using the graphs, it was easy to compare population growth rates in the various sets of nations (Tables 3.1, 3.2). The fastest population growth is taking place in African and Muslim nations (which is why these nations suffer and export so many demographic problems). Growth rates in South America and non-Muslim South-east Asia including India are on the high side of average. East-central Asia including China is on the low side. North America has low growth, Western Europe very low growth, and in Eastern Europe including Eurasia the population is actually shrinking. This picture may soon change, especially in southern Africa, as deaths from rampant HIV/AIDS (section 4.7) accelerate.

The graphs illustrate a vital relationship, in the developing world, between fast population growth, poverty and migration (sections 5.5, 6.4). They demonstrate that the first causes the second, which causes the third. Some religious and/or ethnic groups deliberately

AUSTRALIA   Area 7,682,000 km²

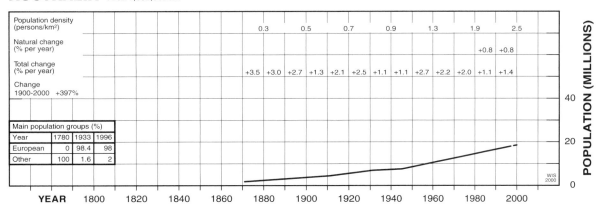

Population density (persons/km²)						0.3	0.5	0.7	0.9	1.3	1.9	2.5	
Natural change (% per year)											+0.8	+0.8	
Total change (% per year)						+3.5	+3.0	+2.7	+1.3	+2.1	+2.5	+1.1	+1.1	+2.7	+2.2	+2.0	+1.1	+1.4
Change 1900-2000 +397%													

Main population groups (%)			
Year	1780	1933	1996
European	0	98.4	98
Other	100	1.6	2

POPULATION (MILLIONS)

YEAR   1800   1820   1840   1860   1880   1900   1920   1940   1960   1980   2000

WIS 2000

The native people of Australia, the Aborigines, arrived from south-east Asia about 60,000 years ago. They spread over the continent and eventually hunted out the naïve megafauna including two-tonne diprotodons and other giant plant-eating marsupials. It is argued (Flannery, 1996) that this act made the dry Australian climate even more arid and the soils even more nutrient-poor, because uneaten vegetation accumulated, fuelling vast bushfires which killed the broadleaved forests that had recycled rainfall into the continental interior by transpiration. Fire-resistant leathery-leaved trees, especially eucalypts, thrived in their place. Rainfall was not recycled, but ran off the bare charred soil, carrying nutrients in the ash out to sea.

The Aborigines, whose nomadic hunter-gatherer lifestyle determined the carrying capacity of their mostly arid homelands, modified by occasional El Niño climatic setbacks, numbered around half a million when Britain established the first European settlement, a penal colony, at Sydney in 1788. Within a few decades their numbers had been greatly reduced by European diseases and repression. Subsequently, immigration has been a major factor in Australian population growth. In 1974 the long-lived 'white only' immigration policy was relaxed.

Although population density is very low, soil erosion, deforestation and salinization of irrigated land are serious local problems. Some authorities believe that long-term carrying capacity has already been reached, and that the snowballing influx of illegal migrants from all over the world should be terminated. In 2001 boats carrying hundreds of smuggled migrants were forcibly prevented from reaching the Australian coast.

Most of the continent is arid or semi-arid. 41 native animal species were wiped out by Europeans and their introduced animals in the 20th century. Censuses: about every tenth year until 1961, then every fifth year.

encourage fast population growth in order to gain numerical advantage over rival groups. This is *aggressive breeding* (section 4.4). The slower breeders, threatened with loss of status and territory, react savagely (see GAZA STRIP, KOSOVO, ISRAEL, MACEDONIA, WEST BANK). Aggressive breeding in a multicultural society is a potent cause of racism (section 5.7).

A common misconception, often wrongly used to 'prove' that population growth is decelerating, derives from the difference between percentage growth rate and numerical growth rate (section 5.1). Over several decades the percentage rate may show significant falls, while over the same period, because the base for the calculation, the actual population, is rising, numerical growth per decade may be constant or even increasing. The graph for BRAZIL from 1950 to 2000 illustrates this perfectly. A similar reservation applies to falling fertility rates.

The genocide and ethnic cleansing in RWANDA (1994), whereby the population was reduced by more than 2 million in 3 months, epitomised thousands of violent world events which, before the WROG period, combined with hunger and disease to keep populations low. In those days, when an aggressor took another country by force, he tended to want the land, not the former inhabitants, most of whom he removed from the scene in a variety of ways. Casual massacres of the losing side were commonplace during the campaigns of successive empire-builders in the Old World, as far back as history goes. For example, the Mongol hordes were inclined to eliminate the populations of cities that resisted their advance. In the New World, invading Europeans were unsympathetic when the native Americans succumbed in millions to aggression, ill-treatment, alcoholism and Old World diseases.

After the WROG period began, it was possible for national populations to expand relatively unrestrained by hunger and disease. Cherishing and saving human life became an

18 – 180 million

BANGLADESH East Pakistan Area 148,400 km²

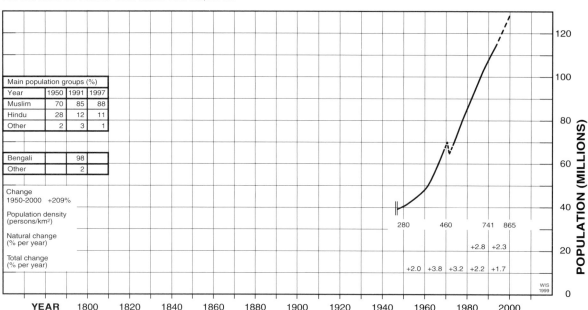

Main population groups (%)			
Year	1950	1991	1997
Muslim	70	85	88
Hindu	28	12	11
Other	2	3	1
Bengali		98	
Other		2	

Change 1950-2000 +209%

Population density (persons/km²)

Natural change (% per year)

Total change (% per year)

The rich alluvial soil of the Ganges/Brahmaputra delta, which makes up most of Bangladesh, was farmed in the first millennia BC and AD by Buddhist and Hindu tribes. It became part of the Muslim Mogul (or Mughal) Empire in the 16th century. When it came under British rule in the 18th century it was the largely Muslim eastern province of Bengal state in British India, prior to the 1947 partition into Hindu India and Muslim Pakistan. East Pakistan, as it then became, split violently from West Pakistan in 1971 because its overwhelmingly Bengali people wanted their own Bengali republic. Millions of Hindus fled to India and up to 3 million people were killed.

Since the 1960s the percentage rate of population growth has decreased dramatically, although the annual numerical increase has stayed almost constant. The enormously high population density, by far the world's highest for a major nation, has forced tens of millions of peasants (who form 85% of the population) to farm land that is often flooded. Half a million people died in tidal floods in the cyclone of 1970, and in many years there is a significant death toll (120,000 in 1991) as the rivers, swollen in the monsoon, overflow

their banks. Even more people have died of poverty. Poor harvests in 1943 and 1974 caused unemployment and high food prices, so that although food was still available those who could not afford it starved to death, 3 million in the former year and 1.5 million in the latter. Fresh water resources are heavily polluted: river water by sewage bacteria and well water by arsenic. That such a crowded, poverty-stricken and disaster-prone country has not suffered disastrous civil strife since 1971 may perhaps be attributed to the ethnic homogeneity of the population (although Buddhist separatists in Chittagong fought a minor war, 1977–98). In 2001 a survey by Transparency International claimed that Bangladesh was the world's most corrupt nation.

Bangladeshi ministers claimed in 2000 that if their country is flooded by rising sea levels due to global warming its people should have the right to emigrate to Western nations who, by excessive fossil fuel consumption, have caused the problem. It is equally arguable that Bangladesh alone is responsible for the huge population increase since independence that has made it so vulnerable.

option. But in some multicultural nations, increasingly from about 1980, population growth has been terminated by extreme violence. One internal group attacks another, massacring it or expelling it as refugees (see RWANDA). It happens when the national population has risen to a level, the *Violent Cutback Level* (*VCL*), at which rivalry for land, resources and power has become so acute that it flashes into genocidal war (section 4.2).

In the 1990s, in the developed world, it was politically correct to speak warmly about "the global village", "globalisation" and "multiculturalism", words evocative of an ideal world in which human communities of different ethnic or cultural backgrounds are expected to live happily together, forgetting past grievances if they exist. Such a scenario tends to be ephemeral, at best, in the real world. I examine it in Chapter 7, based on the historical evidence of 235 graphs and their briefings.

What I find is that dissimilar groups can live peacefully together when population density is low and/or there are ample resources for everyone, as in ISRAEL before World War I. However, if population density rises, and shortages (or even the expectation of shortages) develop, a Darwinian scenario unfolds in which natural selection, the survival of

BRAZIL Area 8,540,000 km²

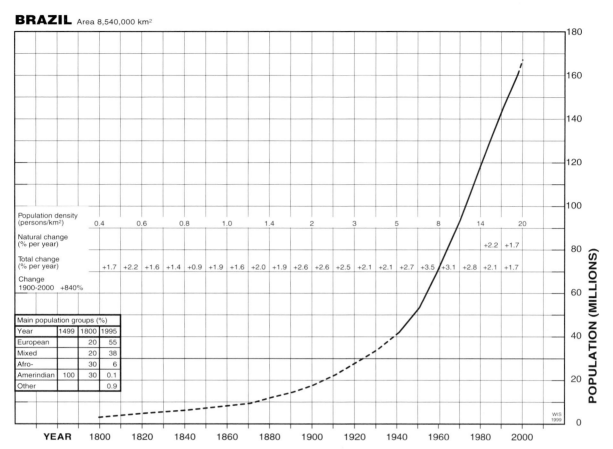

Population density (persons/km²)	0.4	0.6	0.8	1.0	1.4	2	3	5	8	14	20
Natural change (% per year)										+2.2	+1.7
Total change (% per year)	+1.7	+2.2	+1.6	+1.4	+0.9	+1.9	+1.6	+2.0	+1.9	+2.6	+2.6	+2.5	+2.1	+2.1	+2.7	+3.5	+3.1	+2.8	+2.1	+1.7
Change 1900-2000 +840%											

Main population groups (%)			
Year	1499	1800	1995
European		20	55
Mixed		20	38
Afro-		30	6
Amerindian	100	30	0.1
Other			0.9

YEAR 1800 1820 1840 1860 1880 1900 1920 1940 1960 1980 2000

POPULATION (MILLIONS)

18 – 180 million

When Pedro Cabral, a Portuguese sea-captain, landed in Brazil in 1500 its Amerindian population numbered between 3 and 6 millions. They were hunter gatherers, and the concept of working to produce more food than they needed was alien to them. Soon, as they slaved on Portuguese sugar plantations and contracted European diseases, hundreds of thousands of them died. African slaves, more resilient in the service of uncaring masters, were imported to replace them. In 1798 more than half of the 3.3 million population were black and coloured slaves. Brazil became independent in 1822, the world's 5th largest country, half the total area of the South American continent. Coffee plantations boomed in the 19th century. In 1888 slavery was abolished. The land is now cultivated partly by small peasant farmers and partly by workers on huge estates. Brazil is immensely rich in natural resources and is well advanced in manufacturing, but the great majority of the people are very poor. 90% are Catholic.

The tropical rain forest of Brazil's Amazon basin is the largest reservoir of terrestrial biodiversity in the world. So far, about one quarter of the forest has been cleared for agricultural or industrial development, but the rate has decreased since 1995. The deforestation has already caused climate change, reduced rainfall leading to droughts, and huge forest fires. March 1998 saw 104,000 square kilometres of forest ablaze. In 2000 Brazil announced that only 10% of the Amazon rain forest would be saved from destruction.

The population graph shows rapid growth, fuelled by immigration, beginning rather earlier than elsewhere in Latin America. It demonstrates a common misconception: that if the percentage growth rate slows, numerical growth per year decreases also. In Brazil, although the annual percentage increase has halved since it peaked in the 1950s, the graph goes on rising as steeply as ever, because the base of the percentage calculation is rising. The change in population density is a truer gauge of the practical effects of population growth.

the fittest, comes into play (section 5.8). Like-minded individuals maximise their chances of survival by joining together (Unity is Strength) in groups or factions, which compete with growing rancour as the coveted resource depletes and the number of persons coveting it increases. This scenario, if unchecked, ends at VCL, a population crash, and survival of the strongest group – in fact it is the normal course of unconstrained Darwinian evolution.

I have found no precedents wherein an expanding population consisting of rival groups decides voluntarily to stop competing and descend peacefully into poverty and squalor together. In the real world, some groups take steps that bring them out on top, while their rivals fade away.

Probably the most imminent and serious shortage facing humankind today is shortage of oil. The depletion of this finite resource is examined convincingly by Colin Campbell, an

CANADA Area 9,220,000 km²

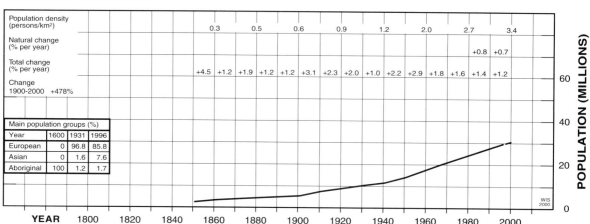

Population density (persons/km²)		0.3	0.5	0.6	0.9	1.2	2.0	2.7	3.4

Natural change (% per year): +0.8 +0.7

Total change (% per year): +4.5 +1.2 +1.9 +1.2 +1.2 +3.1 +2.3 +2.0 +1.0 +2.2 +2.9 +1.8 +1.6 +1.4 +1.2

Change 1900-2000 +478%

Main population groups (%)			
Year	1600	1931	1996
European	0	96.8	85.8
Asian	0	1.6	7.6
Aboriginal	100	1.2	1.7

POPULATION (MILLIONS)

YEAR 1800 1820 1840 1860 1880 1900 1920 1940 1960 1980 2000

WIS 2000

Humans arrived in Canada around 40,000 years ago from Siberia. As they spread over North America much of the indigenous megafauna including mammoth and plains camel was hunted to extinction. Inuit tribes from Siberia populated the northern coasts about 4000 years ago. The first Europeans (other than the short-lived Viking settlement in Vinland around 1000 AD) arrived in 1497, when the indigenous population was about 250,000. The first settlers were French, in 1604. Within a few decades the population of indigenous "Indians" was much reduced by European diseases and alcoholism. English fur traders established the Hudson Bay Company in 1670. Anglo-French rivalry (beginning with the fur trade, which almost wiped out the beaver) led to a succession of territorial wars that ended in Britain's favour with the Treaty of Paris in 1763. French colonists were concentrated in Quebec province, which has threatened several times to secede. The rate of immigration, always high, increased sharply after World War Two. Canada, extremely rich in natural resources and the world's second-largest nation in terms of area, also has one of the Western world's lowest population densities.

oil geologist, in *The Coming Oil Crisis* (1997). Although the search for new oilfields intensifies, the discovery rate has fallen drastically (Figure 7.1). For every one or two barrels discovered, four barrels are used. About half the world's recoverable oil has now been produced and consumed. When supply is unable to meet demand, probably before 2010, the era of *cheap* oil will have ended (section 7.3). There are possible alternatives to oil, but none will replace its potency and convenience (section 9.1).

Between 1860 and 2000, the era of cheap oil, world population was able to expand from 1 billion to 6 billion. In Chapter 9 I try to assess how an overpopulated world will react to the shock of increasingly expensive oil.

In drawing up the graphs I soon encountered a difficulty: many nations, especially in places like Central Europe, have had boundary changes at frequent intervals (Chapter 2). The graphs in such cases do not record the demographic development of a fixed area. For example, neither modern Austria nor modern Hungary are as big as their namesakes were in pre-1914 Austria-Hungary. Relatively few nations have had fixed boundaries for more than 100 years. In much of Africa there were no boundaries at all until late in the 19th century. Because the transience of land frontiers complicates the interpretation of long-term population development (so illuminating in the case of largely insular ENGLAND (LONG TERM)) I have included graphs for certain islands even though they are not independent nations.

As I constructed the graphs, and began to notice similarities and trends, it became obvious to me that politicians, businessmen and media pundits were blissfully unaware of the facts of population growth and its consequences, or else were determined to ignore them in favour of policies based on wishful thinking and political correctness. Some subjects were avoided or misrepresented because of their 'sensitivity' (section 6.7). The convenient but unreliable demographic transition theory (section 5.2) was quoted to assure people that population growth would sort itself out before long. Sustainable development was supposed to take care of global heating and climate change, but I could see that it hadn't a chance, given the exploding world population (section 5.3). Economists and demographers projected current trends for many decades into the 21st century. It seemed to be taken for granted that

COLOMBIA Area 1,141,000 km²

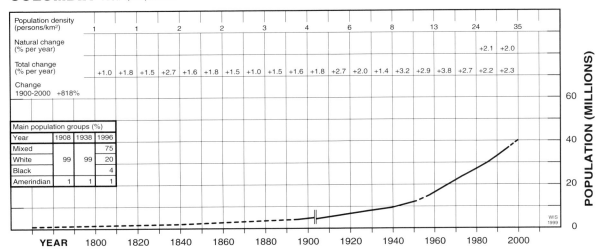

| Population density (persons/km²) | 1 | 1 | 2 | 2 | 3 | 4 | 6 | 8 | 13 | 24 | 35 |

| Natural change (% per year) | +2.1 | +2.0 |

| Total change (% per year) | +1.0 | +1.8 | +1.5 | +2.7 | +1.6 | +1.8 | +1.5 | +1.0 | +1.5 | +1.6 | +1.8 | +2.7 | +2.0 | +1.4 | +3.2 | +2.9 | +3.8 | +2.7 | +2.2 | +2.3 |

Change 1900-2000 +818%

Main population groups (%)			
Year	1908	1938	1996
Mixed			75
White	99	99	20
Black			4
Amerindian	1	1	1

POPULATION (MILLIONS)

WIS 1999

YEAR 1800 1820 1840 1860 1880 1900 1920 1940 1960 1980 2000

18 – 180 million

Colombia was inhabited by Chibcha Indian farmers, subservient to the Incas, when Columbus sighted it in 1499. Spain conquered the territory early in the 16th century and ruled it until 1819 when after a revolution it joined with neighbouring states to form Greater Colombia. This was the first of several short-lived confederacies, the last of which became independent Colombia (including Panama) in 1886. Panama seceded in 1903, helped by the US which at once constructed the Panama Canal. The economy in 2000 depended to a large extent on exports of coffee and oil. Earthquakes in 1994 and 1999 caused widespread damage and thousands were killed.

Colombia's history is one of chronic political and civil strife. The 1899–1903 civil war saw 100,000 deaths, and in another, 1949–57, some 300,000 were killed. Since then several powerful guerrilla groups, Marxist and/or directly linked to the cocaine trade, have been continuously active against the US-backed and financed government forces. In 1998 rebels controlled nearly half the country. However, in spite of almost daily assassinations, disappearances, kidnappings and bombings, population growth is rapid. As elsewhere in Latin America, there is little objection to the 'social cleansing' of vagrants and 'street children'. Censuses: at irregular intervals after 1912. Apparent large ethnic changes in the group table are the result of varying interpretations of 'white'. In 2001 60% of the population was poverty-stricken and 20% was destitute.

business as usual could continue indefinitely (section 8.7).

Most of the social and environmental problems that concern humankind and are addressed in this book are no more than symptoms, the consequences of a single fundamental problem: overpopulation. This is obvious if *reductio ad absurdum analysis* is applied (section 6.1, *climate change*). Trying to treat individual symptoms of overpopulation while ignoring the cause is as misguided as treating headaches with aspirin when they are the symptoms of a brain tumour. So if, for example, we think we can halt climate change by prescribing small reductions in greenhouse gas emissions (the Kyoto protocol), while world population is growing by 80 million every year, we have, as they say, "another think coming".

Humans on 'Spaceship Earth' are already so numerous that the planet cannot cope with their demands on its resources and ecosystems (see WORLD). My conclusion (Chapter 9) that there will be a global population crash, only a few decades from now, echoes the Club of Rome's findings in *The Limits to Growth* (Meadows *et al*, 1972) and the lesson of Easter Island (Bahn & Flenley, 1992; see section 3.8). Three different approaches reaching the same conclusion should not be ignored. Mounting population pressure does have a potential safety valve: recognition and rational analysis of the danger, leading to remedial action (birth control worldwide), but for half a century the valve has been tightly closed by a taboo (section 6.7). The subject is so 'sensitive' that few people are prepared to face it. Birth control is, however, humankind's best hope for a less painful future.

Many scientists believe that life on Earth is experiencing a mass extinction of species (e.g. *The Sixth Extinction* by Leakey and Lewin, 1996). Whereas previous mass extinctions were caused by natural events such as asteroid strike, the current one is being driven by the destruction of wildlife and habitat by proliferating humans. The subject of this book is, therefore, the anatomy of a modern mass extinction.

CONGO DEMOCRATIC REPUBLIC   Zaire, Belgian Congo, Congo Kinshasa   Area 2,345,000 km²

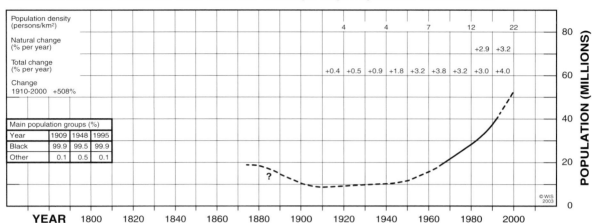

Population density (persons/km²)						4	4	7	12	22
Natural change (% per year)									+2.9	+3.2
Total change (% per year)					+0.4	+0.5	+0.9	+1.8 +3.2 +3.8 +3.2	+3.0	+4.0
Change 1910-2000 +508%										

Main population groups (%)

Year	1909	1948	1995
Black	99.9	99.5	99.9
Other	0.1	0.5	0.1

© WIS 2003

YEAR   1800   1820   1840   1860   1880   1900   1920   1940   1960   1980   2000

POPULATION (MILLIONS) 80 60 40 20 0

Pygmy hunter-gatherers were living in the Congo rainforests when Bantu farmers moved in from the west one or two thousand years ago. The east and west regions of this huge country provided slaves to Arab and European traders respectively for several centuries before slavery was ended in the 19th.

European explorers penetrated the interior in the 1870s and King Leopold 2 of Belgium acquired the 'Congo Free State' as his personal property in 1885. Tribespeople were treated so savagely by his agents that Belgium took on the country as a colony in 1908. It has been alleged that the population halved between 1880 and 1920 (*The Guardian*, 13.7.02). Belgian Congo developed patchily but peacefully until 1960 when, following nationalist riots, Belgium suddenly granted independence. Most of the Belgian residents left for Europe.

Independent Congo (renamed Zaire, 1971–1997) has been chaotic. A civil war in Katanga province, following its attempted secession, was settled in 1963, but under President Mobutu the mineral-rich economy was shamelessly milked to benefit his family and his tribe, the Ngbande. The 1970s saw influxes of refugees from Uganda and invasions by Angolan rebels. Unpaid troops started days of urban rioting in the early 1990s but the showdown came in 1994 when a million refugees from the Rwandan genocide flooded over the eastern borders. Their presence sparked interethnic warfare and rebellion throughout eastern Zaire, led by Laurent Kabila who advanced to Kinshasa and replaced Mobutu as president in 1997. Kabila's equally chaotic regime in the renamed Congo Democratic Republic (or "Congo Kinshasa") has provoked ongoing civil war, involving soldiers (and entrepreneurs seeking plunder, especially minerals) from other African states including Rwanda, Uganda, Angola, Namibia and Zimbabwe, in the north, east and south. Some main towns, such as Kisangani (Stanleyville), are being destroyed. Predatory rebels and tribal warlords now hold vast swathes of Congolese territory, refugees stream back and forth across frontiers, and the fighting caused 2 million deaths between 1998 and 2001. About 2% of the population is HIV-positive.

• • • • • • • • • • • • •

"The basic behaviour mode of the world system is exponential growth of population and capital, followed by collapse"

(The Club of Rome, (Meadows et al) 1972)

• • • • • • • • • • • •

"New occasions teach new duties, time makes ancient good uncouth.
They must upward still, and onward, who would keep abreast of truth"

(Songs of Praise)

Chapter 2

Constructing the Graphs

The graphs are plotted using data from a wide variety of sources. Some of the oldest data sets are the most complete, being the findings of detailed censuses or 'enumerations' that were initiated around the beginning of the 19th century by the United States (1790) and certain North European nations (e.g. Britain, 1801) and their colonies. They have been repeated almost every subsequent decade. Occasional population estimates for other countries appeared in the 1800s, and they became more common, punctuated at intervals by censuses, in the 20th century.

Some data sources, such as *The Statesman's Yearbook* and some encyclopedias, usually distinguish between a population estimate and a census result, and give the year in which it was made. Others simply give a figure, not necessarily the latest (as it may be repeated year after year while the estimates in other sources are changing). Thus, different sources often disagree as to the population of a particular nation in a particular year. Mid-year or end-year figures may be quoted. In recent years there have been considerable discrepancies between the up-to-date estimates made by different authorities, such as national governments, the United Nations, and the Population Reference Bureau. This applies especially to countries where there has been civil strife, entry or exit of refugees, or famine, and to countries where there has been no census for many years. Some estimates include refugees as part of the population, others exclude them, and some apparently ignore them.

Population estimates are just that, estimates. Where a previous census figure is available they may be calculated on the basis of assumed changes (births, deaths and migration) during the intervening years. Not infrequently the assumed changes are themselves adjusted in subsequent years. The next census enables previous estimates to be corrected, and provides a new base for calculating the next series of annual estimates.

Population censuses, ideally, record the presence of every individual resident in the country on a chosen date, together with a mass of data about him or her. The head of every household completes a questionnaire covering every person resident in that dwelling place on the census date. In practice, even where this procedure is followed (usually in developed nations), the scope for error or falsification is wide. In any survey people may answer questions in ways which, they believe, are advantageous to them. Just as elections may be rigged, census figures may be adjusted by officials or governments to achieve political ends. (For example, the Turkish census of October 2000 found an improbably large population increase of 8.4 million (13.4%) since the previous census in 1997, which was said by some observers to result from inflation of the figures by provincial authorities in order to maximise grants from central government.) Some demographers argue that modern sampling methodology can provide more accurate population figures than censuses, at a small fraction of the cost.

For a year or two after a census the "raw" figures are usually quoted. Later they are replaced by adjusted figures, when checks have been made with other data sets such as medical and employment registers and migration records. But the final figure can never be precise, and

EGYPT Area 997,800 km²

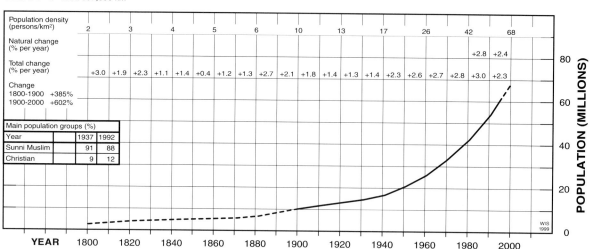

| Population density (persons/km²) | 2 | 3 | 4 | 5 | 6 | 10 | 13 | 17 | 26 | 42 | 68 |

Natural change (% per year): +2.8 | +2.4

Total change (% per year): +3.0 | +1.9 | +2.3 | +1.1 | +1.4 | +0.4 | +1.2 | +1.3 | +2.7 | +2.1 | +1.8 | +1.4 | +1.3 | +1.4 | +2.3 | +2.6 | +2.7 | +2.8 | +3.0 | +2.3

Change
1800-1900 +385%
1900-2000 +602%

Main population groups (%)		
Year	1937	1992
Sunni Muslim	91	88
Christian	9	12

YEAR 1800 1820 1840 1860 1880 1900 1920 1940 1960 1980 2000

POPULATION (MILLIONS) 0 20 40 60 80

WIS 1999

Agriculture developed in Egypt's Nile Valley nearly 8000 years ago, only slightly later than its world premiere in south-western Asia. Natural irrigation and fertilisation of the soil by annual floods generated abundant crops, allowing the early establishment of rich and powerful civilisations. By 5000 years ago Egypt was a single nation-state with the resources to construct enormous religious monuments that still stand today. Even so, by modern standards the population was tiny.

Egypt was a granary of the Roman Empire. Islam was introduced in the 7th century by Arab conquest. In the 16th century the Ottoman Turks invaded and ruled until 1882. Finally, after half a century as a British protectorate, modern Egypt became independent in 1936. In 1958 Egypt merged with Syria to form the United Arab Republic. It was hoped that more Arab nations would join, but internal disagreements broke up the union after only 3 years. In the 1990s Muslim fundamentalists began to harass the moderate government by terrorist activities directed at tourism and at Coptic Christian communities in the south.

All of Egypt is desert except the Nile valley, the Nile delta (vulnerable to sea-level rise), and a few oases, amounting to 3.5% of the whole country. Virtually all the population occupies the agricultural land, at a density of 1900 per square kilometre. The creation of Lake Nasser, behind the Aswan High Dam, in 1971 has increased the options for cultivation (by irrigation) but the annual fertilisation of the land by flood-borne silt, which now settles out in the lake, has ceased. Agriculture, tourism and oil from recently-discovered fields are the main supports to the economy, which also receives as much US aid as does Israel. Before about 1940 the high birth rate was almost equalled by the high death rate, but then the introduction of modern medicines and agrochemicals began the DC population surge.

when it applies to a nation with tribes in a remote hinterland, or urban shanty towns concealing illegal immigrants, or regions made inaccessible by war or terrorism, the room for error is large. Nevertheless, a census-based final figure must be considered more reliable than an estimate.

When all the figures for estimates and censuses are plotted, a fuzzy band of dots appears. Deciding the final graph is a matter of interpretation; it tends to be, broadly, a median line through the dots, heavily weighted towards census evidence where available. Allowance is made for recorded sudden events such as flows of migrants or refugees, or deaths in wars, terrorism, famines, volcanic eruptions (see ICELAND, MARTINIQUE) or similar disasters.

In the graphs, a continuous line indicates relative reliability, as where there are censuses. A broken line is based on estimates, but may discount census figures that seem unreasonable (see GUATEMALA). The double vertical lines mark discontinuities, where frontiers were changed and the area of the country increased or decreased, by conquest (e.g. CHILE, 1884), or by unification (e.g. YEMEN, 1990), or partition (e.g. CZECHOSLOVAKIA, 1992).

How accurate are the graphs? Reality cannot be ascertained. I can only suggest that continuous lines may have up to 10% error, and dashed lines up to 25%. Normally, given that populations change gradually, except during dramatic events, errors should be much less than these extremes, especially in developed nations. (The most recent population estimates, for 2000 and a few preceding years, have not, of course, been adjusted.)

Accompanying the graphs are horizontal *data rows*. The "Population Density" row is

ENGLAND Area 131,000 km²

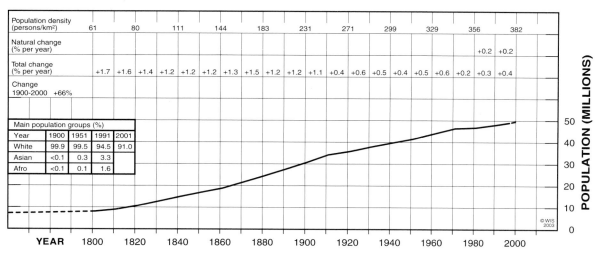

Population density (persons/km²)	61	80	111	144	183	231	271	299	329	356	382

Natural change (% per year): +0.2 +0.2

Total change (% per year): +1.7 +1.6 +1.4 +1.2 +1.2 +1.2 +1.3 +1.5 +1.2 +1.2 +1.1 +0.4 +0.6 +0.5 +0.4 +0.5 +0.6 +0.2 +0.3 +0.4

Change 1900-2000 +66%

Main population groups (%)

Year	1900	1951	1991	2001
White	99.9	99.5	94.5	91.0
Asian	<0.1	0.3	3.3	
Afro	<0.1	0.1	1.6	

POPULATION (MILLIONS): 50 40 30 20 10 0

© WIS 2003

YEAR 1800 1820 1840 1860 1880 1900 1920 1940 1960 1980 2000

England's population history is uniquely long. England has been a fixed geographical entity since the Roman conquest of 43 AD, and a single political unit since the Danish conquest of 1016. There exists enough historical and archaeological data to construct a graph of population change in England over the last 2 millennia (see ENGLAND LONG TERM).

The graph above records population change during England's DC surge which began about 1750 with the Industrial Revolution. Numerical growth was fastest in the 19th century (23 million), but was nearly as fast in the 20th (20 million). The first kink in the graph reflects the loss of men and their potential offspring in World War One; the second kink, in the 1970s, reflects a fall in the birth rate apparently linked to social developments such as increased use of the contraceptive pill and the advent of legal abortion in 1968. The 1980s and 90s saw a pronounced rise in immigration, much of it illegal, to the point where it made up more than half the total population increase of c. 200,000 per year in the 1990s. In 1999 local men attacked asylum-seekers in the port of Dover. Illegal immigration is dangerous; in 2000 58 immigrants died of asphyxiation hiding in a container truck.

Censuses: every tenth year from 1801, except 1941 (during World War Two). From 1998 coastal properties perceived to be at risk from sea level rise (greenhouse effect) have attracted raised insurance premiums. See also: UNITED KINGDOM.

constructed by dividing the population on a given date (read off the graph) by the area of the country at that time. Figures are given every twentieth year.

The "Natural Change" row in most cases is derived from the World Population Data Sheet issued annually by the Population Reference Bureau, by averaging the yearly estimates of 'natural increase' (births minus deaths) for the decades 1981–89 and 1990–99. A very few nations (e.g. YUGOSLAVIA, ENGLAND) have natural change data for earlier decades, quoted from studies by historians and statisticians. More data could be obtained by lengthy research.

The "Total Change" data row is derived directly from the graph, by calculating the average annual percentage population change from the beginning to the end of each decade (except in decades which have seen area changes). Total Change differs from Natural Change by including changes due to migration. However, unless the difference between them in any one decade is large, it may merely reflect imprecision in estimates or censuses.

The final short data row, "Change", gives percentage population change from the beginning to the end of the longest continuous period since 1900 during which population totals have not been affected by changes in the land area. It is derived directly from the graph.

The "Main Population Groups" tables are derived mainly from census data in the *Statesman's Yearbook*, though the introductory sections of travel guidebooks are often helpful. They are included because rivalry between ethnic and/or religious groups is a potent cause of civil war, especially where poverty is worsening, or one group is gaining strength at the expense of another (e.g. KOSOVO).

Below each graph is a *briefing*: a very short history of the country with emphasis on events relevant to population change and development, the origins of ethnic and religious groups and the conflicts they have triggered, and the environmental changes, such as desertification, caused by and affecting the evolution of ancient and modern civilizations.

18 – 180 million

Briefing data is drawn from the same yearbooks, encyclopedias, travel guidebooks, etc., as the population data (see SOURCES AND REFERENCES). Perhaps the most informative single publication is Ponting's *Green History of the World* (1991).

The graphs are presented in 4 groups, from very large nations (populations more than 180 million) to small nations (populations less than 1.8 million). In each group all the graphs have the same vertical scale (representing population), which differs by a factor of ten from the vertical scale of the next group. This is advantageous because the graphs in any one group are directly comparable, thus differences and similarities between nations, especially in growth rate, are visually obvious. This would not be the case if, for example, scales were varied to make each graph fit the page. The graph of Russian Empire population in RUSSIA (18–180 million) is repeated in USSR (more than 180 million), illustrating the factor of ten scale difference. If all the graphs from the Falkland Islands to China were at the same scale, most of them would be so close to the bottom of the page that all detail would be lost.

I have been asked why I didn't simply request the statistical services of every nation to provide details of its population history, instead of poring over old books in libraries. Apart from practical questions (do they all have such histories, and would they reply?), the fact that the data in the old books is *contemporary* removes any possibility that it might subsequently have been manipulated for political or other purposes. The same argument applies when the data is drawn from a wide range of sources. The great disadvantage of studying old books is that one is so easily diverted down fascinating sidelines, such as the campaigns of the Mongol hordes, or proposals to develop atmospheric railways. Some of these are relevant to, and appear in, the briefings.

• • • • • • • • • • • •

"The further back you look, the further forward you can see"

(Churchill)

Chapter 3

Two Centuries of Surging Populations, Worldwide

Historical analysis of population change in a nation is helped by referring to the relevant graph and briefing (named in capitals in the text, thus: RWANDA). This chapter looks at population development in the main categories of nations: Developed World, Sub-Saharan Africa, etc. (In Tables 3.1 and 3.2 the 'Nation Sets' are grouped according to different criteria.)

Before about 1750, all nations had populations that were very small by modern standards. They were controlled by the *carrying capacity* of the land (section 4.1). This was, effectively, the number of people that could survive on the food that peasant farmers, who formed the bulk of the population, could produce using manual labour (at best, horse- or ox-ploughs), animal or human dung ('night soil') as fertilizer, and hand-weeding of crops. It was true organic farming. Carrying capacity was reached when malnutrition raised the death rate to equal the birth rate. This rarely happened. It only took a plague, or a period of war and lawlessness, for the population to shrink by as much as half (see ENGLAND LONG TERM). In any geographical region, carrying capacity has fluctuated over the centuries, as agriculture replaced hunter-gathering, as forests were felled and deserts irrigated, or as fertile land was desertified (see INDIA, IRAQ). Sometimes conquered lands supplied food to their rulers, as in the grain-growing Roman provinces of North Africa.

Everything changed with the advent of the Industrial Revolution, which began in the mid-18th century in Western Europe. New technologies increased agricultural productivity. Populations were no longer held at carrying capacity by starvation and malnutrition. Death control (DC) became possible, and DC population surges (section 4.3) took off in Europe and North America.

3.1 The Developed World (omitting ex-communist Europe)

DC surges in most of Europe have been fairly restrained, with annual population growth in the more industrialised northern nations (see UNITED KINGDOM, GERMANY) around 1% to 1.5% in the 1800s and less than 1% in the 1900s. In southern Europe where industrialisation was more gradual, they have mostly been less than 1%. Developed nations outside Europe (e.g. AUSTRALIA, USA) experienced much faster growth, often 2% to 3% per year at first, thanks to the attraction of their vast empty spaces to migrants from the crowded countries. JAPAN, with TAIWAN, South KOREA and SINGAPORE the developed nations of Asia, industrialised late but fast. Most of western Europe experienced low to very low natural growth rates in the late 20th century, but they were offset by increasing immigration from developing nations (section 5.2).

It is arguable that the main aggressor nations in World War Two (essentially a war of the developed world) were driven by population pressure to attempt territorial conquest. If the factors: population size (potential strength) and population density (potential overcrowding) in

ENGLAND: LONG TERM POPULATION CHANGE

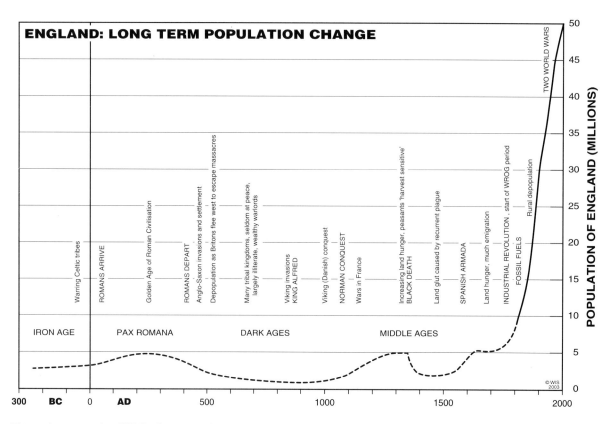

This graph is precise after 1801 (the first census of a series taken every tenth year except 1941) but becomes less precise as it is traced backwards in time. That admitted, it is still accurate enough to be meaningful. It follows the evolution of the world's first industrial nation from the end of its prehistory to its present position as a leading Western power. Population changes in England, and their causes, have great relevance to other nations worldwide.

Most of the graphs in this book begin during or just before a nation's DC population surge (section 4.3) that followed the introduction of improved farming and health-care techniques. This graph for England, which would extend back over 8 pages at the standard time-scale, illustrates the main types of change that probably affected all countries during the ages of low population density before their DC surges.

Roman invaders in 43 AD found England inhabited by Celtic Iron Age tribes with no written language and usually at war with each other. The Romans imposed a peaceful ordered society, initiating a weak DC surge. Population rose to near 5 million, which was England's carrying capacity at that time, the maximum number that Romano-British farmers could feed. In the 4th century AD Rome's declining power coincided with incursions by barbarians from northern Europe. When the Romans finally departed in 409, waves of these Angle and Saxon invaders arrived on the east coast and drove westward. Genetic studies show that they 'cleansed' England of most of its native Celts. By slaughter and displacement over 2 centuries the population was reduced to little more than a million. England's 'Dark Ages' had begun.

England was now ruled by tribal warlords who, when not disposing of the Celts, fought each other to gain land and loot. Population remained low as farmers went to war and crops were plundered. By the 8th century the dozen or so kingdoms, influenced now by Christianity, were beginning to merge. A more peaceful society might have developed had not new waves of piratical invaders, the Danes or Vikings, landed on England's shores. Their armies roamed the country, harassing the Anglo-Saxons and settling conquered territory. Eventually, in 1016, the Dane Cnut, or Canute, became king of all England. The population, still mainly Anglo-Saxon, was not 'cleansed', and began to rise as it enjoyed relative peace after 6 centuries of ethnic and tribal warfare.

Population was about 1.5 million when the Norman Conquest of 1066 initiated new wars, but they were mainly in France. Englishmen, who now had to be cherished as the providers of wealth to an organised nation, tilled the soil and multiplied. Marginal land, neglected since Roman times, was cultivated again as land-hunger developed. By 1300 population was back at 5 million, national carrying capacity, with further growth prevented by the starvation and disease that followed every bad harvest.

Out of Asia came a plague, the Black Death, which between 1348 and 1400 reduced the population to about 2 million. Major wars in France and at home (Wars of the Roses), and recurrences of plague, kept the population low until the mid-1500s. Then, with peace and stability restored, there was another surge. By 1650 population was back at carrying capacity, still 5 million. There was less starvation, because Britain now had colonies to which hungry people could emigrate.

In the mid-18th century, after 100 years at carrying capacity, England became the first country to experience the great population surge that has since spread to the whole world as the global "population explosion". Scientific and technical innovation, the Industrial Revolution, ended the recurring food shortages that, by starvation and malnutrition-related diseases, had kept population growth in check. Death Control (DC) was now possible, and would soon become the norm. Human life, previously so expendable, would soon be treated as sacred.

England's DC surge has continued unabated for two and a half centuries. The population is now increasing by one million in five years. Less than half of the increase is natural change, the excess of births over deaths, the remainder being the excess of immigrants over emigrants. The latter shows every sign of accelerating as the years pass, unless Draconian preventative measures are taken. In 2000, business interests were actively promoting immigration, alleging that England needs a young workforce to support the naturally ageing population, but opponents argued that with a population density of 382 England was already unbearably congested and the natural environment was deteriorating. With ethnic minorities comprising more than 7% of the English population, and increasing faster than the native majority, racial tension was endemic in many towns and cities.

Table 3.1 Population Growth, Main Sets of Nations, 1970–2000
(in millions, totals affected by rounding)

Nation sets (see Appendix 1)	Population increase 1970–80	Population increase 1980–90	Population increase 1990–2000	Population increase 1970–2000	% increase 1970–2000 and (rank)	% of world increase 1970–2000 and (rank)
Developed World incl. ex–USSR, excl. Muslim nations (52 nations)	91	73	51	215	20.5 (6)	8.8 (6)
South and Central America (25 nations)	77	76	85	238	86.9 (4)	9.8 (5)
Sub–Saharan Africa excl. Muslim nations (33 nations)	63	85	94	242	131.3 (1)	9.9 (4)
Central to South–east Asia, excl. Muslim and developed nations (13 nations)	191	213	233	637	88.5 (3)	26.1 (1)
East Asia excl. developed nations (3 nations)	152	160	162	474	57.0 (5)	19.4 (3)
All Muslim nations (49 nations)	172	208	254	635	111.0 (2)	26.0 (2)
Ex–USSR (non–Muslim) and East European Soviet satellites (15 nations, repeats from rows above)	24	20	– 7	37	12.5 (7)	1.5 (7)

Table 3.2 Changing Percentage Of World Population, Main Sets Of Nations, 1970–2000
(totals affected by rounding)

Nation sets (as in Table 3.1)	% of world population in 1970 (3632 million)	% of world population in 2000 (6074 million)
Developed World	28.9	20.8
South and Central America	7.5	8.4
Sub-Saharan Africa	5.1	7.0
West-central to South-east Asia	19.8	22.3
East Asia	22.9	21.5
Muslim	15.7	19.9
Ex-USSR and satellites	8.3	5.6

1940, and population growth rate (potential urgency) in the preceding decade, are multiplied together, an *aggressivity index* is obtained for the main warring nations as follows: *Japan 17, Germany 7, Italy 7, UK 5, Netherlands 3, Poland 2, Belgium 1, USA 1, USSR 1, France 0.* Unlike most of the other nations, Japan and Germany felt population pressure acutely because they had no large overseas territories to absorb emigrants. Their advanced technology enabled them to plan for major wars of conquest.

Tables 3.1 and 3.2 show how drastically population growth in the developed world (including part of the ex-communist block) has slowed, compared to any other set of nations, in both numerical and percentage terms. It has decelerated progressively in each of the last three decades. Currently, more than half of the annual increase is caused by immigration plus the high birth rate among recent immigrants.

ETHIOPIA   Abyssinia   Area 1,100,000 km² (excluding Eritrea)

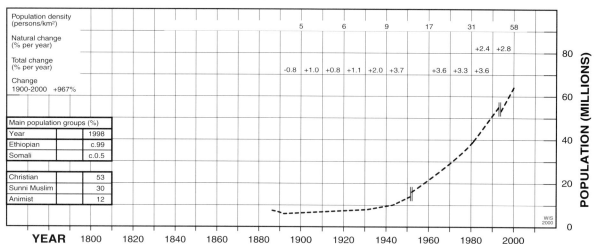

Ethiopia's history as a state goes back several millennia. The Queen of Sheba, who supposedly visited Israel around 1000 BC, may have been Ethiopian. The country was converted to Christianity in the 4th century AD, and Christianity co-existed more or less amicably with Islam after the 7th century, when Arabs occupied parts of Axum, a Christian kingdom in north Ethiopia. Deforestation impoverished the soil and Axum migrated south to become Abyssinia, which resisted Ottoman Muslim expansionism from the 16th century, disintegrated in the 18th, and was reunited under Emperor Menelik 2 from the 1880s. In spite of a harsh famine, 1888–92, that reduced the population by almost one third, Menelik fought off invasion by Italy in the 1890s and expanded Abyssinia southward at the expense of the Somalis. Italy returned and conquered Abyssinia in 1935 but was expelled in World War Two.

In 1952 Eritrea federated with Ethiopia and was forcibly annexed by Ethiopian emperor Haile Selassie in 1962, provoking 30

years of draining civil war which overflowed into Tigray province. Selassie's feudal regime was overthrown in 1974 and replaced by a chaotic military dictatorship which, thanks to support from the USSR and Cuba, was able to repel attacks by Somalia and various rebel groups. Droughts and famines in the 1980s (the great famine of 1984 caused up to a million deaths) further weakened the regime until in 1991, following withdrawal of Soviet support, it gave way to the rebels. Eritrea managed to secede in 1993. The new multiparty government had not repaired the shattered economy (still largely dependent on peasant farming) when it fought a savage frontier war with Eritrea (1999–2000), coinciding with a new drought and famine. Unless Ethiopia can escape from the combination of droughts, famines, destructive wars and rapid population growth, a poverty-stricken future is assured. Now, even when harvests are good, 5 million Ethiopians need food aid.

There are some 70 ethnic groups and sub-groups. About 5% of the population was HIV-positive in 2000.

3.2   South America

South American nations, European colonies for several centuries and still retaining a significant European element in their populations, began their DC surges towards the end of the 19th or early in the 20th centuries. As a general rule, the more European the population (e.g. URUGUAY), the earlier the DC surge began. Once initiated, the surge was fast, 2% to 4% per year. In most cases the last few decades of the 20th century show a slackening in percentage growth, which does not prevent numerical growth continuing unabated (see Table 3.1, which includes Central America). As in most of the New World, population densities (PDs) in South America are still very low compared to the Old World, the highest being ECUADOR (PD 46). In consequence, few if any wars have been caused by population pressure. In COLOMBIA (PD 34) the chronic civil strife is generated by warlords who profit from drug trafficking. Poverty is worsening in almost all nations as population growth outpaces economic growth (the Micawberish Rule, section 6.4).

3.3   Sub-Saharan Africa

The DC surge arrived earlier in southern Africa than in the tropics, thanks to earlier and more intense European settlement. In SOUTH AFRICA it began before 1900; in the Rhodesias (ZAMBIA and ZIMBABWE) and Portuguese colonies (ANGOLA and MOZAMBIQUE) and KENYA around 1910–1920, but in most of Central and West Africa it was delayed until 1930–50, even as late as 1960 in LIBERIA. NIGERIA, GHANA and

FRANCE with Alsace-Lorraine. Area 549,000 km²

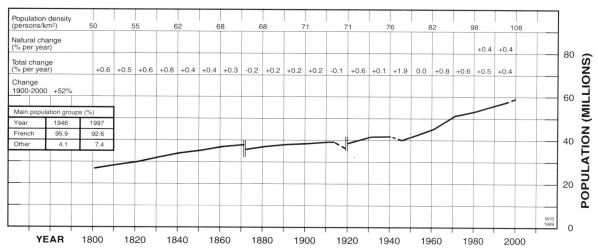

Population density (persons/km²)	50	55	62	68	68	71	71	76	82	98	108
Natural change (% per year)										+0.4	+0.4

Total change (% per year): +0.6 +0.5 +0.6 +0.6 +0.4 +0.4 +0.3 -0.2 +0.2 +0.2 +0.2 -0.1 +0.6 +0.1 +1.9 0.0 +0.8 +0.6 +0.5 +0.4

Change 1900-2000 +52%

Main population groups (%)		
Year	1946	1997
French	95.9	92.6
Other	4.1	7.4

POPULATION (MILLIONS): 80, 60, 40, 20, 0

WIS 1999

YEAR 1800 1820 1840 1860 1880 1900 1920 1940 1960 1980 2000

Cave paintings from 35,000 to 20,000 years old, stone monuments 14,000 years younger, and intricate Iron Age jewellery, tell of continuous occupation of France through recent prehistory. Romans conquered it (Gaul) in the 1st century BC and introduced Christianity in the 4th century AD, but were driven out by Germanic tribes in the 5th. Charlemagne briefly reunited France in the 8th and 9th centuries. Following William of Normandy's conquest of England in 1066 successive royal marriages perpetuated English claims to parts of France that were only extinguished by the Hundred Years War (1337–1453).

Until the 18th century France was at carrying capacity, its largely peasant population controlled by repeated famines and plagues (one third of the population died in the Black Death, 1348–50) until a DC surge began with the Industrial Revolution in the mid-18th century. In 1700 the population was 21 million (four times that of England); by 1790 it had risen to 28 million. The French Revolution of 1789 eliminated royalty, established a republic, and unleashed conquering French armies all over Europe under Napoleon Bonaparte. After the Napoleonic Wars which ended in 1815 at Waterloo the only significant changes in France's borders were the loss of Alsace-Lorraine after the Franco-Prussian War of 1870–71 and its recovery in 1919 after World War One. The loss of French population was considerable in both world wars.

France before 1950 could well have been cited as an example of the successful working-out of the 'demographic transition' theory, which proposes that when a nation becomes prosperous its birth rate naturally falls and population growth ceases. From about 1870 to 1950 the birth rate was lower than in most European nations, but an unexpected 'baby boom' began in the late 1940s. It has been maintained by the large numbers of immigrants from North Africa. In 2002, 8% of the population was Muslim. Tension between the Jewish and Muslim minorities, heightened by the Israeli-Palestinian *intifada*, escalated into frequent attacks on synagogues and Jewish-owned property.

Censuses: every 5 to 10 years after 1860; at longer intervals before 1860. About 220,000 abortions are performed each year. France lacks large reserves of fossil fuels; in 1996 most of its electricity was generated in 54 nuclear power stations.

CAMEROON, in 1910–20, were exceptions probably linked to British administration. Before the surge proper there seems to have been a gentle population rise, perhaps the effect of colonially-imposed peace preceding the more significant death control brought about by improved agriculture and health care. (In Angola in the 1950s the entire population, black and white, of tsetse fly country was regularly inoculated against sleeping sickness by mobile 'pentamadina' brigades.)

Since the DC surge began in sub-Saharan nations they have typically continued to grow at 2% to 5%, rarely 6%, per year. This is faster than in any other set of nations (Table 3.1), but its impact on growth worldwide is quite small, because sub-Saharan Africa's total population (excluding Muslim nations) is the smallest of any set. In many nations percentage growth has slackened in recent decades, but the graphs, which represent numerical growth, have continued to rise steeply. Ominously, in the 1990s, four nations experienced such catastrophic setbacks that a jagged peak and trough appears in their graphs.

In two, BURUNDI (1993) and RWANDA (1994), the population reached VCL (Violent Cutback Level, section 4.2). Their population densities were extremely high, 220 and 290 respectively, the people were almost all peasant farmers, and rivalry for land was intense between two tribal groups. Eventually one group plotted a coup to destroy or drive out the other, and the outcome was a drastic population decrease. Both nations had overshot their VCLs. Intertribal massacres had begun 20 to 30 years before the final showdowns, but were

18 – 180 million

GERMANY Area 541,000 km² to 1914; 471,000 km² to 1939; 355,000 km² after 1945

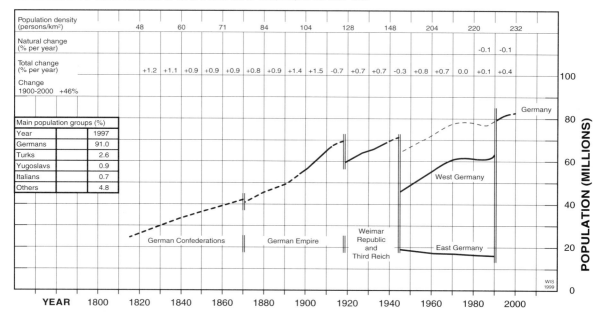

The Germanic tribes of Middle Europe repelled Roman armies early in the first millennium AD and turned the tables in the 5th century by sacking Rome itself. Charlemagne united them by conquest in the 8th and 9th centuries to form part of his Holy Roman Empire, within which many individual German princedoms flourished and became semi-independent. For centuries the princes formed and broke alliances and fought civil wars. After the Thirty Years War (1618–48) Prussia became the dominant kingdom. In 1815, following the Napoleonic Wars, 39 states joined in the German Confederation, which fought internal wars before defeating France in 1871.

Bismark, prime minister of Prussia, masterminded the birth of the German Empire in 1871, taking Alsace-Lorraine from France. Austria stayed independent, so the Empire began with a small population fall. Loss of territory after both world wars caused further falls. The division of Germany into Russian and Western occupation zones in 1945 was formalised in 1949 by the creation of West Germany (Federal Republic) and East Germany (Democratic Republic). Population growth in the former during its 'economic miracle' was augmented by influxes of foreign workers, mainly Turks but partly from the relatively stagnant communist regime in the east until the communists built the Berlin Wall to stop emigration in 1961. Population stability was achieved in the 1970s and 80s with, in the West, immigration making up for the low birth rate. Reunification in 1990 triggered new waves of immigration which, as the natural and total change percentages show, were the sole cause of rapid population growth in the 1990s.

Of the major European nations, Germany's population growth rate was highest in the decades preceding World Wars One and Two (section 3.1). Only the UK approached the German growth rate, but British population pressure had a convenient safety valve: emigration to Canada, Australia and elsewhere in the British Empire. The Nazi leader Adolf Hitler exploited the German longing for *lebensraum* (living space) in planning World War Two (1939–45), which began with Nazi Germany's attempt to enslave Europe. The Nazis exterminated 6 million Jews, and other races and groups regarded as inferior, in the 'Holocaust', by direct murder or by working them to death in 'concentration camps' like Auschwitz and Birkenau.

quashed by external powers including the UN. When the rival groups became strong enough, the peacekeepers were brushed aside. Tribal violence is now ongoing in both nations, restricting population growth, but large-scale massacres have been inhibited, so far, by more peacekeepers.

In LIBERIA (1990) and SIERRA LEONE (1997) the conflicts are less readily explained. Population density is low and there is land to spare. The Liberian civil war began as a struggle for power between rival tribes, which broadened until law and order broke down and warlords seized control of their own regions. When the conflict crossed the border into Sierra Leone the diamond mines were targeted, and again warlords seized power and wealth for themselves. Both nations seem to have regressed into tribalism. Whether their populations have reached VCL will be determined by whether they increase. If they do not, due to unending civil conflict, they are by definition at VCL.

The human race has been evolving in Africa for several million years, but the sub-Saharan population remained low until the 20th century AD. Endemic hunger and disease, tribal warfare and in-tribe killings (dictated by the chief or the witch-doctor to serve their ends) suppressed growth. The centuries of European and Arab slavery may have marginally

GHANA Gold Coast Area 238,500 km²

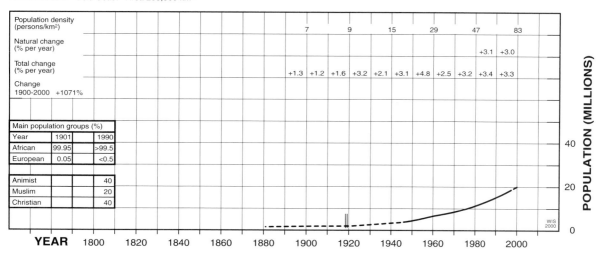

Population density (persons/km²)						7	9	15	29	47	83	
Natural change (% per year)										+3.1	+3.0	
Total change (% per year)						+1.3	+1.2	+1.6	+3.2	+2.1	+3.1	+4.8	+2.5	+3.2	+3.4	+3.3
Change 1900-2000 +1071%												

Main population groups (%)		
Year	1901	1990
African	99.95	>99.5
European	0.05	<0.5
Animist		40
Muslim		20
Christian		40

POPULATION (MILLIONS)
40
20
0

YEAR 1800 1820 1840 1860 1880 1900 1920 1940 1960 1980 2000

WIS 2000

The coastal region of Ghana was sparsely populated when Portuguese navigators reached it in 1472, but the profitable trade with European merchants seeking gold and slaves soon fostered the growth of rich Fante (coastal) and Ashanti (inland) kingdoms. Britain annexed the coast as a colony in 1874 and, having subdued the Ashanti, declared a protectorate over the whole Gold Coast in 1901. British Togoland joined it administratively in 1919.

The economy prospered, based on cocoa and gold. Independence came in 1957 and the country was renamed Ghana (after an ancient empire far to the northwest). Kwame Nkrumah's promising presidency became corrupt and was ended by a coup in 1966. Further coups severely damaged the economy until 1979 when Jerry Rawlings seized power. The country is now enjoying stability and some economic recovery, but still depends on foreign aid. Most of the people are peasant farmers, and problems of deforestation and soil erosion are widespread. There are about 75 ethnic groups, mostly Muslim in the north. In 2000, women were still being tested for witchcraft, and punished if found guilty. Censuses: every 10–20 years from 1948.

18 – 180 million

lowered populations, but the unfortunates who were exported were often prisoners of tribal wars or fetish victims who would not have survived otherwise (section 8.6).

European colonisation, the 'scramble for Africa', took off in the later 19th century. The colonial powers quickly enforced tribal peace, but the agro-medical revolution that began the DC surges arrived several decades later. In the 1950s and 1960s the whites were forced to grant independence to the increasingly powerful black populations (many of which have now increased nearly ten-fold in less than a century, section 6.6). Before long, unprecedented population pressures provoked violent ethnic and/or tribal competition for natural resources such as fertile land in Burundi, Rwanda and Somalia, or oil and/or diamonds in ANGOLA and Sierra Leone. This reversion to old-fashioned tribalism was probably inevitable, given the outpacing of economic development by population growth, accentuating poverty and disorganisation (section 6.4). On the other hand the HIV/AIDS pandemic, which struck about 1980, has drastically reduced life expectancy in southern and eastern Africa (Section 4.7). This has yet to show up in the graphs (but recent population estimates for many sub-Saharan nations must be suspect, as in SOMALIA).

3.4 West-central to South-east Asia

The Hindu, Muslim and Buddhist nations of this region, from Kazakhstan through India to Indonesia, are experiencing population growth almost as fast as sub-Saharan Africa (Table 3.1). After centuries and millennia during which Asian civilizations were the most advanced in the world, populations were low by modern standards until some degree of death control became possible. This was through contact with Europeans or the spin-off from their DC technologies, even before they established their colonies and enforced internal peace. The populations of most countries were already growing when they were first estimated, but the main DC surges, marked by a pronounced acceleration in the annual growth rate, took off at various times from about 1900 to as late as 1950 (INDONESIA).

IRAN   Persia   Area 1,648,000 km²

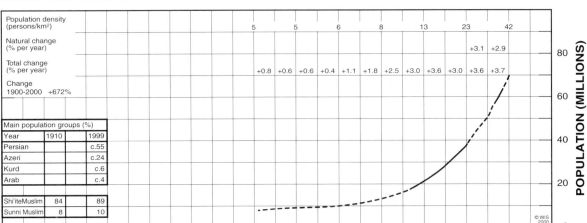

Population density (persons/km²)						5	5	6	8	13	23	42		
Natural change (% per year)											+3.1	+2.9		
Total change (% per year)						+0.8	+0.6	+0.6	+0.4	+1.1	+1.8	+2.5	+3.0	+3.6	+3.0	+3.6	+3.7
Change 1900-2000 +672%														

Main population groups (%)		
Year	1910	1999
Persian		c.55
Azeri		c.24
Kurd		c.6
Arab		c.4
Shi'iteMuslim	84	89
Sunni Muslim	8	10

YEAR   1800   1820   1840   1860   1880   1900   1920   1940   1960   1980   2000

POPULATION (MILLIONS): 0 – 20 – 40 – 60 – 80

© WIS 2000

Western Iran was part of the 'fertile crescent' extending through Syria and down to Palestine in which population pressure forced the transition from hunter-gathering to agriculture, some 10,000 years ago. Aryans, ancestors of the Persians, came from the north. Villages developed, then city-states and minor empires like that of the Medes, but the first great Persian empire was founded by Cyrus the Great in the 6th century BC, extending from Egypt to India at its peak. It declined after the emperor Xerxes' defeat by the Greeks at Marathon in 490 BC. A Sassanian Persian empire arose which held off the Romans but was overwhelmed by Arab armies spreading Islam in the 640s AD. Persian culture was emerging from 600 years of Arab domination when Genghis Khan's Mongols conquered and ravaged the country early in the 13th century. Mongol rule lasted for 2 centuries. In 1501 Sheikh Ismail took power and proclaimed himself the first Shah of Persia, now a Shi'ite Muslim nation. The Shahs were constantly at war with their neighbours, but could not prevent the growth of Russian (north) and British (south) influence in the 19th century.

The discovery of huge oilfields about 1900 intensified Great Power rivalry. Britain began their exploitation and defended them against the Turks in World War One. Persia became Iran in 1935.

Again in World War Two Britain and Russia occupied Iran to protect the oilfields. The oil industry was nationalised in 1951 and the Shah, continuing the trend to Westernisation begun between the wars, presided over growing prosperity. Resentment of his dictatorial and corrupt rule was fanned by Muslim fundamentalists to ignite civil war. He fled in 1979 and Iran abruptly became an Islamic republic, led by the fundamentalist Ayatollah Khomeini, who outlawed Western ways. Political opposition was ruthlessly suppressed, the economy weakened, and a war with Iraq, 1980–90, cost nearly a million lives. Moves towards a more liberal regime were blocked by clerics in 2000.

Iran is host to more than 2 million refugees, mostly from Afghanistan and Iraq. The country still has, in its oil and gas reserves, the potential for several decades of prosperity. The fast population growth that had begun as an oil-based DC surge in the 1920s was at first encouraged by the Islamic state, but the policy was reversed in the 1990s when the government, realising the difficulty of hoisting such a large mainly backward population out of poverty, began providing free family planning services for all. Even so, with 60% of the population younger than 20, a big increase is inevitable.

The region includes nations with enormous populations and high population densities. INDIA alone has 1.4 times as many people as the whole of Europe (including European Russia) and is six times as densely populated. BANGLADESH is the world's most densely populated major nation (PD 865 in 2000), so crowded that millions of people have to inhabit flood plains and coastal lowlands liable to catastrophic inundations, when tens or hundreds of thousands die. Their populations, like those of the other great, fast-growing and densely populated nations including PAKISTAN, the PHILIPPINES, VIETNAM and INDONESIA (the world's fourth largest population) have either tripled or quadrupled since they expelled their colonial masters soon after World War Two. Pakistan's population has nearly quintupled. Continued growth at this rate is likely to bring VCL devastation to parts of the region and, perhaps, to the planet.

Most of the medium-sized nations (e.g. THAILAND, MYANMAR) and the smaller ones (e.g. NEPAL, MALAYSIA) have grown as fast or nearly so. A few of their governments have begun to encourage small families, as did SINGAPORE until the late 1990s when it decided (with a PD of 6100) that larger ones were desirable. The mainly Sunni Muslim ex-Soviet nations of Central Asia have significantly large numbers of Russian and other east European settlers, but only KYRGYZSTAN, with the most (27%) non-Muslims, has a growth rate less than 2% per year.

IRAQ Mesopotamia Area 438,300 km²

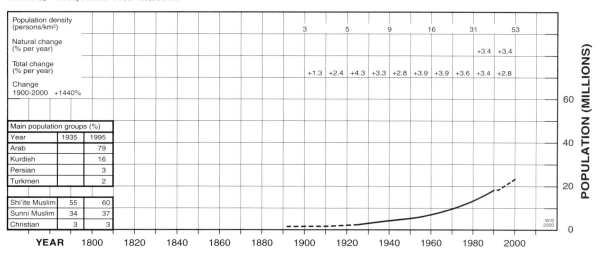

Population density (persons/km²)						3	5	9	16	31	53
Natural change (% per year)										+3.4	+3.4
Total change (% per year)						+1.3 +2.4 +4.3 +3.3 +2.8 +3.9 +3.9 +3.6 +3.4 +2.8					
Change 1900-2000 +1440%											

Main population groups (%)

Year	1935	1995
Arab		79
Kurdish		16
Persian		3
Turkmen		2
Shi'ite Muslim	55	60
Sunni Muslim	34	37
Christian	3	3

YEAR 1800 1820 1840 1860 1880 1900 1920 1940 1960 1980 2000

POPULATION (MILLIONS)

The Garden of Eden, where Adam dwelt with Eve, was supposedly located in southern Iraq at the confluence of the Tigris and Euphrates rivers. It was in that region, some 10,000 years ago in the 'fertile crescent' of ancient Mesopotamia, that humankind learned to farm. City-states arose, then the Sumerian, Assyrian and Babylonian empires, but the land was slowly degraded by intensive irrigation and much of it is now desert. Romans, Persians, Tatars and Mongols came and went before, in the 16th century, Mesopotamia was conquered by the Ottoman Turks. After World War One three Ottoman provinces, Kurdish, Sunni Muslim and Shi'ite Muslim, merged to create a poor but independent Arab kingdom, which was transformed, in the 1920s and 30s, by the discovery and exploitation of vast oil resources (only Saudi Arabia has more).

Iraq was developing into a rich advanced Arab society when, in 1958, King Faisal was assassinated. In the ensuing succession of military dictatorships Iraq managed to fight or otherwise alienate all her neighbours except Jordan. Power was seized by the Sunni Muslim ethnic minority, the Ba'ath Party led by Saddam Hussein, which savagely suppressed its ethnic and religious rivals, the Kurds in the north and the Shi'ite Muslims in the south, going so far as to physically drain the vast southern marshes in which the Shi'ite 'Marsh Arabs' lived. The war with Iran (1980–90) caused huge loss of life on both sides.

UN sanctions, imposed following the 1990–91 Gulf War in which Iraq invaded and briefly occupied Kuwait (losing about 50,000 soldiers and civilians), have ensured a much reduced quality of life for most of the people. Iraq is a once-prosperous nation brought low by ethnic and religious ambition; even so its population growth is extremely fast. The population fall, 1991–92, reflects the exodus of persecuted Kurds and Shi'ites to Turkey and Iran. There was intermittent in-fighting between rival Kurd groups in the 1990s. Iraq's growing population increasingly depends on water from the Tigris and Euphrates rivers, which enter the country from Turkey and Syria, states which also need more water. Dangerous shortages are developing. Capital punishment is commonplace; several hundred executions took place in 2000. Censuses: roughly every 10 years from 1935.

Unlike the developed world, which has seen no major conflict since World War Two and has prospered in consequence, most nations in the region are, or have recently been, impoverished by ethnic or religious violence. The dispute between Hindu India and Muslim Pakistan over Kashmir has flared up at intervals since partition in 1947. Muslim fundamentalists conquered most of AFGHANISTAN in the 1990s but were ejected in 2001, others fought a civil war in TAJIKISTAN in the 1990s, and yet others are attacking Christian and Chinese minorities in INDONESIA and the PHILIPPINES. Expansionist communism spawned guerrilla war in MALAYSIA (1948–60), a war of conquest in VIETNAM (1959–75), ideological genocide in CAMBODIA (1975–79) and several other conflicts.

The jagged peak and trough terminating a long population rise, that can be indicative of a population reaching Violent Cutback Level (VCL), appears in the graphs for AFGHANISTAN, TAJIKISTAN and EAST TIMOR. Afghanistan traditionally suffers from droughts, famines and civil wars, but the population cutback that began in 1979 with the Soviet invasion was largely due to the flight of several million refugees, many of whom have since returned. The graph continues to rise, but it seems likely that massive cutback by hunger or violence is being temporarily held at bay by peacekeepers and foreign aid. In Tajikistan the trough marks a temporary outflow of refugees from the civil war, but again, foreign aid is vital to combat drought and famine. The three troughs in the East Timor graph since 1941 correspond with interventions by foreign powers, respectively Japan, Indonesia and the

18 – 180 million

ITALY Area 301,300 km² after 1919.

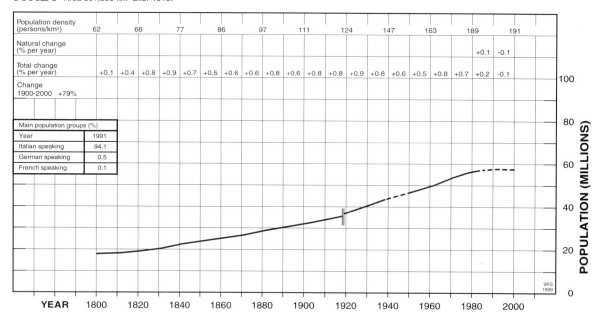

Population density (persons/km²)	62		66		77		86		97		111		124		147		163		189		191
Natural change (% per year)																		+0.1	-0.1	
Total change (% per year)	+0.1	+0.4	+0.8	+0.9	+0.7	+0.5	+0.6	+0.6	+0.8	+0.6	+0.8	+0.8	+0.9	+0.8	+0.6	+0.5	+0.8	+0.7	+0.2	-0.1
Change 1900-2000 +79%																				

Main population groups (%)	
Year	1991
Italian speaking	94.1
German speaking	0.5
French speaking	0.1

The Italian peninsula has given birth to great empires and cultures over two millennia. The fall of the Roman Empire, caused in part by deforestation, soil loss and desertification of Rome's North African grain-growing provinces, was followed by a succession of lesser empires and rich militant city-states. They spawned literary and artistic geniuses like Dante and Giotto, then by the 15th century Renaissance, artists like Botticelli, Leonardo da Vinci and Michelangelo. Modern united Italy dates from 1861, except for Istria and the Trentino which were gained from Austria after World War One and are still partly German-speaking.

The graph before 1861 relates to land approximately within the present frontiers. For nearly 2 centuries Italy's population grew slowly and steadily, European-style, until the 1980s when the graph levelled off and then declined slightly. Many Italians now view children as an economic burden, with women preferring work over having a family. The total fertility rate is well below replacement level. However, Italy is the target destination or way-station for great numbers of refugees and economic migrants from the Balkans, especially chaotic Albania and Kosovo, and Africa, desirous of settling in Europe. In 1997 there were 1.25 million legal and illegal immigrants resident or temporarily resident in Italy.

United Nations. Given independence and peace, this fertile island should be able to support an increasing population for a while.

HIV/AIDS is known to be taking a hold in some countries. More than 1% of the population of MYANMAR (Burma) was HIV-positive in 1999. 'Safe sex' practices were promoted in THAILAND throughout the 1990s, but the disease is widespread. In other nations, such as India, it is believed to be a serious problem that must be faced in the near future.

The region is an object lesson as regards the damage to national standards of living that ethnic or religious conflict, coupled with rapid population growth, causes. The resultant poverty and backwardness on a huge scale have held up economic development. People have banded together in rival ethnic or religious groups, assuming that groups have more power and influence than individuals, which has led to further conflicts and crises. It is a self-perpetuating cycle of ignorance and misery that persuades growing numbers of people to emigrate in search of a better life. Usually they carry their ethnic and religious prejudices with them, to the detriment of more relaxed host nations.

Table 3.1 shows that this set of nations, even excluding the Muslim ones, has contributed more to world population growth than any other during the last three decades. Its growth rate is only the third fastest in percentage terms, but the enormous population base, three-quarters of it in India, produces correspondingly large numerical growth.

JAPAN Area 377,800 km²

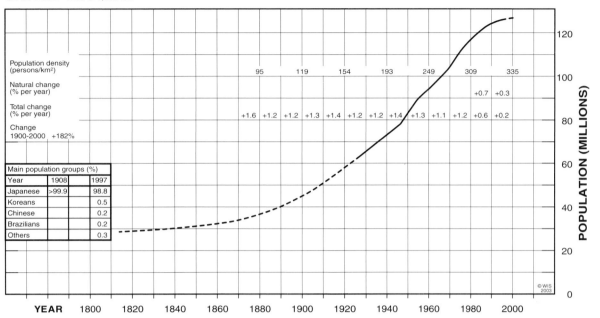

Population density (persons/km²)		95	119	154	193	249	309	335
Natural change (% per year)							+0.7	+0.3
Total change (% per year)	+1.6 +1.2 +1.2 +1.3 +1.4 +1.2 +1.2 +1.4	+1.3	+1.1	+1.2	+0.6	+0.2		
Change 1900-2000 +182%								

Main population groups (%)

Year	1908	1997
Japanese	>99.9	98.8
Koreans		0.5
Chinese		0.2
Brazilians		0.2
Others		0.3

© WIS 2003

POPULATION (MILLIONS)

120 — 100 — 80 — 60 — 40 — 20 — 0

YEAR 1800 1820 1840 1860 1880 1900 1920 1940 1960 1980 2000

The Japanese archipelago of 4 large islands and about 1000 smaller ones is mountainous, with volcanoes and frequent earthquakes. Early in the first millennium AD one of the many tribes living in the archipelago, the Yamato, subdued the others, establishing an empire from which the present ruling family is descended. Buddhism became the state religion in the 7th century, combining with traditional Shintoism. Military power was exercised by Shogun warlords who (helped by a typhoon) repelled Mongol invasions in the 13th century. Until the 19th century Japan developed in isolation, with borders largely closed, but not without chronic internal strife between local lords and religious movements including Christians.

In 1853 a US naval force demanded the relaxation of trade barriers. Japan quickly Westernised and became a military power, driving the Chinese from Korea in 1895, the Russians from Manchuria and Korea in 1905, and annexing Korea in 1910. Between the world wars Japan occupied Manchuria as well as Pacific islands that had been German. In World War Two Japan conquered Southeast Asia and many Pacific islands, but was retreating and fighting bloody defensive battles when the war was suddenly ended by the atomic destruction of Hiroshima and Nagasaki. Occupied by the US, and pledging itself to peace, post-war Japan concentrated on industry and commerce, achieving a dominant position in world trade until the 1990s when mounting financial problems culminated in a deep recession. In 2002 Japan had the world's highest level of public debt: 140% of GDP. Japan is blamed by other nations for excessive over-fishing in the Pacific, in the name of 'scientific research'.

Japan's population is concentrated in a narrow coastal strip between mountains and sea, making it particularly vulnerable to global heating and its consequence, sea level rise. There are advantages, therefore, in the fact that the birth rate has drastically reduced in recent decades. The average Japanese woman now has only 1.4 babies. Business and government view the low birth rate as a crisis. Until 1999, when the contraceptive pill was legally approved, abortion was the main practice controlling births. Censuses: frequent from 1930.

3.5 East Asia

CHINA, with its millennia of advanced civilisation, had a uniquely large population (c. 450 million) when it was first credibly estimated in the 1930s. The DC surge began around 1900 and it accelerated until the 1960s when the population was increasing by some 150 million (three Englands) per decade. Marxist Chinese leaders initially believed that socialism would solve overpopulation problems, but by the late 1970s the threat to China's future prosperity was so evident that they introduced the Draconian 'one child per family' policy. Since then the percentage growth rate has fallen but numerical growth per decade has hardly changed.

Although the majority of Chinese are peasant farmers, the national talent for commerce has nurtured thriving industrial cities, especially near the south-eastern coasts where Western influence is strong (e.g. HONG KONG). This ability has been exploited in neighbouring countries which have escaped from Chinese communist rule or influence. TAIWAN, industrialised after World War Two, enjoys great prosperity. Its population density has risen to 605, the world's second highest for a major nation, and its growth rate is falling numerically as well as percentage-wise.

KENYA   Area 571,400 km² (excluding lakes)

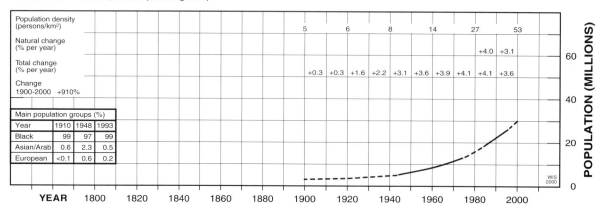

	5	6	8	14	27	53
Population density (persons/km²)						
Natural change (% per year)					+4.0	+3.1
Total change (% per year)	+0.3 +0.3 +1.6 +2.2 +3.1 +3.6 +3.9 +4.1 +4.1 +3.6					
Change 1900-2000 +910%						

Main population groups (%)			
Year	1910	1948	1993
Black	99	97	99
Asian/Arab	0.6	2.3	0.5
European	<0.1	0.6	0.2

Some of the world's first hominids lived in Kenya 2.5 million years ago. Their fossil remains, found near Lake Turkana, helped formulate the belief that humankind originated in Africa. Within the last 5000 years Cushitic and Nilotic pastoralists (cattle-herders) arrived in Kenya from the north and Bantu farmers came from the west. Tribal disputes, often violent, for land and dominance, have continued ever since. Arab traders and slavers settled along the coast, introducing Islam, from the 8th century. Portuguese adventurers conquered the coastal strip in the 1520s and ran a harsh colonial regime for more than a century, until British, Dutch and Arab merchants replaced them. Britain established an East African Protectorate in 1895. White settlers displaced Nilotic Masai and Bantu Kikuyu from their highland territories in central Kenya, triggering resentment that culminated in the Kikuyu-led Mau Mau uprising (1952–60). Independence followed in 1963. Stable Kikuyu-run governments under Jomo Kenyatta and Daniel arap Moi bought out European farms to resettle the booming black population. However, the first multi-party election in 1992 unleashed serious ethnic violence between Kikuyus (22% of the population) and the other tribal groups. Tension continues.

Although development has been mainly peaceful, local tribal fighting causes many deaths. The Kikuyu tribe has maintained its power by dubious means and the economy, dominated by agriculture and wildlife tourism, has deteriorated as population growth outstrips national productivity. In 2000, after a 3-year drought, 4 million Masai were surviving on food handouts. A hopeful sign is that the 3.1% average natural population growth per year in the 1990s conceals a rapid fall to 2% in 1998 and 99, but there is growing immigration from strife-torn neighbour states. If government figures suggesting that 25% of Kenyans were HIV-positive in 2000 are correct, population will soon peak and then decline. In 2001 the International Monetary Fund suspended aid to Kenya because of widespread corruption.

SOUTH KOREA's development mirrors Taiwan's, with a very high population density of 472 and a slightly declining numerical growth rate, but the population of NORTH KOREA, which remained communist, under an eccentric dictator, and concentrated on military strength rather than commerce, grew even faster in percentage terms until the 1990s. Then, the collectivised agriculture was affected by drought and began a disastrous decline, and famine halted population growth. MONGOLIA has remained pastoral, with a late DC surge that is still advancing, made possible by foreign aid.

JAPAN, which has always been culturally distinct from China, was the first East Asian country to industrialise, achieving great prosperity in the late 20th century when the low fertility rate and minimal immigration caused a slump in population growth.

Although this set of nations (even when Japan, Taiwan and South Korea are removed to the developed world) has the world's second largest population, its percentage growth rate is the lowest, apart from the developed world (Table 3.1). Uniquely, except for the developed world, its numerical growth rate is hardly increasing.

3.6 South-west Asia and North Africa

The nations of this region, from Iran to Mauritania, are with few exceptions Arab and Muslim, with populations growing as fast as, or faster than (2% to 6% per year) any other group. The exceptions are ISRAEL (mostly neither Arab nor Muslim), LEBANON (mostly Arab but 38% Christian), TURKEY and IRAN (Muslim but non-Arab), and Sahelian nations from MAURITANIA to SUDAN, which are largely or wholly Muslim but ethnically part- or non-Arab. In the 7th and 8th centuries AD the Arabs poured out of their homeland, the Arabian peninsula, in a prolonged campaign of conquest to spread the Muslim faith, which the prophet Muhammad had just founded. Much of the Arab empire fell to the Ottoman

KOREA   Corea   Areas: North Korea 122,800 km², South Korea 99,400 km², Total 222,200 km²

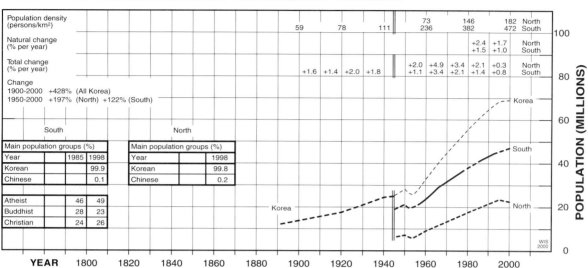

Population density (persons/km²)		59	78	111	73 / 236	146 / 382	182 / 472	North / South		
Natural change (% per year)						+2.4 / +1.5	+1.7 / +1.0	North / South		
Total change (% per year)		+1.6	+1.4 +2.0 +1.8		+2.0 / +1.1	+4.9 / +3.4	+3.4 / +2.1	+2.1 / +1.4	+0.3 / +0.8	North / South

Change
1900-2000 +428% (All Korea)
1950-2000 +197% (North) +122% (South)

Main population groups (%)		
Year	1985	1998
Korean		99.9
Chinese		0.1
Atheist	46	49
Buddhist	28	23
Christian	24	26

Main population groups (%)	
Year	1998
Korean	99.8
Chinese	0.2

The Korean nation was founded in 2333 BC, according to tradition, but the region was home to many tribal dynasties until early in the first millennium AD when 3 kingdoms competed for supremacy. In the 7th century the Buddhist Silla kingdom absorbed the others, unifying the peninsula, and was succeeded by the Koryo dynasty in the 9th century. Mongols invaded and devastated the country in 1231 but were driven out in 1392 with support from the Chinese Ming dynasty, which replaced Buddhism by Confucianism. Japanese invaders in the 1590s destroyed most of Korea's temples and palaces. Manchu Chinese were the next overlords, and Korea closed its borders to foreigners until Japan, following its defeat of China and Russia, annexed it in 1910 and treated it as a colony.

After World War Two, Korea was divided into a Soviet-backed north and a US-backed south. The Korean War (1950–53) was an attempt by the North to conquer the South, which was rescued by United Nations forces. Subsequently the South developed into a powerful densely populated industrial nation, its security guaranteed by the US in face of threatened attack by the eccentric closed Marxist society of the North, which developed its armed forces at the expense of its people. Since 1995 the North has suffered from severe famines exacerbated by floods and droughts and only foreign aid prevents mass starvation (in 1999 it denied allegations by the South that 3 million had died in recent famines). Thousands of North Koreans migrate illegally into China. The ethnic homogeneity of both Koreas is notable. In the South, up to 20,000 female babies are aborted each year.

Turks, also Muslim, in the 16th century. The Arab provinces of the Ottoman Empire mostly lay quiescent until they were liberated by the defeat of Turkey in World War One.

The region was always in touch with developments in Europe, and populations were already rising when the first estimates were made around 1900, but the main DC surge in most nations began in the mid-20th century with the new agricultural and medical technologies. Some populations, mainly along the Gulf (e.g. KUWAIT, UNITED ARAB EMIRATES) expanded astonishingly fast (up to 33% per year) as foreign workers came in following the discovery of productive oilfields.

Recent decades have seen a resumption of the messianic fervour to spread the Muslim faith, or particular versions of it, that characterised the 7th and 8th centuries. (Similar religious zealotry pervaded Christian nations in medieval times, driving the Crusades, the Reformation and the Spanish Inquisition.) Oil wealth allied with messianic zeal, fast population growth and Arab solidarity make Muslim expansionism a potent and disruptive force. Most Muslim nations in the region have avoided prolonged violent conflict, but a few (e.g. IRAQ, YEMEN) have embarked on a series of internal and external wars that have kept their populations in poverty and distress.

The establishment of Israel in 1948, and its expansion in 1967, occupying lands that had been Arab for 12 centuries, including many localities traditionally holy to both Muslims and Jews, has kindled such intense racial hatred that both sides are intent on gaining short-term advantage by means that, unless checked, are bound to end in catastrophe. The Arab populations of Israel's occupied territories, WEST BANK and GAZA, are expanding as fast as possible ("our children are our weapons", section 4.4) and some Jewish sects are doing the

18 – 180 million

MALAYSIA   (Includes Singapore before 1965)   Area 329,760 km²

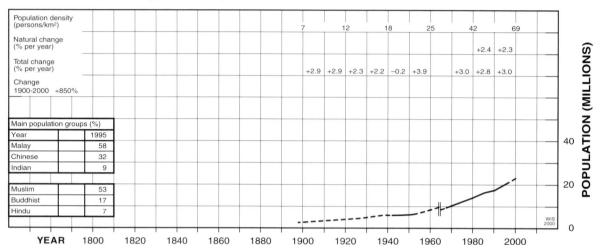

Population density (persons/km²)					7	12	18	25	42	69		
Natural change (% per year)									+2.4	+2.3		
Total change (% per year)					+2.9	+2.9	+2.3	+2.2	−0.2	+3.9		+3.0	+2.8	+3.0	
Change 1900-2000 +850%												

Main population groups (%)		
Year		1995
Malay		58
Chinese		32
Indian		9
Muslim		53
Buddhist		17
Hindu		7

POPULATION (MILLIONS) — 40, 20, 0

WIS 2000

YEAR   1800   1820   1840   1860   1880   1900   1920   1940   1960   1980   2000

Peninsular Malaya and Singapore were part of a Sumatra-based Buddhist empire from the 8th century AD to the 13th, when Siam conquered most of the peninsula. Arab traders brought Islam to the region when they established the port of Malacca about 1400. The first European traders were Portuguese and Dutch, but three British trading posts combined to form the Straits Settlements in 1826 and Britain declared protectorates over other peninsular Malay states between 1874 and 1914. Many thousand Chinese and Indians immigrated to work the tin mines and rubber plantations.

After Japanese occupation in World War Two, Britain united the states of Malaya as a colony and suppressed a communist guerrilla revolt (1948–60). The independent Federation of Malaysia was created in 1963 by the union of Malaya and Singapore with the sparsely populated North Borneo regions of Sabah (a British protectorate since 1888) and oil-rich Sarawak (which had belonged to the British Brooke family from 1840 to 1946). Singapore seceded in 1965. Malaysia suppressed an Indonesian-backed 'confrontation' (1963–66) and, with Malays dominant in government and Chinese in business, enjoyed growing prosperity until the temporary setback of the Asian recession in 1997, when tens of thousands of illegal immigrants, mostly Indonesian, were forcibly repatriated. Malaysia has become an important industrial nation that is also rich in oil, gas and minerals and in cash crops such as palm oil, cocoa and rubber. Rainforests cover more than half the land area but are being exploited unsustainably. Censuses: about every tenth year from 1970.

same, on the premise that an increasing population equates to increasing power. Muslim fertility in Palestine was about twice Jewish fertility as long ago as 1936–45 (Witherick, 1990, Table 2.1). Given the fanaticism and lack of forethought exhibited by both religious/ethnic factions, an immense disaster in the near future seems inevitable.

Table 3.1 shows how the Muslim world, i.e. the Muslim nations of this region combined with those of South-central and South-east Asia and parts of West and East Africa, was responsible for the second-fastest percentage growth rate, and the second-largest contribution to world population growth, between 1970 and 2000. Numerical growth, decade on decade, was increasing faster than any other. Most Muslim nations are already very densely populated, either absolutely (e.g BANGLADESH) or in relation to their agriculturally productive (non-desert) areas (e.g. EGYPT). It is not surprising, therefore, that extreme poverty is widespread and worsening fast (section 6.4, the *Micawberish Rule*), and that the lives of so many Muslims are dominated by resentment and aggression kindled by their self-generated plight. The *aggressivity index* (section 3.1) of some Muslim nations must be high, but, lacking the means to wage wars of conquest, their frustrations find outlets in emigration, terrorism and sabotage.

3.7 Central America

From giant Mexico to little Belize, the eight continental Central American nations share a common history with the large islands of the Caribbean: Cuba, Hispaniola (Haiti plus Dominican Republic), Jamaica and Puerto Rico. All were inhabited by Amerindians, some in advanced civilisations, but when colonised by Spain in the 15th and 16th centuries most, sometimes all (in the islands), of the indigenous population died of enslavement and disease. African slaves were imported to replace them. Mixed populations developed, except where

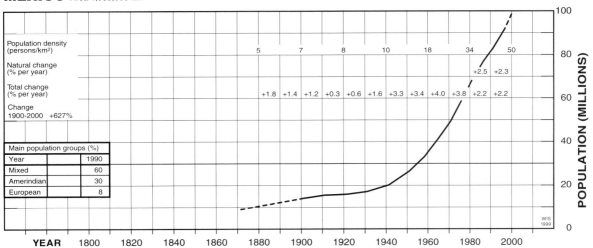

MEXICO Area 1,965,000 km²

Agriculture developed in south Mexico and adjacent countries about 2000 years later than its initiation in Mesopotamia. About 4000 years ago the productivity of small plots of ground growing peppers, gourds, beans and tomatoes became sufficient to support non-nomadic life in villages. Early forms of maize came 1000 years later. From these small beginnings there developed the great successive Olmec, Maya and Aztec civilisations of Central America, the last of which disintegrated when Cortes and his conquistadores arrived in 1519. The European diseases they carried reduced the indigenous population from about 20 million to about one million in less than a century.

After 3 centuries of Spanish rule Mexico became independent in 1821. Much territory in the north was lost in a disastrous war with the US (1846–48); it became California and New Mexico. There followed a long turbulent period of dictatorships, coups and peasant revolts. A population boom that began in the 1930s is

ongoing, causing widespread poverty as it outpaces the availability of new resources (much of northern Mexico is desert). The situation is alleviated to some extent by the exploitation of large offshore oil reserves discovered in 1976, and by rapid expansion of manufacturing and tourism. There is continuous, large-scale, mainly illegal emigration to the USA (in 2000 some 7 million Mexicans were living there) and an ongoing low-key revolt by 'Zapatista' Mayan Indians hungry for land in the southern state of Chiapas. In 2000 the Montes Azules biosphere reserve of tropical rainforest in the south was being invaded, deforested and settled by landless peasant farmers. 95% of the population is Christian, mostly Catholic, and 75% is living in poverty, compared to 49% in 1981.

Mexico City, the world's fourth largest city in 1995, with a population of 16 million, is so prone to atmospheric pollution that, at times, industrial activity and the use of private cars have to be drastically curtailed. Censuses: roughly every tenth year after 1900.

the colonists were expelled at an early date (see HAITI).

In the mainland nations a slow population rise began late in the 19th century, but the main DC surge was delayed until the 1940s or 1950s, except in PANAMA (1930) where US influence was strong thanks to American control of the Canal Zone. In the islands, which were more suited to settlement and development, and now have no Amerindian survivors, the DC surge began about 1880 and has continued ever since. In consequence their population densities are much higher than those of the mainland nations, with the notable exception of EL SALVADOR which has acknowledged its overpopulation problem since the 1970s.

The region's populations are still growing rapidly, except in communist CUBA, where growth has slowed markedly since the cessation of Soviet aid. In most nations population growth has far outstripped the growth of the national economy, causing severe poverty (sometimes exacerbated by natural disasters like Hurricane Mitch in HONDURAS), and a widespread determination to emigrate, especially to the USA.

3.8 The small islands: Oceania, the Caribbean, etc.

Small island nations tend to have a distinctive pattern of population growth that reflects their limited land area, lacking an empty hinterland to absorb a growing number of people. The initially rapid DC surge slackens its gradient when all the fertile land has been taken into cultivation. The annual population surplus must then emigrate, unless some natural resource (e.g. the phosphate rock of NAURU) or commercial activity (e.g. tourism in the MALDIVES, or industry in MALTA, or the military base in GUAM) earns enough foreign

18 – 180 million

MOROCCO (excluding Western Sahara) Area 458,700 km²

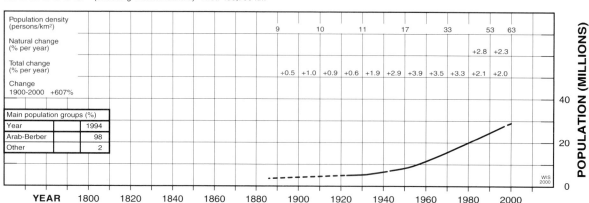

Population density (persons/km²)						9	10	11	17	33		53	63		
Natural change (% per year)												+2.8	+2.3		
Total change (% per year)						+0.5	+1.0	+0.9	+0.6	+1.9	+2.9	+3.9	+3.5	+3.3	+2.1	+2.0
Change 1900–2000 +607%															

Main population groups (%)		
Year		1994
Arab-Berber		98
Other		2

Since the first settlement of Morocco by Berber tribes who may have lived in the Sahara before it became a desert, the long succession of invaders has included Phoenicians, Romans, Vandals, Byzantines, Arabs, Portuguese, Spanish and French. The Arabs converted the population to Islam; the French and Spanish built modern cities and ports. The country was ruled by France and Spain jointly from 1904 until nationalist pressure, with violence, led to independence in 1957. The economy is mainly agricultural but tourism is growing in importance.

As in all North African Muslim states, population growth since World War Two has been explosive. In 1994, 55% of the population were under 21 years old, 40% were 'absolutely poor', and 70% were illiterate. Millions of young children go into industry or domestic service after minimal schooling. Deforestation followed by intensive agriculture to grow fruit and cereals is causing serious soil erosion and desertification. Not surprisingly there is great pressure to emigrate to Europe. The sea crossing between Morocco and Spain is so short that the north Moroccan coast has become a major embarkation point for illegal migration to Europe from all over Africa. Censuses: occasional from 1926.

exchange to support the excess population. Alternatively, an over-large population may survive on foreign aid (e.g. AMERICAN SAMOA).

The first European visitors to some Pacific islands estimated their populations and described their living conditions (in certain islands, e.g. HAWAII in 1778, SOLOMON ISLANDS c. 1790, FIJI in 1774, the population had reached VCL and was busy with communal slaughter) prior to the inevitable catastrophic decline that followed the arrival of Old World diseases and alcoholism. The Hawaii graph probably mirrors the population crashes triggered by Spanish colonisation of Central America nearly three centuries earlier. Pacific island populations seem to have slowly declined, after the initial crash, until DC surges began (from the late 19th to the mid-20th century, depending on the impact of Western technology).

The Polynesian inhabitants of tiny isolated Easter Island in the south Pacific "carried out for us the experiment of permitting unrestricted population growth, profligate use of resources, destruction of the environment and boundless confidence in their religion to take care of the future. The result was an ecological disaster leading to a population crash". Bahn & Flenley in *Easter Island, Earth Island* (1992) go on to argue that what happened to the Polynesians will be repeated on a vast scale on 'Earth Island' unless humankind learns 'the Lesson of Easter Island'.

Relatively detailed population records are available for a few island groups, like the AZORES or CAPE VERDE, that have a long history of European occupation. They show how, given the small total population, emigration in hard times can cause significant troughs and peaks in the graphs. The "hard times" can be weather-related, as in droughts, or may simply reflect the perception that a more agreeable life than the constant drudgery of subsistence farming may be available elsewhere. Some categories of natural disaster are severe enough to greatly reduce an island population, such as volcanic eruptions (e.g. MARTINIQUE, MONTSERRAT, ICELAND) or epidemic disease (e.g. FIJI). The Fiji graph exhibits a recent VCL-type peak and trough, the result of inter-ethnic violence in 1987 between aboriginal Fijians and the descendants of Indian plantation workers, which caused

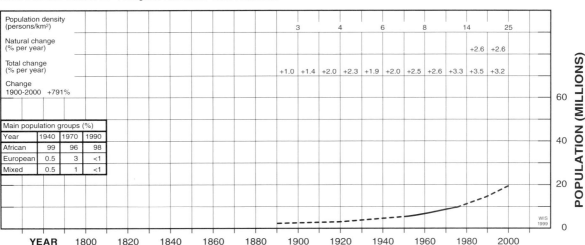

MOZAMBIQUE Portuguese East Africa Area 799,000 km²

Mozambique was settled by Bantu tribes from the north and west about 2000 years ago. They drove out or exterminated the hunter-gatherer Bushmen. Arab traders established coastal settlements in the centuries before the first European, the Portuguese navigator Vasco da Gama, arrived in 1498. Portugal displaced the Arabs and gradually subjugated the tribal kingdoms, developing an extensive slave trade which was outlawed in 1752. Mozambique was declared a Portuguese colony in the same year, but Portuguese control was not complete until the 1900s.

In 1951 the country became a Portuguese overseas province. Six decades of peaceful development ended in 1964 when Frelimo guerrillas began a war for independence. Portugal withdrew from all its African territories in 1975, when most Europeans left Mozambique and independence was achieved, with a Marxist government. Two years later the Renamo guerrillas, backed by South Africa, began a devastating civil war that lasted until 1992 when a peace treaty was signed soon after Marxism was rejected. During the civil war a million people were either killed or died in famines. Some 1.5 million refugees fled the country, but most had returned by 2000. In the late 1990s the largely agricultural economy was recovering, helped by foreign aid, but in 2000 floods of unprecedented severity caused a major setback, rendering a million people homeless. There are ten main black ethnic groups.

massive Indian emigration. Another anti-Indian coup in 2000 failed but showed that a potential VCL scenario exists, generated by ethnic rivalry in spite of the low population density.

In most islands, ongoing advances in agricultural technology are still slowly raising the basic carrying capacity, so even where there are no significant sources of foreign exchange, not all of the annual population increase is forced to emigrate.

3.9 Ex-communist Europe

Distinct from all other regions has been population change in republics of the ex – Soviet Union and in the nations of eastern Europe that were Soviet satellites. From World War Two until 1991 growth was steady and mostly slow, but the collapse of the Soviet monolith in that year triggered remarkable changes (Table 3.1). In the Slav nations of the north and west population growth suddenly ceased, and in many cases a decline began, particularly marked in Russia itself. By contrast, populations of the ex-USSR Muslim nations in the south continued their rapid expansion.

The nations of eastern Europe have been so remodelled over the years, by military conquest and political bargaining, that many of them cannot be recognised as geographical entities before World War One, or even, in some cases, World War Two. Some of them had previously existed as names (e.g. POLAND, HUNGARY), but their frontiers were so different to today's that there is no continuity in their population development.

Individual socialist republics within the Soviet Union were established in the years following World War One, and they persisted, with mostly minor border changes, up to and beyond the 1991 Soviet collapse. Internally, however, populations were subjected to massive changes by the policy of 'Russianisation': the deportation of native peoples and the

MYANMAR　Burma　Area 676,600 km²

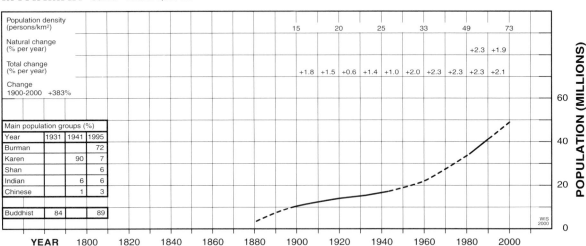

Population density (persons/km²)		15	20	25	33	49	73
Natural change (% per year)						+2.3	+1.9
Total change (% per year)	+1.8 +1.5 +0.6 +1.4 +1.0 +2.0	+2.3	+2.3	+2.3	+2.1		
Change 1900-2000　+383%							

Main population groups (%)

Year	1931	1941	1995
Burman			72
Karen		90	7
Shan			6
Indian		6	6
Chinese		1	3
Buddhist	84		89

POPULATION (MILLIONS): 60　40　20　0

YEAR　1800　1820　1840　1860　1880　1900　1920　1940　1960　1980　2000

WIS 2000

Myanmar (Burma until 1989) was largely covered by rain forest when Mongoloid Burman tribes moved into the mountainous north from Tibet and China in the early centuries AD. By 1060 King Anawrahta had consolidated the First Burmese Empire and was building a huge complex of Buddhist temples at his capital, Bagan. Kublai Khan's Mongol hordes sacked Bagan in 1287, and there followed several centuries of chronic warfare between tribal kingdoms. Britain fought the First Burma War (1824–26) to stop Burmese incursions into India, and after 2 more wars, 1852 and 1886, the whole of Burma was annexed to British India (causing the steep population rise, 1880–1900). Britain brought in Chinese and Indian labour to plunder the vast teak forests and hugely expand rice production for export. Dissident tribes constantly harassed the British, and since the Japanese occupation in World War Two and independence in 1948 they have never been subdued by the central governments. A military coup in 1962 gave birth to an oppressive socialist state run by the military which devastated the previously prosperous economy and has survived several coups and a failed election in 1989. The economy is still mainly agricultural, with rice the most important export. Timber and oil are important also. There are 8 main ethnic groups. 3.5% of adults were infected with HIV/AIDS in 2001. Censuses: every tenth year from 1901 to 1941, then occasional.

importation of tens or hundreds of thousands of Russians to replace them. In some cases Russians became a large ethnic minority (e.g. LATVIA, ESTONIA), a source of great friction after Soviet domination ended in 1991. In KAZAKHSTAN, Russian settlers actually outnumbered the native Kazakhs until the high Kazakh birth rate restored the traditional majority.

The boundaries of Soviet satellite nations were more ephemeral (e.g. POLAND, CZECH REPUBLIC AND SLOVAKIA, ROMANIA). YUGOSLAVIA affords a case-history of idealistic unification that ended in tears. Eight small nations or provinces merged in 1918 to form a Slav superstate (section 6.3), but economic inequalities and the disproportionate expansion of Muslim populations tested the multicultural regime to breaking-point, which was reached in the 1990s. Slav-Muslim rivalry causes conflict elsewhere, (as in Chechnya, whose population probably reached VCL in 1994), exacerbated by the contrast between Slav population decrease and Muslim growth.

This being a developed region, DC surges were already in progress when most graphs begin. Percentage growth has normally been low (around 1% per year or less, as in the rest of Europe) except in less developed and/or mainly Muslim nations where an extra surge reaching 2% to 3% per year followed World War Two (e.g. ARMENIA, ALBANIA, KOSOVO).

Why have so many ex-communist Slav nations experienced population declines since the 1991 Soviet collapse? In Russia it has coincided with a fall in life expectancy, attributed by some to excessive drinking and smoking. One may speculate that there is an emotional link with the failure of the ideology that guided national development for 70 years, and with the decline in the economy, and in national self-esteem, that has continued and worsened in the post-communist years. In 2002, however, it seemed possible that, with a new less corrupt regime in Russia, optimism was returning.

NEPAL Area 140,800 km²

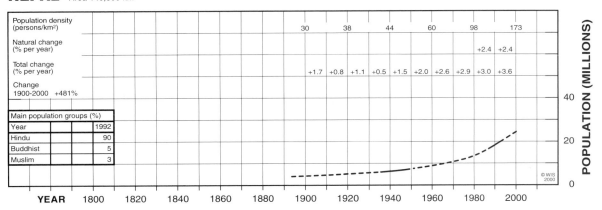

Population density (persons/km²)						30	38	44	60	98	173		
Natural change (% per year)										+2.4	+2.4		
Total change (% per year)						+1.7	+0.8	+1.1	+0.5	+1.5	+2.0	+2.6	+2.9	+3.0	+3.6
Change 1900-2000 +481%													

Main population groups (%)		
Year		1992
Hindu		90
Buddhist		5
Muslim		3

POPULATION (MILLIONS)
40
20
0

YEAR 1800 1820 1840 1860 1880 1900 1920 1940 1960 1980 2000

© WIS 2000

The South Himalayan kingdom of Nepal includes many of the world's highest mountains including Mount Everest. Over millennia it was settled by Caucasian Indian migrants from the south (including the Gurkhas) and Mongoloid Tibetans from the north (including Sherpas). Prince Siddhartha Gautama, the founder of Buddhism, was born in south Nepal in 563 BC and the country became largely Buddhist for a few centuries until Hinduism returned. From about 1200 AD the ruling Malla dynasty achieved great prosperity, the arts flourished and the frontiers expanded south to the river Ganges and east to include Sikkim, but internal conflicts between vassal city states and tribal groups were endemic and erupted into violence whenever Malla domination faltered. In 1768 the Gurkha Shah dynasty replaced the Mallas. Defeats by China in Tibet in 1792, and by British India in 1816, reduced Nepal to roughly its present size

The kingdom stayed independent throughout the colonial period by isolating itself from the outside world. Not until the 1950s did the frontier controls begin to relax (explaining the late and slow development of the DC population surge). A form of democratic government has developed characterised by sometimes violent party politics and considerable corruption. The economy, based on subsistence farming and supported by tourism, remains fragile, depending on foreign aid. The widespread extreme poverty is aggravated by rapid population growth which forces the subdivision of land holdings and the deforestation of mountainsides, resulting in soil erosion on a massive scale. More than 100,000 Nepalese were HIV-positive in 1998. A Maoist uprising began in 1996. By 2002 the rebels had killed c. 7000 people, controlling one third of the country.

18 – 180 million

Using these generalised histories as a basis, it is possible to project population change for perhaps a decade into the future, *assuming* that the driving forces remain more or less the same. But during that decade powerful opposing forces, that are being generated by excessive human pressure on Earth's ecosystems and natural resources, will come into play (see Chapter 6 and WORLD). They are already acting, ignored by the great majority of people and their rulers. They will exert a restrictive influence on nations in general, and a catastrophic influence on those nations which are already, for reasons such as unsustainably high population density, particularly vulnerable.

• • • • • • • • • • • • •

"The critical point in [world] population growth is approaching, if it has not already been reached."
(The Club of Rome, 1972)

NIGERIA Area 923,800 km²

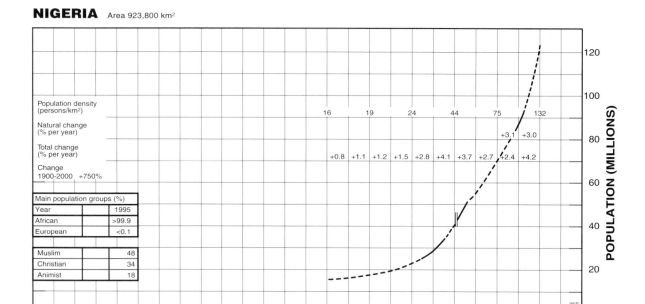

Population density (persons/km²)										16	19	24	44	75	132	
Natural change (% per year)														+3.1	+3.0	
Total change (% per year)										+0.8	+1.1	+1.2	+1.5	+2.8	+4.1	+3.7	+2.7	+2.4	+4.2
Change 1900–2000	+750%															

Main population groups (%)		
Year		1995
African		>99.9
European		<0.1
Muslim		48
Christian		34
Animist		18

YEAR   1800   1820   1840   1860   1880   1900   1920   1940   1960   1980   2000

POPULATION (MILLIONS)

WIS 2000

In the dry savannahs of northern Nigeria Islamic Hausa kingdoms waxed and waned in the 11th and later centuries AD, while in the southern rainforests the Yoruba peoples established their own empires, constructing the huge Eredo earthworks near Lagos. The first European visitors were Portuguese navigators in 1470. Large-scale trading for slaves developed and millions were shipped to the Americas before abolition in the 19th century. Britain proclaimed a protectorate over Lagos and the coast in the 1860s and pushed northward to include the Muslim emirates by 1900. In 1914 the British Colony and Protectorate of Nigeria was declared. Development and population growth accelerated after World War Two as death control (DC), thanks to agricultural and medical improvements, took hold. The birth rate continued high. Rich oilfields were found in the Niger Delta region in 1958.

Independence in 1960, and the accession of British North Cameroon in 1961, paved the way for violent conflict between the main ethnic groups: Hausa (Muslim), Ibo (Christian) and Yoruba (mixed). In 1967 the Ibo southeast attempted to secede as so-called Biafra. In the 3-year civil war that followed, more than a million Biafrans died, mainly of starvation, before the Ibos acknowledged defeat. The 1970s, 80s and 90s saw a succession of civilian governments, all with an ethnic or regional bias, brought down by military coups. There was a border war with Cameroon in 1994.

Most Nigerians are peasant farmers, but the country is a net importer of food. Droughts and floods complicate crop growing. Vast oil revenues have not discouraged the spread of profligacy, corruption and abuse of human rights. Tribal violence is ongoing. Proposals made in 2000 to minimise corruption by introducing Islamic 'Sharia' law (including amputations for thieves, lashings for single mothers, and segregation of the sexes in schools and on public transport) to new areas of the Muslim north initiated Muslim-Christian riots, with several hundred deaths in 2000 and more in 2001. 5% of the adult population was HIV-positive in 1999. There are about 250 ethnic groups and sub-groups.

• • • • • • • • • • • • •

"The population of this country is stable and has been stable for a long time"
(Jonathan Dimbleby, February 2000)

Chapter 4

Natural Limits and Controls on Population Growth

Most of the circumstances in which populations grow, shrink or stabilise can be illustrated by referring to 3 graphs: ENGLAND LONG TERM, WORLD and RWANDA. Some of the concepts were familiar to Thomas Malthus in 1798. This chapter lists, defines and discusses them.

4.1 Carrying Capacity

The simplest definition of national carrying capacity is "the maximum number of people a country can support using its own indigenous resources". England, as the ENGLAND LONG TERM graph shows, could support (i.e. feed) no more than 5 million people in the centuries before 1750. In those days most Englishmen were agricultural workers whose weaker family members survived when harvests were good, but died of hunger and disease when they were bad. The 5 million plateau could only be reached when conditions for survival were optimal, i.e. England was at peace, free of plagues, and all cultivable land was in use (as in the 3rd to 4th, 13th to 14th, and 17th to 18th centuries).

Life was very hard when England was at carrying capacity. The food-producers were stretched to the limit, cultivating the most unrewarding terrain such as steep hillsides or land liable to flood. Accident or illness affecting the main breadwinner usually meant deaths in the family. But for those who were left after warfare (Dark Ages) or plague (Black Death) had cut England's population by half, the living was easier because the survivors could abandon marginal land and grow their crops on productive ground.

A country has reached or passed its carrying capacity "if its people need to eat all the food produced in good years and cannot reserve enough for use in bad ones" (Duguid, 2002).

Carrying capacity in the sense of a country's maximum population is not, however, a number fixed for ever. It varies with time, as resources are expanded or lost by human ingenuity or fecklessness. The Industrial Revolution of the mid-18th century began the mechanisation of agriculture, powered by fossil fuels, with crop rotation and imported fertilizers improving soil fertility. Chemicals were developed to deal with weeds, pests and diseases, allowing higher-yielding strains of food plants and animals to be introduced. All this happened slowly, step by step, which is why DC population surges (section 4.3) in the industrialised world seldom exceeded 1% per year, compared to up to 5% per year in undeveloped nations to which European colonists brought 150 years' worth of improvements all at once.

In England, an area of land that in 1700 AD produced food for 10 people could by 1900 feed 50. Carrying capacity had multiplied 5 times. And there was more. Industrialised England was selling manufactured goods to the world, and the money thus earned could be used to buy food produced abroad. If food availability was the criterion, England's carrying capacity was almost limitless.

The Optimum Population Trust (OPT) contrasts carrying capacity with *optimum population*,

PAKISTAN (excluding Kashmir) Area 945,000 km² (including Bangladesh) to 1971, then 796,000 km²

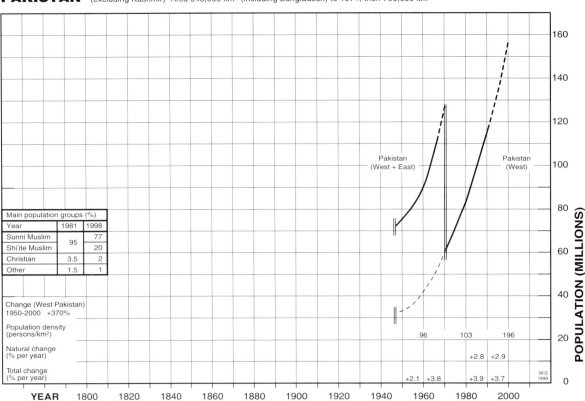

Main population groups (%)		
Year	1981	1998
Sunni Muslim	95	77
Shi'ite Muslim		20
Christian	3.5	2
Other	1.5	1

Change (West Pakistan)
1950-2000 +370%

Population density (persons/km²)	96	103	196
Natural change (% per year)		+2.8	+2.9
Total change (% per year)	+2.1	+3.8	+3.9	+3.7

Labels on chart: Pakistan (West + East); Pakistan (West); POPULATION (MILLIONS); WIS 1999

The Indus valley, a fertile plain between high mountains and arid highlands, that bisects Pakistan, was cultivated by some of the world's first farmers 5000 years ago. Hindu and Muslim civilisations followed. In the 1840s the area became part of the British Indian Empire. By the 1930s the dominantly Muslim population was campaigning for independence as a Muslim state to be called Pakistan ('Pure Nation') and this was achieved in 1947 when the British departed. Accompanied by much slaughter, millions of Hindus and Sikhs left the new Pakistan for India, and millions of Muslims arrived from there. The new state comprised West Pakistan (now Pakistan) and East Pakistan (now Bangladesh). East Pakistan was mainly Muslim but its population was overwhelmingly of Bengali origin and craved separation from the ethnically distinct West Pakistan. In 1971 after a bloody civil war East Pakistan achieved its goal and became Bangladesh.

In Pakistan since 1971 civilian and military governments have alternated amid charges of corruption. Tension between Sunni and Shi'ite Muslims, Christians and Ahmadis has caused many killings. Insulting the prophet Mohammed is punishable by death. In 1990 there were about 5 million Afghan refugees, including active rebels, in camps along the Northwest Frontier, reduced to about 2 million in 2001.

Pakistan's population growth rate since 1947 is almost unequalled. The official annual figures for natural change (those usually quoted by the media) in the 1980s and 90s, 2.8% and 2.9% respectively, are almost 1% less than the figures for annual total change calculated from the reported population increase over the period. Female infanticide is commonplace. Pakistan's problems of severe and growing poverty, illiteracy (only c. 5% of women are literate), unstable governments, a huge foreign debt, religious conflicts, agriculture plagued by droughts and salinization of irrigated areas, and chronic intermittent warfare with India over Kashmir, would be bad enough without an annual population increase of nearly 4%. Violent cutback (VCL) in the near future is a real possibility.

which they define as "the population that is most likely to make the option of a good quality of life available to everyone, now and in the future". Unfortunately, the populations of many nations have grown so fast, in recent decades, that the OPT's definition of *absolute over-population* applies to them: "a population already too large for any foreseeable possibility of acquiring the resources necessary for a decent quality of life".

In their latter definition the OPT may be too pessimistic. It is all too foreseeable that in the not-so-distant future national populations may plummet, as happened in England in the 5th and 14th centuries (see ENGLAND LONG TERM), through conventional or other forms of warfare, through new diseases, or even by accident (section 9.4). The survivors, with more national resources per person, should then be able to improve their living conditions. But an alternative scenario is equally possible, even probable: if climate change is real and intensifying (section 6.1), carrying capacities will fall drastically as large regions stop growing food due to sea level rise (e.g.

PERU   Area 1,285,000 km² (after 1941)

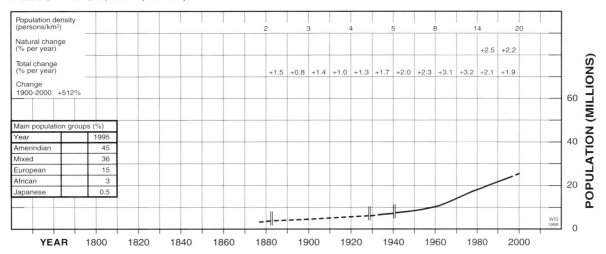

Population density (persons/km²)							2	3	4	5	8	14	20	
Natural change (% per year)												+2.5	+2.2	
Total change (% per year)						+1.5	+0.8	+1.4	+1.0	+1.3	+1.7	+2.0	+2.3	+3.1	+3.2	+2.1	+1.9
Change 1900-2000 +512%														

Main population groups (%)		
Year		1995
Amerindian		45
Mixed		36
European		15
African		3
Japanese		0.5

WIS 1999

YEAR   1800   1820   1840   1860   1880   1900   1920   1940   1960   1980   2000

POPULATION (MILLIONS)   60   40   20   0

18 – 180 million

Peru straddles the cordilleras of the Andes between the arid Pacific coastal plain and the Amazonian tropical rainforest. Aggressive Wari and Tiwanaku empires developed in the first millennium AD, typified by irrigated farming that supported great religious centres in which violence and ritual sacrifice operated. The Inca civilization followed them and was in full flower when Spanish conquistadores attacked and destroyed it in 1533. Five thousand years of agriculture had developed maize, potatoes and other wild plants into important crops. Three quarters of the Amerindian Inca population died from exploitation and disease following the European conquest, so African slaves, and subsequently Chinese labourers, were brought in to work plantations and mines.

Simon Bolivar led the fight for independence from Spain,

achieved in 1824. Since then, rule by elected governments has alternated with coups and long periods of military dictatorship. Large areas in the south were lost in the 'Pacific War' with Chile (1879–83), but part was regained in 1929. Amazonian lands were won from Ecuador in 1941, and the border dispute dragged on until 1998. All these areas were sparsely populated. Maoist 'Shining Path' guerrillas were active in the 1980s and 90s, causing c. 30,000 deaths. Peasant farming, mining, fishing and tourism (especially to visit the Inca ruins) are mainstays of the rather weak economy. A slow DC population surge began in the 1930s, accelerating in the 1960s. Growth rates from the 1980s were affected by education in family planning and government-sponsored sterilisation of poor women (c. 100,000 in 1998).

BANGLADESH, EGYPT), or desertification (e.g. USA, SPAIN), or even excessive cold (e.g. north-west Europe). A further alternative is certainly real: a rich nation that has enjoyed a good quality of life by buying food and other necessities from abroad may become so impoverished (e.g. by war or by political or economic incompetence) that it is thrown back on its own territorial resources. Most European nations, if this happened, would find their mineral and energy resources gravely depleted and much of their best agricultural land built upon. Their own intrinsic carrying capacities having plummeted, their populations would follow suit. The same applies to the many overcrowded nations that now survive on foreign aid, if this ceased.

Carrying capacity is not a mere theoretical concept. Populations experiencing DC surges have been able to disregard it for up to 250 years (the WROG period, section 4.5) as they increase food availability by exploiting the planet's finite resources, but that carefree lifestyle cannot last for ever. The signs are that it is nearing its end (see WORLD). Already the populations of some developing nations (e.g. ETHIOPIA, North KOREA) would be cut back by starvation, in the famines caused by natural droughts, were they not given charitable aid.

4.2 Violent Cutback Level (VCL)

The population of a nation or region has reached VCL when more of its people die and /or emigrate than are born and/or immigrate, in an ambience of violence and hatred. Further population growth is impossible. A VCL scenario is typically associated with genocide and/or ethnic cleansing, as distinct from the carrying capacity scenario where population increase is ended by non-violent starvation or diseases of malnutrition. The violence is usually a struggle for power and possessions between rival population groups or factions, which are normally ethnic, tribal, religious or political, less often based on differences in class or wealth. Total elimination of the opposition may well be the goal.

PHILIPPINES   Area 300,000 km²

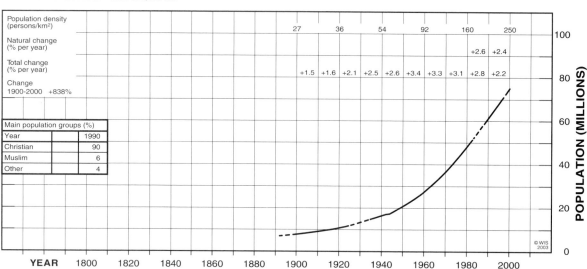

Population density (persons/km²)							27	36	54	92	160	250	
Natural change (% per year)											+2.6	+2.4	
Total change (% per year)							+1.5	+1.6	+2.1	+2.5	+2.6	+3.4	+3.3	+3.1	+2.8	+2.2
Change 1900-2000 +838%													

Main population groups (%)	
Year	1990
Christian	90
Muslim	6
Other	4

This archipelago of 7000 islands, mountainous with active volcanoes, was populated over many millennia by waves of Malayan and Mongoloid people, crossing land bridges or island-hopping from mainland south-east Asia. It was colonised by Spain in 1565 and named in honour of the Spanish king. The USA acquired it in 1898 after the Spanish-American War. At once the Filipinos rose in revolt, but were put down with 200,000 killed. Independence was achieved in 1946. Subsequent administrations have been accused of corruption on a massive scale. Since the 1970s Muslim separatists have been waging a guerrilla war in the southern island of Mindanao.

As in most of southeast Asia, the population doubled three times in the 20th century. Half the people are engaged in agriculture and logging, activities which have reduced the natural forest cover from about 90% when the Spanish arrived to about 7% in 1990. Soil erosion has greatly increased, and agricultural productivity decreased, in consequence. Less than 10% of the coral reefs are undamaged. The Philippines' exceptionally exuberant biodiversity is sliding into mass extinction.

In modern times the best example of a population reaching its VCL is that of RWANDA. This small East African country is home to Hutu subsistence farmers and Tutsi pastoralists, tribes with a long history of antipathy because the Tutsi minority has dominated the Hutu majority. The first German colonists, around 1900, found a low population density of around 40 persons per square kilometre, and no shortage of cultivable land. By 1959 the DC surge was in full swing, the population had more than doubled and serious tribal warfare began. Many Tutsis were killed or expelled to Ugandan refugee camps, where their numbers grew rapidly and hatred festered. In 1990 they fought their way back into part of Rwanda. By 1994 Rwanda's population density had risen to 290 (Africa's highest), land-hunger dominated the national consciousness, tribal killings happened every day, and the Hutus plotted to eliminate their rivals. The genocide attempt failed, but a million Rwandans were killed (of both tribes), and 2 million, mostly Hutus, were driven out of the country. The resident population fell by 3 million in less than 3 months, initiating the tell-tale peak and trough on the graph.

Since 1994 some of the refugees have returned and the population is rising again. Renewed genocide has been prevented, so far, by a UN force, but small-scale tribal killings are commonplace. Rwanda's population is fated to oscillate about its VCL of roughly 6 million, unless either

   a) invincible peacekeepers or unlimited aid prevent renewed genocide, or
   b) tribal rivalry ceases, through partition or complete genocide, or
   c) economic development provides the means to import ample food supplies, or
   d) natural deaths persistently exceed natural births, through disease (e.g. AIDS) or (wonder of wonders) family planning.

As the WORLD graph shows, the Industrial Revolution ushered in a period of rapid population growth, expressed as DC surges, because food surpluses were suddenly available.

POLAND Area 389,000 km² to 1939: 312,500 km² after 1945

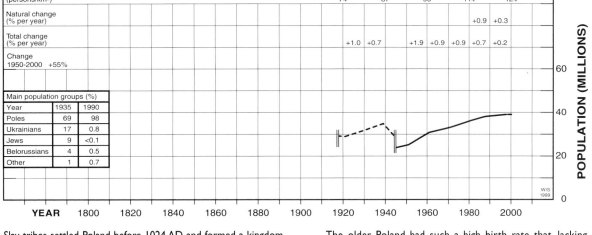

Population density (persons/km²)								74	87	95	114	124	
Natural change (% per year)											+0.9	+0.3	
Total change (% per year)								+1.0	+0.7	+1.9	+0.9	+0.9 +0.7 +0.2	

Change 1950-2000 +55%

Main population groups (%)		
Year	1935	1990
Poles	69	98
Ukrainians	17	0.8
Jews	9	<0.1
Belorussians	4	0.5
Other	1	0.7

YEAR 1800 1820 1840 1860 1880 1900 1920 1940 1960 1980 2000

POPULATION (MILLIONS) 60 40 20 0

WIS 1999

Slav tribes settled Poland before 1024 AD and formed a kingdom, powerful enough to halt the Mongol hordes' advance into Europe in 1241. By 1466 the Polish kings ruled territory from the Baltic Sea to the Black Sea. Union with Greater Lithuania in the 16th century created Europe's largest state, but a long series of wars against Sweden, Russia, Prussia and Turkey weakened the nation (one third of the population died in the 1655–60 Swedish war). In 1795 Poland vanished from the maps, partitioned between Austria, Prussia and Russia. A Polish state was re-established after World War One, incorporating parts of Lithuania, Belarussia and Ukraine; but after World War Two a new (communist) Poland was created, 250 kms further west, by loss of the eastern territories and the incorporation of German Silesia, from which about 7 million Germans were expelled to achieve "historically acceptable" boundaries.

The older Poland had such a high birth rate that, lacking significant industry, the standard of living drastically declined and some 200,000 Poles emigrated each year. The Soviets deported 1.7 million Poles to labour camps between 1939 and 1941; one million of them died. More than 6 million Poles were killed in World War Two, many of them Jews (see the table) in the Auschwitz and other extermination camps. New Poland threw off communist rule in 1989, but unlike most other ex-communist Slav states its population is still growing, if only just. Almost all the population is Christian. New Poland's industry, developed Soviet-style, has been so environmentally unfriendly that by 1990 pollution had damaged three quarters of all forests, many water supplies were unusable and historic cities were crumbling, matters which did not help the important tourist industry. Censuses: roughly every tenth year after 1931.

This period of *Weak Restraints On Growth* (*WROG*, section 4.5) is a historical anomaly, because previous periods of WROG (rhymes with dog, or, if it seems apt, rogue) had only occurred briefly, when new lands empty of humans were discovered and settled. However, since the late 1970s populations have begun to reach VCL in growing numbers, indicating that the slack is being taken up and the WROG period is coming to an end. Nations, and rival groups within them, will soon return to their old habits of competing violently to take possession of territory, food, water, oil or whatever. Everyone alive today has grown up with WROG and we mostly assume that weak restraints on growth are normal. People are totally unprepared for the vicious inter-group competition for resources that held populations down before the Industrial Revolution, and will strike afresh every time a population reaches the critical (VCL) level.

What were the controls that restricted world population growth in the centuries and millennia before the Industrial Revolution? Mankind's Darwinian acquisitiveness (using force to acquire other peoples' possessions, i.e. wars of conquest) is the main thread of recorded history. Violent death, often but not always in a VCL situation, was a commonplace population limiter. When inter-group bickering was inhibited, by strong civil government or by external force (as in European colonies, or in England during the 'Pax Romana') VCL scenarios were prevented. If population growth was restricted it was by carrying capacity. Epidemic disease, of course, could strike at any time.

Governments have not yet realised the futility of intervention in an VCL situation. In Kosovo, in 1999, NATO used massive force to end the violent showdown between Serbs and Albanians. It had been building up for 5 decades as the Albanians worked to "Albanianise" the country (i.e. take it from the Serbs) by maintaining a high, typically Muslim, birth rate, a

18 – 180 million

ROMANIA Area 237,500 km² (130,000 km² to 1918, 310,000 km² to 1940)

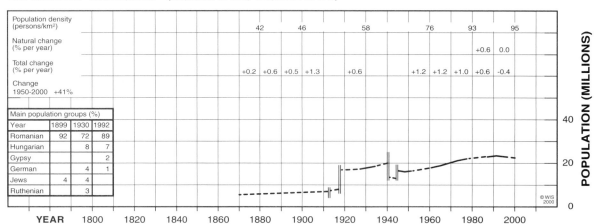

Population density (persons/km²)	42	46		58		76	93		95	
Natural change (% per year)							+0.6	0.0		
Total change (% per year)	+0.2	+0.6	+0.5	+1.3	+0.6	+1.2	+1.2	+1.0	+0.6	-0.4
Change 1950-2000 +41%										

Main population groups (%)			
Year	1899	1930	1992
Romanian	92	72	89
Hungarian		8	7
Gypsy			2
German		4	1
Jews	4	4	
Ruthenian		3	

POPULATION (MILLIONS)

YEAR   1800   1820   1840   1860   1880   1900   1920   1940   1960   1980   2000

©WIS 2000

The kingdom of the Dacia, tribes of Greek origin, included most of modern Romania. It was conquered by Rome in the 2nd century AD, became a province of the Empire, and converted to Christianity. Rome withdrew in the 3rd century, leaving the country open to 1700 years of invasion, slaughter and pillage by Goths, Huns, Slavs, Magyars, Austrians and Turks (among others). Transylvania in the west had become Magyar (Hungarian) by the 13th century but Moldavia (north) and partly Slav Wallachia (south) formed a proto-Romanian nation that resisted the Mongols in the 1240s. Transylvania fell to the Ottoman Turks in the 15th and 16th centuries but Romania's brutal resistance (by princes such as Vlad the Impaler, a Wallachian who "rarely ate without a Turk writhing on a stake in front of him") ensured its continued autonomy, though paying tribute to the Ottomans. Massacre followed massacre as Romanian peasants rose against the Turks, the Hungarian and Saxon nobles, and the Austrian Habsburgs who conquered Transylvania in 1687. Ottoman domination of Moldavia and Wallachia ended with the Russo-Turkish war of 1828–29. The state of Romania (without Transylvania) was declared in 1862.

Romania fought with the Allies in World War One and was rewarded in 1918 by the return of Transylvania from Austria-Hungary and Bessarabia from Russia. After German occupation in World War Two, when 400,000 Jews and 40,000 Gypsies were deported and murdered, and 700,000 Romanian soldiers died, Bessarabia was ceded to Russia. Romania became a Soviet satellite state, subject to the usual industrialisation and collectivisation of agriculture, but managed to distance itself from the USSR in foreign policy. Under Nicolae Ceausescu (1965–89) Romania received much foreign aid but succumbed to chronic corruption and inept administration that caused food and fuel shortages. In 1989 Ceausescu was overthrown in a violent popular uprising and executed. His anti-birth control policy promoting large families was reversed, and the population, though not mainly Slav, began a gentle decline. The new government, effectively neo-communist, gave way to a democratic reforming one in 1997 but Romania remains poor, suffering especially from the imminent exhaustion of the Ploiesti oilfields which began production in the 1850s and were the largest European producer between the World Wars. Some ethnic minorities, in particular the Gypsies with their large families, are unpopular and persecuted. Romania has the world's largest Gypsy population. Almost everyone is Christian. Censuses: roughly every tenth year from 1930.

procedure that may be termed *aggressive breeding* (section 4.4, and see KOSOVO). Having stopped the overt fighting, NATO naively assumed that Serbs and Albanians would soon see the error of their ways and settle down peacefully together. Gradually it is dawning on NATO that if and when their 50,000 peacekeepers leave Kosovo, the local residents will resume their battle (ending, probably, with the expulsion of the remaining Serbs) because the mixed population is above its VCL.

The population of a well-off homogeneous nation has a high VCL by definition, because there is no tendency for it to split into factions competing for resources. If, however, the nation becomes multicultural, a strong divisive tendency appears, causing the VCL to fall. In the first case the VCL could be 20 million, in the second case, 10 million. So even if a nation's population has been static, it can suffer violent cutback by the introduction of religious or ethnic diversity and/or decreasing wealth. VCL may be compared to the pressure at which a balloon bursts; if the rubber has flaws it fails at a lower pressure than if it is homogeneous.

Once a national population has increased and/or diversified to the point that internal groups are constantly quarrelling, with violent incidents, there is no easy way back. Truces may be declared, or promises made, but unless the underlying cause of dissension, population pressure, can be relaxed in some way the fine words usually prove to be no better than appeasement, a variation in the ongoing struggle for advantage (see IRELAND AND

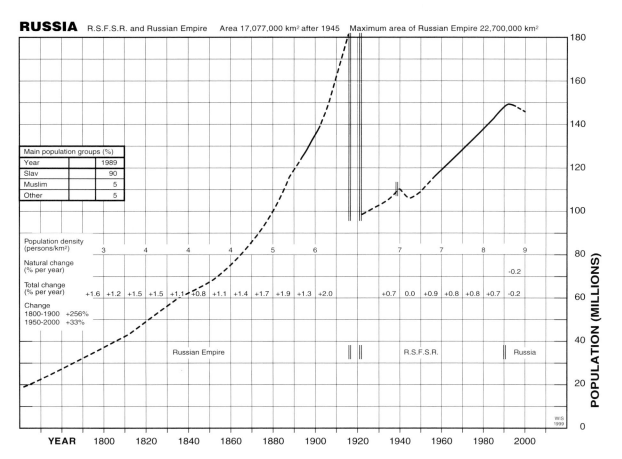

RUSSIA   R.S.F.S.R. and Russian Empire   Area 17,077,000 km² after 1945   Maximum area of Russian Empire 22,700,000 km²

Main population groups (%)

Year		1989
Slav		90
Muslim		5
Other		5

Population density (persons/km²): 3   4   4   4   5   6   7   7   8   9

Natural change (% per year): -0.2

Total change (% per year): +1.6   +1.2   +1.5   +1.5   +1.1   +0.8   +1.1   +1.4   +1.7   +1.9   +1.3   +2.0   +0.7   0.0   +0.9   +0.8   +0.8   +0.7   -0.2

Change
1800-1900 +256%
1950-2000 +33%

Russian Empire   R.S.F.S.R.   Russia

YEAR   1800   1820   1840   1860   1880   1900   1920   1940   1960   1980   2000

POPULATION (MILLIONS)

18 – 180 million

Russia's steppes, forests and tundra, extending across 11 time zones from Europe through Siberia to the Pacific Ocean, are so vast that the world's sixth most populous nation has one of the lowest population densities. The Slav principalities of European Russia endured invasions of Vikings, Lithuanians and Poles between the 9th and 17th centuries, but the most devastating attackers were Mongol hordes from the Far East who destroyed every town between Kiev and Moscow and held sway over much of southern Russia from the 13th to the 15th centuries. In the 16th century Ivan the Terrible sent the first waves of settlers and fur traders eastward into Siberia; after only 200 years they had reached the Pacific coast and crossed into Alaska.

The Romanov Tsars ruled Russia from 1613 to 1917 and extended the Empire by conquest to the borders of Turkey, Afghanistan and China. Gains of new territories included their populations, largely explaining the rapid population growth of the Russian Empire. Lenin's Bolshevik coup of 1917 initiated seven decades of communism. Russia, as the Russian Soviet Federal Socialist Republic (RSFSR), was the all-powerful leader of up to 16 communist republics, the fragmented Empire, which became the Soviet Union (USSR) in 1922. The communist system was very inefficient, except in war. In spite of its enormous resources of land, minerals, oil, gas and forests, the USSR collapsed in a welter of corruption and bankruptcy in 1991. Russia reappeared at the head of the Commonwealth of Independent States (CIS) to which most of the ex-Soviet nations belong.

The early years of the USSR saw communist dogma, obsessed with productivity, put into practice at the expense of those it was supposed to serve. Millions of peasants starved to death during the collectivization of agriculture before World War Two, when the harvests were seized to fund imports of machinery and to feed industrial workers in cities.

The end of communism coincided with an unexpected development. According to demographic transition theory (section 5.2), population growth will cease as nations become more affluent, but Russia and other Slav nations of the ex-USSR are experiencing a population downturn linked to economic setbacks, poverty, corruption and 'psycho-social depression'. With poor medical facilities, fear of childbirth is common among Russian women. Alcohol poisoning and suicide are major killers of men. Male life expectancy had fallen to 60 by 1999. In contrast, population growth continues in CIS Muslim nations. Growing Slav-Muslim hostility has been evident in Russia's recent attempts to suppress militant Islam in the Caucasus and Balkans. The ongoing separatist revolt in Muslim Chechnya, begun in 1996, is particularly brutal, with 60,000 deaths in 1996 alone and 200,000 refugees by 1999.

In 2001 half a million Russians were HIV-positive. The disease is expected to worsen, given the high incidence of drug addiction. By the late 1990s, settlers in many parts of Siberia were experiencing ground subsidence problems as global heating began to melt the permafrost.

NORTHERN IRELAND). The only *painless* way to reduce population pressure is to have fewer babies, and it is a desperately slow-acting remedy.

People and governments cling to unrealistic politically-correct beliefs, such as the idea that ethnic groups competing for territory or resources can be converted, (e.g. by "education, education, education") to multiculturalism. The formative years of these people were within

SAUDI ARABIA   Area 2,200,000 km²

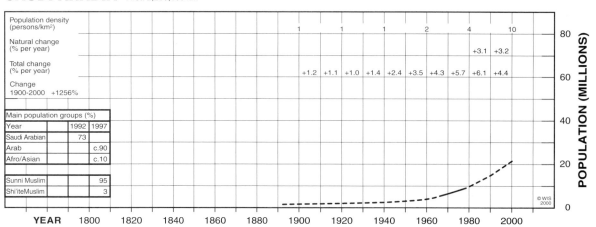

Population density (persons/km²)								1	1	1	2	4	10	
Natural change (% per year)												+3.1	+3.2	
Total change (% per year)								+1.2	+1.1	+1.0	+1.4	+2.4	+3.5	+4.3	+5.7	+6.1	+4.4
Change 1900–2000 +1256%														

Main population groups (%)		
Year	1992	1997
Saudi Arabian	73	
Arab		c.90
Afro/Asian		c.10
Sunni Muslim		95
Shi'iteMuslim		3

© WIS 2000

POPULATION (MILLIONS): 80 — 60 — 40 — 20 — 0

YEAR: 1800 1820 1840 1860 1880 1900 1920 1940 1960 1980 2000

In 1902 Ibn Saud of the Al-Saud family began a military campaign to conquer the Arabian regions of Hejaz, Hasa and Nejd and drive out the Ottoman Turks. He succeeded, with British help in World War One, in 1925, and named his new kingdom Saudi Arabia. The Sauds have been in power ever since. Two-thirds of their country is desert. They had established a comparable empire previously, in the 18th and early 19th centuries, but were ousted by the Turks. Even earlier, after the prophet Mohammed (born in the Hejaz at Mecca) had converted the Arabians to Islam in the 7th century, evangelising Arab armies had spread the Muslim faith from Spain to India, but after the 13th century the Muslim Empire fragmented and Arabia was dominated by Egypt and then Turkey.

From the 1940s the Saudi kings have presided over the exploitation of vast oil resources (one sixth of the world total) that exist beneath the Arabian (Persian) Gulf coastal region. Oil-related industries have attracted a flood of foreign workers, contributing to the DC population surge (each Saudi woman has 6 children on average) that the oil wealth made possible. However, dissident groups objecting to the monarchy's links with the West organised several terrorist incidents in the 1990s.

the WROG period, so they naturally assume that resources will always be adequate. But feckless mankind's response to two and a half centuries of unprecedented plenty has been the DC population surge. The ratio of population to resources is now returning to what was normal before the Industrial Revolution, and even organisations as powerful as NATO must begin to realise that it is beyond their power to prevent the two sides in a VCL situation "getting on with it".

4.3 The DC (Death Control) Surge

In the pre-industrial world, death control was seldom an option. Premature death resulted from war, banditry, starvation or disease, afflictions which the peasant farmers who comprised the great bulk of humankind knew all too well. One or more afflictions could strike a peasant and his family at any moment, and he was powerless to resist. Only the uppermost classes could control them to some extent. It was a great incentive to become a priest, merchant, landowner, military leader or doctor, positions jealously guarded within families, once achieved. In primitive tribes the chiefs and witch-doctors enjoyed similar advantages.

Change (well illustrated by ENGLAND LONG TERM) began in Europe in the mid-18th century. Clever scientists and engineers worked out procedures and invented technologies that revolutionised agriculture. Food surpluses were produced. Then in the 19th century disease became more controllable when Pasteur, Lister and others discovered the importance of antisepsis and hygiene. These advances did not reduce the incidence of war, but death control (DC) had reached the point where populations were beginning their current explosions.

DC population surges were the knee-jerk response of humans, prolific breeders, to an unfamiliar environment of relative plenty in which more babies survived and people lived longer. The surge began in Europe and North America and spread in fits and starts to the rest of the world, along with trade and colonialism, which needed a large and healthy work force.

In retrospect, the world would have been infinitely better served if birth control had

SOUTH AFRICA   Area 1,224,700 km²

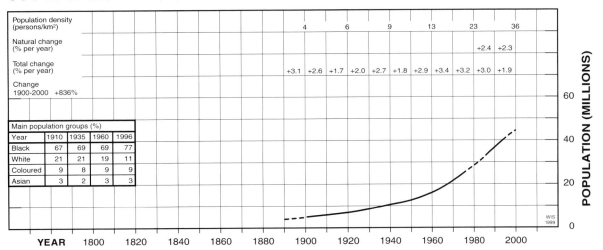

Population density (persons/km²)								4	6	9	13	23	36
Natural change (% per year)												+2.4	+2.3
Total change (% per year)						+3.1	+2.6	+1.7	+2.0	+2.7	+1.8	+2.9	+3.4	+3.2	+3.0	+1.9
Change 1900-2000 +836%													

Main population groups (%)				
Year	1910	1935	1960	1996
Black	67	69	69	77
White	21	21	19	11
Coloured	9	8	9	9
Asian	3	2	3	3

POPULATION (MILLIONS)

YEAR   1800   1820   1840   1860   1880   1900   1920   1940   1960   1980   2000

Bantu tribes were migrating south through South Africa, displacing the local Khoi and San peoples (Hottentots and Bushmen) at the same time as Dutch farmers (Boers) were expanding their settlement at Cape Town, established in 1652. (Europeans at first paid little attention to the Cape of Good Hope after Portuguese explorers rounded it in 1488.) 19th century wars between Bantu, Boers and British, who bought the Cape from the Netherlands in 1814, are well documented (e.g. Knight, 1994). Genocidal strife broke out when powerful Bantu leaders attacked adjacent tribes to win lands and cattle. The death toll in the 1820s *mfecane* (crushing) by Shaka Zulu of his Nwandwe and other neighbours, and dissidents within his own tribe, is thought to have been about 200,000. Elsewhere in Africa, before the slave trade was abolished, tribal leaders preferred to enslave the defeated and barter them for European goods.

In the 1840s, following their Great Trek north, Boer farmers established the Transvaal and Orange Free State republics which were conquered and annexed by the British in the Second Boer War (1899–1902). By this time the black tribes had been forcibly subjugated by British and Boers, notably in the Zulu War of 1879. The Union of South Africa was created in 1910 as a self-governing British dominion in which the Boers, or Afrikaners, were the ruling group.

White minority rule, threatened by the growing black majority, developed the notorious 'apartheid' (racial segregation) policy which held back the blacks and coloureds from 1948 to 1991, when it succumbed to international disapproval and internal violence. Subsequent black governments, in spite of brilliant leadership by Nelson Mandela (1994–99), have been unable to curb the continuing violence which has included 30,000 to 40,000 murders annually. In 2001 racial tension between whites and non-whites, and between the latter and immigrant blacks (up to 180,000 annually), was still a chronic problem.

During the 1990s the HIV/AIDS epidemic became a major concern. By 2001 13% of the adult population was infected and life expectancy was falling drastically. Yet in 2000 only 8% of men were using condoms. The economy, originally dependent on gold and diamond exports, has evolved to include manufacturing and financial services. Censuses: at intervals of 5 to 21 years from 1904.

been available, and used. What actually happened was that population growth kept pace with the increasing availability of food. Agriculture became more intensive, and people went on cultivating marginal land, felling forests, draining marshes and irrigating deserts to plant crops. Over much of the planet there was little or no sustained improvement in resources per person, so people stayed poor, and there were vastly greater numbers of them. The opportunity to maintain a low population enjoying a high standard of living was missed.

4.4   Aggressive Breeding

An alternative to the above title is *competitive breeding* (Parsons, 1998). When a national population consists of two or more groups, it is often the case that families in one group consistently have more children than the other groups (see HUNGARY). In consequence the size of that group increases at the expense of the others. In democratic societies the power and influence of the enlarging group increases proportionately. Eventually it may become the national majority and, in a lawful election, take over the country.

Western European governments are discovering with anxiety that some sections of their populations view the non-native minorities (approaching 10% in several countries in 2002) as a threat. Racial tension is generated which could be avoided if the minorities were not perceived to be aggressively expansionist. In such circumstances, all the precedents are *against* the peaceful establishment of a multicultural society, and *for* the progressive worsening of race relations.

18 – 180 million

SPAIN including Balearic and Canary Islands. Area 504,750 km²

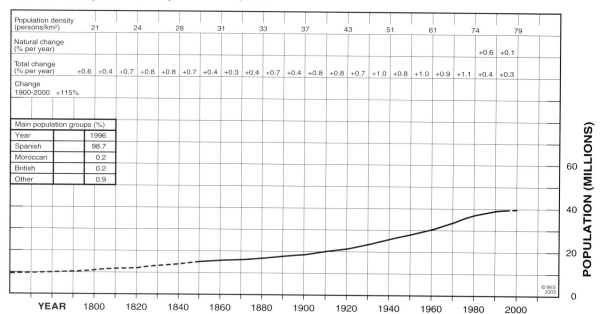

Stone Age dwellers in Spain, some of whom created remarkable cave paintings, merged over many millennia BC with Iberian tribes from North Africa and, in the 1st millennium BC, with Celts from France. The country was a major source of grain, minerals and olive oil to the Roman Empire, but was conquered in the 5th century AD by Vandals and then Visigoths who established Christianity. Spain was over-run by 'Moors' (Muslim Berbers from North Africa) in the 8th century. Impressive monuments to Moorish civilisation and culture, such as the Alhambra in Granada, remain. Christians slowly fought their way back from the north until the last Muslim stronghold, Andalusia, was cleared in 1492. In the same year Columbus crossed the Atlantic, opening the door to Spanish conquest of huge regions of central and southern America.

Exploitation of the New World's riches made Spain the wealthiest nation in Europe, but by the late 17th century most of the money had been spent in fruitless wars attempting to dominate Italy, the Netherlands, England and France. By the early 19th century the main American colonies had achieved independence, and Spain became a relatively poor largely agricultural country until tourism and manufacturing blossomed after World War Two.

For several decades the ethnically distinct Basque peoples of northern Spain have been agitating, with terrorist murders, for independence. The 1990s saw a rapid increase in illegal immigration from North Africa, across the 13-kilometre wide Straits of Gibraltar (as many drown as reach Spain, it is said) leading to growing ethnic intolerance in Spain between natives and immigrants. The graph shows the steady slow growth typical of most western European nations, although the main DC surge came later than in northern Europe. The near-cessation of growth in the 1990s has been attributed to prosperity triggering the 'demographic transition' (section 5.2). Censuses: about every tenth year from 1860.

The rise of the fertile minority is usually slow, but it is seldom unnoticed. Where tension already exists between groups, it can be singled out as a cause of resentment, even outrage, if it is perceived to be a deliberate policy, bringing advantage to the fast breeders and disadvantage to the other groups. Extreme examples of aggressive breeding and its consequences are found in GAZA, WEST BANK, ISRAEL, MACEDONIA and KOSOVO. People who feel threatened by aggressive breeders tend to react savagely because, as Euripides remarked in the 5th century BC "There is no greater sorrow on earth than the loss of one's native land".

In Israel's occupied territories, the West Bank and Gaza, families of 5 to 14 children are common among the Arab Muslim population, although their parents can seldom support them without charitable aid. The population of Gaza has more than quintupled since 1950. The purpose, often stated ("our children are our weapons") before the current *intifada*, is to gain numerical advantage vis-a-vis the Israelis, achieving status and sympathy in the eyes of a democratic and compassionate world. Some Jewish sects have equally large families, and successive Israeli governments have sought immigration from the worldwide Jewish diaspora. Race hatred is intense, the population is far above VCL, and a violent massive readjustment has only been prevented, so far, by Israel's overwhelming military might. The Israelis are the "invincible peacekeepers" (section 4.2.a).

Having many children is considered advantageous when, during conflicts, their

SRI LANKA Ceylon Area 65,600 km²

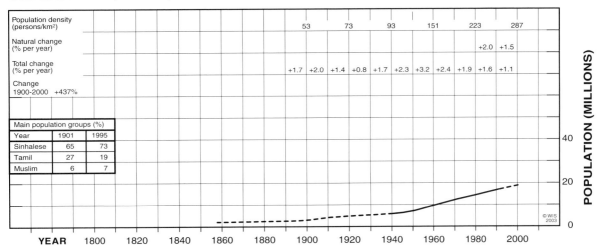

Population density (persons/km²)							53	73	93	151	223	287	
Natural change (% per year)											+2.0	+1.5	
Total change (% per year)							+1.7	+2.0	+1.4	+0.8	+1.7	+2.3	+3.2	+2.4	+1.9	+1.6	+1.1
Change 1900-2000 +437%													

Main population groups (%)		
Year	1901	1995
Sinhalese	65	73
Tamil	27	19
Muslim	6	7

YEAR 1800 1820 1840 1860 1880 1900 1920 1940 1960 1980 2000

POPULATION (MILLIONS)

© WIS 2003

Sinhalese tribes from north-east India established themselves on the island of Sri Lanka, subjugating the Vedda aboriginals, and adopting Buddhism, about 2500 years ago. Muslim Tamils came from southern India some 1500 years later. From the 15th century, Chinese, Portuguese, Dutch and finally British administrations were imposed. In 1948 the island achieved independence as Ceylon; the name was changed to Sri Lanka in 1972. The economy is based on agriculture (rice, tea, rubber and coconut products), manufacturing and tourism, though the latter has declined due to the civil war. The extensive tropical forests are being steadily depleted by the expansion of agriculture, resulting in serious soil erosion.

The DC population surge began about 1940. Although the percentage growth rate has greatly declined since then, the numerical growth rate has remained almost constant. There has been considerable emigration of the Tamil minority, including 200,000 to India in the 1980s. Traditional ethnic hostility between Tamils and Sinhalese, suppressed during British rule, intensified after independence as the population density became uncomfortably high. Prohibition of Tamil as an official language in schools was a flash point. Civil war broke out in 1983 with the Tamil Tiger guerrillas fighting to establish their own state in the northeast of the island. An Indian peacekeeping force was ineffective and was withdrawn in 1990. By 1999 the war had caused c. 100,000 deaths and about 300,000 Tamils had left the country as refugees or asylum-seekers. The population is still growing, so has not reached VCL. In 2001 the government reversed the 'small families' policy of the 1970s by urging the Sinhalese majority to have more babies, to strengthen the army against the Tamils. Peace talks, involving Tamil autonomy, were under way in 2002. Censuses: irregular from 1946. In 1995 69% of the population was Buddhist, 15% Hindu, 7% Muslim and 7% Christian.

18 – 180 million

sufferings are publicised and the enemy's heartlessness can be blamed.

In Kosovo the Albanian Muslims outnumbered the Serbs 3 to 1 at the end of World War Two, but by 1998, thanks to the Albanian policy summarised by Vickers (1998) "Having a large number of children... was seen as ensuring for Kosovo an Albanian as opposed to a Serb future", the proportion was 12 to 1. Serb rage boiled over in 1998 and Serbia, led by President Milosevic, began an VCL-style ethnic cleansing of Kosovan Albanians that was halted and reversed by NATO intervention. A letter from a Serb woman to Tam Dalyell MP (*Hansard*, 19 April 1999) put the Serb point of view (condensed and updated by me): "You share your house with another young couple. Children are born. You have a few, they have many. They tell you the house should be theirs, because they are the majority. They try to expel you, but you are strong and start pushing them out. They call in NATO, which beats you up. The other family then returns and expels you". President Clinton of the USA was backing the Albanian aggressive breeders when he said (13 May 1999) that the Kosovan situation "is an affront to humanity everywhere".

In Macedonia, adjoining Kosovo, the Albanian minority increased from 17% to 23% (though they claim 40%) between 1948 and 1994, despite much emigration. "We will beat you in the beds" they tell the Slav majority (*National Geographic*, March 1996).

In BOSNIA the high Muslim birth rate helped to foment the racial hatred that fragmented the country in 1992–95, but religions other than Islam practise aggressive breeding. In Northern IRELAND, Catholic and Protestant factions both tend to have large families, but the Catholics are gaining and may be the majority within two decades. "We're outbreeding the [Protestants]", Irish nationalists boast (*The Guardian*, 27 August 2001).

SUDAN   Anglo-Egyptian Sudan   Area 2,505,800 km²

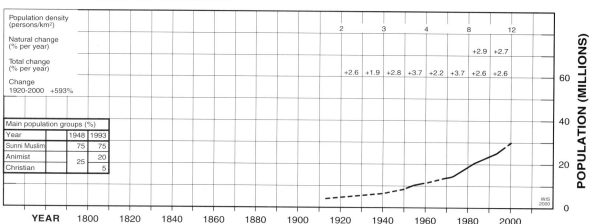

Population density (persons/km²)									2	3	4	8	12		
Natural change (% per year)													+2.9	+2.7	
Total change (% per year)									+2.6	+1.9	+2.8	+3.7	+2.2	+3.7	+2.6	+2.6
Change 1920–2000 +593%															

Main population groups (%)

Year	1948	1993
Sunni Muslim	75	75
Animist	25	20
Christian		5

POPULATION (MILLIONS)

60

40

20

0

WIS 2000

YEAR   1800   1820   1840   1860   1880   1900   1920   1940   1960   1980   2000

Prehistoric inhabitants of Sudan's Nile Valley, as long as 14,000 years ago, were killing each other with stone weapons, possibly competing for territory on the narrow fertile floodplain (*National Geographic*, July 2000). Egypt colonised Nubia (the deserts and Nile Valley in northern Sudan) in the first millennium BC. Soon after, the Nubians conquered Egypt. In the 4th century AD Nubia fell to Axum (part of Ethiopia) and was Christianised, only to be over-run by Muslim Arabs 800 years later. When, in the 1820s, the Arabs began to carry Islam into the savannahs and forests of southern Sudan, with its humid equatorial climate, slavery, pillage and disease devastated the black African population.

British interest in the area provoked the successful Mahdi revolt (1881–85), but Anglo-Egyptian forces defeated the Mahdi in 1898 and established the Anglo-Egyptian Sudan, essentially a British colony. Development progressed peacefully until 1955 (the year

before independence) when civil war broke out between the Muslim north and the black African south, which had hoped to secede. Since then, the country has been impoverished by almost continuous conflict, in which the North (itself divided) has failed to subdue the two rebel groups in the South (which depend on UN food aid) in spite of repeated droughts, famines, systematic destruction of villages and vast flows of refugees to and from Ethiopia, Chad and Uganda. The war has cost at least 2 million lives. It intensified in the 1990s when oil was discovered in the South and the government tried to consolidate its hold over the oilfields by evicting their black population. Oil exports began in 1999.

There are several hundred ethnic groups and sub-groups in the Sudan. Censuses were held in 1956, 73 and 93 but the results are of dubious value. Lack of medical care has allowed sleeping sickness to return on a large scale.

In the USA, in 2002, the Hispanic minority (13% of the population) overtook the black minority (12.7%), partly by its high birth rate and partly by immigration (*The Guardian*, 23 January 2003).

The Maoris of NEW ZEALAND, who were relegated from sole inhabitants to a small minority by tribal fighting and European immigration, are now rapidly gaining ground, like the Mormons of Utah, USA. When I asked the Luxembourg Embassy in London to explain the accelerating population growth of LUXEMBOURG in the 1990s, a spokesperson linked it to the high fertility rate of immigrant foreign workers and their families. Most Western European nations have immigrant minorities with higher birth rates than their hosts, as do states such as California and Florida in the USA. In Britain in the 1980s, Pakistani and Bangladeshi women had two or three times as many children as white or Caribbean women, but the difference lessened in the 1990s (Berthoud, 2001).

It is in the context of economic migration and asylum-seeking that a high immigrant birth rate is most likely to arouse resentment, a suspicion that the newcomers are taking advantage of their hosts' hospitality, even trying to change the culture of their adopted country to suit themselves. In the UK in December 2000, members of the Rastafarian cult argued that their children should be allowed to wear 'dreadlocks' in school (where traditional hairstyles were the rule), and that drug dealing should be permitted, because it was part of their religion. "Britain's ethnic minority voters have the power to swing 100 seats in the [2001 general] election, according to research by Operation Black Vote", reported *The Guardian* newspaper on 21 May 2001. In the USA, a leader of the 7 million Muslims claimed (BBC World Service, 30 August 2000) that the Muslim vote would decide the Presidential election.

When highly fertile Albanian Kosovans are persecuted by resentful Serbs, or

TAIWAN Formosa Area 36,180 km²

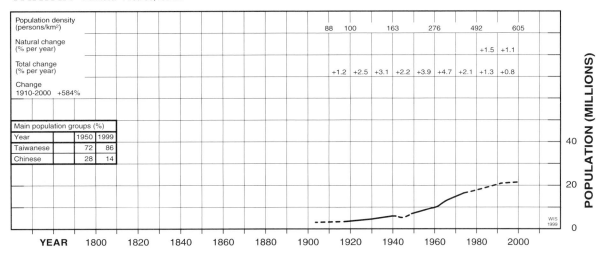

Population density (persons/km²)							88	100	163	276	492	605	
Natural change (% per year)											+1.5	+1.1	
Total change (% per year)							+1.2	+2.5	+3.1	+2.2	+3.9	+4.7	+2.1	+1.3	+0.8

Change 1910-2000 +584%

Main population groups (%)		
Year	1950	1999
Taiwanese	72	86
Chinese	28	14

POPULATION (MILLIONS)

40

20

0

YEAR 1800 1820 1840 1860 1880 1900 1920 1940 1960 1980 2000

WIS 1999

Taiwan's original population of Malayan origin was diluted by Chinese immigration after the 14th century. The island escaped European colonisation apart from four decades of Dutch control in the 1600s. In 1683 it was annexed by China, who ceded it to Japan in 1895. After World War Two it returned to Nationalist China. Two million nationalists fled to Taiwan in the late 1940s to escape the communist revolution in China, which now claims sovereignty over Taiwan and threatens to invade. Even so, commercial cooperation between the two nations is close.

Rapid population growth after World War Two reflected the booming industrialised economy, boosted by US aid. More recently there has been pressure to reduce the birth rate. Female infanticide, by abortion, is widely practised. Taiwan's population density is the second highest in the world for a major nation, so high that the annihilation of several thousand Taiwanese and the destruction of 100,000 homes in the 1999 earthquake, an event normally classed as natural, could also be described as an anthropogenic (man-made) disaster. Taoism and Buddhism have most followers, but about 5% of the population is Christian, and 1% Muslim.

18 – 180 million

Palestinians by Israelis, they are often referred to as "innocent civilians", but in a world mortally threatened by human overpopulation, aggressive breeding with intent to achieve racial or religious advancement must be regarded as antisocial and bound to cause trouble.

4.5 The Period of Weak Restraints On Growth (WROG)

This is the period of about 250 years between the mid-18th century AD and the present. Before it, for most of the time, restraints on population growth were very strong. Populations pressed vainly against immovable ceilings held down by finite food supplies. Surplus people were eliminated by starvation, disease and warfare, as nations, tribes and individuals competed with each other for land and scarce resources. The Darwinian struggle to survive pervaded society at all levels. Suspicion, fear and conflict between nations, races and creeds were normal and natural.

Then, as the Industrial Revolution began its spate of new procedures, technologies and discoveries that augmented the supply of food and improved the treatment of disease, the ceilings disintegrated, initiating the period of WROG. DC population surges began. To a degree without precedent in human history, populations could expand rapidly without needing to take resources from other populations (although Darwinism still ruled at Imperial, empire-building, levels). Long episodes of tolerance and peace became possible. Punishments became less severe (section 8.1). Charity, generosity towards disadvantaged people, which in the pre-WROG Darwinian world tended to be rare and the prerogative of saintly persons (like, for example, Good King Wenceslas), could now be widely practised. Slavery was abolished.

Humankind was slow to understand that a fundamental change had happened. When Malthus wrote his *Essay* in 1798 his thinking was still influenced by the pre-industrial past. But as the DC surges continued and spread to more and more nations the old regime was gradually forgotten. By the late 20th century, when world population had exploded from 600

TANZANIA   Tanganyika plus Zanzibar   Area 886,000 km² (excluding lakes)

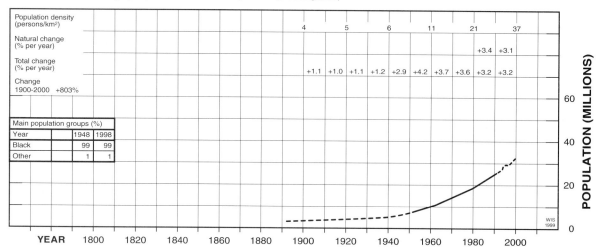

		1948	1998
Main population groups (%)			
Year		1948	1998
Black		99	99
Other		1	1

Nilotic peoples (the Masai) were moving south into Tanzania while Bantu were arriving from the west, one to two thousand years ago. The first millennium AD saw Arab merchants establishing coastal trading posts, which developed into sultanates from which they sent expeditions inland to obtain slaves and ivory. British influence ended the slave trade in the 1870s and the small island of Zanzibar became a British protectorate in 1890, but the mainland (Tanganyika) was allotted to Germany. Several native revolts were bloodily suppressed by the Germans.

Tanganyika was mandated to Britain after World War One. It was agriculturally unexciting, and development was slow but peaceful until independence was achieved under Julius Nyerere in 1961. In Zanzibar an African revolt against the ruling Arab minority ended in massacre or expulsion of most Arabs, and union with Tanganyika, creating Tanzania, in 1964. President Nyerere autocratically began a socialist experiment based on collective agriculture, which largely failed, and he almost bankrupted the country by sending armed forces to take sides in civil wars in Angola, Mozambique, Rhodesia, Uganda and elsewhere. He retired in 1985 and multiparty politics began in 1992. The weak economy is based mainly on peasant farming, with specialist export crops such as cloves and sisal, but wildlife tourism is important. In 2000 there was economic deprivation and separatist agitation in Zanzibar.

As elsewhere in East Africa, periodic droughts are causing food shortages that become more serious as the population rapidly expands. In 1992, 35% of Tanzania's population was Muslim (mainly in Zanzibar and along the coast) and 32% Christian. The blip in the graph (1994–96) represents the influx and later expulsion of half a million refugees from Rwanda.

million in 1750 to 5 billion, WROG conditions were universally thought of as normal. Neo-Malthusians like Paul and Anne Ehrlich who predicted coming disaster in *The Population Bomb* (1968) were derided as doom and gloom merchants.

Under WROG conditions humankind can rapidly create a large population, which cannot be reduced equally rapidly, humanely, because of the long human life span. The attitudes and actions (tolerance and love) that lead to overpopulation are politically correct, unlike those that are caused by overpopulation (conflict and destruction) which are condemned, at least in the developed world. Unfortunately, a nation whose DC surge has continued throughout its WROG period is likely to find itself overpopulated when, for whatever reason, the period ends. If its population is then above VCL, conflict and destruction are inevitable.

By the 1990s it was clear that the worldwide WROG period was drawing to a close, and strong restraints were on the way back. In some nations, growth was terminated as populations reached their VCLs (see WORLD). In others, population cutback was delayed by peacekeepers, or by charitable aid.

Global resources and ecosystems were beginning to fail (see WORLD). Marine fish stocks were so depleted that annual catches passed their peak in 1989. Climate change, the result of excessive pollution of the atmosphere by humans, was beginning. Holes in the atmospheric ozone layer over the Poles were developing for the same reason. The demand for oil would soon exceed supply. It turns out that the weakest restraint of all seems to have been human intelligence (section 8.9).

Guardian (17 July 2001) points out that in a "truly democratic" world government, Britain would have no more votes than Ethiopia, and might not like it. Rightly so, given that ETHIOPIA would have won its voting power not by successfully running a country, but merely by having lots of babies very quickly and irresponsibly. Throughout sub-Saharan Africa the DC population surges which according to Monbiot should entitle entire nations to great power in a democratic world assembly have, in practice, condemned them to failed economies and chaotic governments, in accordance with the Micawberish Rule (section 6.4). "Ten thousand puppy dogs can't be wrong" said a British advertisement in the 1980s (extolling the tensile strength of a toilet paper), but responsible and effective democratic government depends on more than sheer numbers of voters.

One of the gravest threats posed by the *lots of babies* procedure is to developed nations. It is not straightforward. The babies have been and are being born elsewhere, mainly in developing countries of Africa and Asia which are too poor to support them. When the babies reach adulthood they realise their plight and, perfectly naturally, are tempted to emigrate in search of a more agreeable life. Their target has to be the West, where by a significant coincidence *multiculturalism* has become a popular concept among liberal philosophers and human rights activists, who have promoted it so forcibly that most politicians have accepted

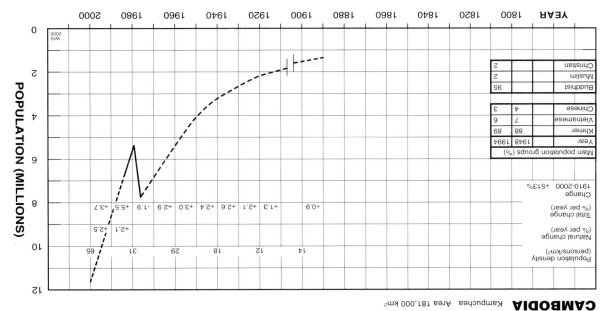

CAMBODIA Kampuchea Area 181,000 km²

Population density (persons/km²)						
Natural change (% per year)						
Total change (% per year)						

Change 1910-2000 +513%

Main population groups (%)		
Year	1948	1994
Khmer	88	89
Vietnamese	7	6
Chinese	4	3
Buddhist	95	
Muslim	2	
Christian	2	

From the 9th to the 14th century the Khmer Empire, which controlled the trade routes between India and China, extended far beyond the boundaries of modern Cambodia. Its rulers were building the monumental temples of Angkor that later were lost in the jungle. The Khmers had absorbed the older Hindu-Buddhist Fou-Nan kingdom, but their empire based on worship of the god-king declined with the advance of peace-loving Buddhism. Its remnants were saved after centuries of Thai and Vietnamese onslaughts by the French, who declared a protectorate in 1863. France ruled Cambodia as part of French Indo-China, adding parts of Laos and Thailand in 1904-07, but after Japanese occupation in World War Two separatist guerrilla activity increased and independence was granted in 1953.

Cambodia then became the theatre of seemingly endless conflict between the national army of King Sihanouk and communist guerrillas backed by China and North Vietnam, intensifying when US and South Vietnamese forces invaded half way through the Vietnamese War (1964–75). In 1975 the communist Khmer Rouge guerrillas were victorious. At once their leader, Pol Pot, began the amazingly brutal transformation of Cambodia into a Maoist peasant-dominated agrarian state. Educated people and minorities were targeted as 'parasites', forced to cultivate the land, starved, or simply executed. Between one and three million are said to have died between 1975 and 1979, when Vietnam invaded and drove the Khmer Rouge into remote forests where they continued guerrilla attacks. Peace was agreed in 1991 and King Sihanouk returned, but the Khmer Rouge remained active and killing civilians on a diminishing scale as its fighters defected until it finally capitulated in 1999. Much of the original rain forest has been destroyed by logging, legal and illegal.

The graph, guided by a census in 1981, indicates a population decline of 2.4 million during the Pol Pot reign of terror, but the subsequent population recovery is so steep that the decline may represent more refugees, who subsequently returned, than deaths. The figures may, of course, be unreliable. 4% of adults were HIV-positive in 2001.

the people', as in Britain when the majority has been shown to favour capital punishment).

However, there is a procedure whereby a minority group, perhaps driven by ethnic or religious ambition, can work its way to the top in a democracy by perfectly legitimate means. It takes time, usually several generations, but the majority and the power it confers, once achieved, can be permanent. The procedure is: *have lots of babies*. Sometimes it just happens, reflecting the group's normal lifestyle (see HUNGARY), but sometimes it is a deliberate policy (see KOSOVO, MACEDONIA), when it equates with *aggressive breeding* (section 4.4).

If the group having lots of babies is already the dominant group in a population, the procedure presents no threat to civil stability. But if a fast-growing majority happens to be governed by a slow-growing, stable or even shrinking minority (e.g. SOUTH AFRICA, KOSOVO), the latter can only keep power by stratagems that defy democracy. In South Africa the stratagem was racist *apartheid*, which withheld the vote from the black majority and aroused such condemnation, worldwide, that the white government capitulated and transferred power by democratic elections in 1991. Draconian policing had prevented serious uprisings against the white minority during the apartheid period.

There were 3.0 million Muslim Arabs in the West Bank and the Gaza Strip in 2000, only half as many as Israel's population (6.2 million, including about a million Arabs), but they were increasing twice as fast. If current growth rates could persist they would outnumber the Israelis in 2024, when both populations would reach about 12 million. With population densities extremely high already (Israel 296, Gaza/West Bank 479) and natural resources especially fresh water in short supply, there could hardly be a more unwise policy than having lots of babies, but religious and ethnic ambition seems oblivious to common sense.

Politically-correct pundits are apt to deplore the dominance of developed nations with small populations in global assemblies like the United Nations. George Monbiot in *The*

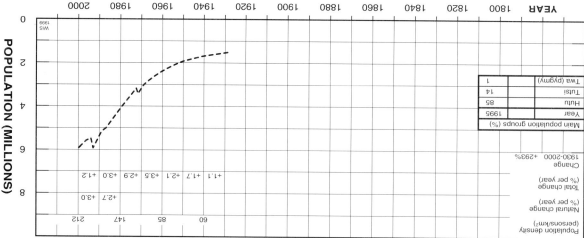

BURUNDI Area 27,800 km²

Main population groups (%)	
Year	1995
Hutu	85
Tutsi	14
Twa (pygmy)	1

Change 1930-2000 +293%

Population density (persons/km²)	60	85	147	212			
Natural change (% per year)			+2.7	+3.0			
Total change (% per year)	+1.1	+1.7	+2.1	+3.5	+2.9	+3.0	+1.2

YEAR: 1800 1820 1840 1860 1880 1900 1920 1940 1960 1980 2000

POPULATION (MILLIONS): 0 2 4 6 8

SWS 1999

The original inhabitants, Twa hunter-gatherers, were displaced by Bantu farmers, the Hutu, early in the second millennium AD. By the 17th century they were dominated by Nilotic Tutsi, cattle-owners who had migrated from lands further east. European colonisation began in 1890 with German military occupation, when the area known (with Rwanda) as Ruanda-Urundi formed part of German East Africa. After World War One Ruanda-Urundi became a Belgian mandate.

Independence came in 1962 with separation from Rwanda. Population density had probably doubled since pre-colonial days. At once, violent rivalry erupted between the ruling Tutsi minority and the Hutu majority. In 1972 the king, a Tutsi, was assassinated. Racial conflict ensued, in which 150,000 Hutus were massacred and 100,000 more fled to Tanzania and elsewhere. There were further uprisings in 1988 and 1993-94, causing 200,000 deaths and the departure of some 800,000 refugees. Since then the country has been anarchic with almost daily ambushes and killings, especially of Hutus in and around the many internal and Rwandan refugee camps, resulting in roughly 1000 deaths per month. A small UN presence does little beyond distributing food and medical aid. 400,000 refugees were camped in Tanzania in 2000.

Recent estimates vary widely, but it would seem that Burundi's population reached Violent Cutback Level (VCL) when its density approached 220. The deciding factors are tribal hatred and the intense competition for agricultural land, given that 95% of the people are subsistence farmers.

UGANDA   Area 197,100 km² (excluding lakes)

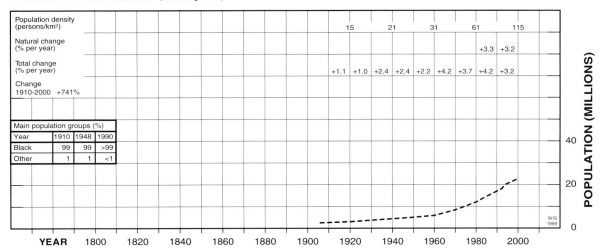

Population density (persons/km²)									15		21		31		61		115	
Natural change (% per year)															+3.3		+3.2	
Total change (% per year)									+1.1	+1.0	+2.4	+2.4	+2.2	+4.2	+3.7	+4.2	+3.2	
Change 1910-2000 +741%																		

Main population groups (%)

Year	1910	1948	1990
Black	99	99	>99
Other	1	1	<1

YEAR   1800   1820   1840   1860   1880   1900   1920   1940   1960   1980   2000

POPULATION (MILLIONS)   40   20   0

WIS 1999

Uganda was home to many Nilotic and Bantu tribal kingdoms during the second millennium AD, including the 15th century Bunyoro which gave way in the 18th century to the powerful Buganda state, ruled by its Kabaka. Not until the later 19th century did foreigners: Arab traders and British explorers, penetrate the country, bringing Islam and Christianity. A religious war soon followed. Britain took the opportunity to help the Bugandans to win, and declared a protectorate over Uganda in 1894.

Uganda did not attract many European settlers and the nation moved smoothly to independence in 1962. Tribal strife followed in which the Bugandan Kabaka was deposed by Milton Obote who as President was deposed in turn by Idi Amin. Amin's reign of corruption and terror (1971–79) saw 300,000 Ugandans selected, on tribal, ethnic or class bases, for elimination. The Asian population of some 60,000 was expelled and foreign assets were seized.

Uganda fell into chaos and Amin fled abroad after unsuccessfully invading Tanzania. Obote returned, but financial and political bankruptcy continued. In 1986, after a civil war, Obote was ousted and Yoweri Museveni was installed as leader.

Guerrilla activity continues, especially in border regions, e.g. close to Sudan where refugees have settled. As elsewhere in Africa, children are kidnapped, recruited to rebel armies and desensitised by being forced to commit atrocities. Uganda has recovered economically to some extent, but foreign aid comprised half the budget in 2001. Several hundred thousand refugees have returned, including some of the expelled Asians. 1994 saw a large influx of refugees from Rwanda. Life expectancy in Uganda fell from 47 years in 1985 to 41 in 1998, due to the HIV/AIDS pandemic. In 2000 Ugandan troops were fighting in eastern Congo, with dubious objectives.

18 – 180 million

anthropogenic disasters. The events themselves will be bad enough, but the cost of repairing the damage will eventually cripple a world made poor by overcrowding. The ultimate anthropogenic disaster may be caused by global heating. If and when sea level rises, and a billion-plus people are flooded out of their coral island and coastal lowland homes, the consequences will be very interesting – in the sense of the old Chinese curse: "May you live in interesting times".

4.7   HIV / AIDS

This pandemic disease, first recognised as late as 1983 and possibly non-existent in humans before the 1950s (it may have jumped the species barrier from chimpanzees about then), is likely to kill more people in less time than even the Black Death (see WORLD). There is as yet no cure. By 2000 some 58 million people had been infected, worldwide, 22 million of whom had died: figures which in purely numerical terms are fairly insignificant when compared to the world population increase of about 80 million *each year*, but the annual infection rate appears to be expanding exponentially (5.3 million in 2000). If so, it could overtake the population increase by 2020, when the total number of cases since 1980 would have been around 500 million, compared to world population growth of between 2 and 4 billion over the same period. Thereafter, it could cause the first actual shrinkage of world population since the 14th century.

A potentially limiting factor is that, unlike most diseases, it is for the most part an affliction acquired *voluntarily*, in the sense that it could be avoided by taking appropriate precautions, i.e. practising 'safe sex' or shunning used needles for drug injection. The

UKRAINE   Area 604,000 km² (after 1954)

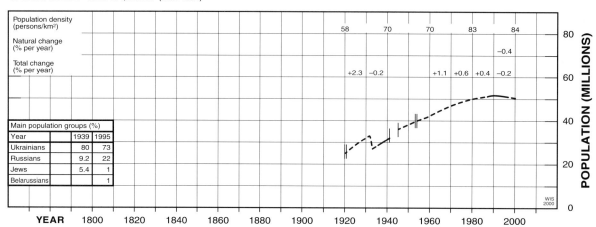

			58	70	70	83	84
Population density (persons/km²)							
Natural change (% per year)						–0.4	
Total change (% per year)			+2.3	–0.2	+1.1	+0.6	+0.4	–0.2

Main population groups (%)

Year	1939	1995
Ukrainians	80	73
Russians	9.2	22
Jews	5.4	1
Belarussians		1

YEAR 1800 1820 1840 1860 1880 1900 1920 1940 1960 1980 2000

POPULATION (MILLIONS): 80, 60, 40, 20, 0

WIS 2000

The first Russian state, centred on Kiev, was established by Rus (Viking) chiefs who overcame warring Slav tribes in the 9th century AD. Kievan Russia (Ukraine) survived until the 13th century when its vast steppes were over-run by Mongol hordes. Later, Ukraine was incorporated into the empires of Poland and Lithuania. After the 16th century Cossack outlaws won much of Ukraine from the Poles and allied themselves with Tsarist Russia. On the partition of Poland in 1793 Ukraine became part of the Russian Empire, except for Galicia in the west which joined Austria.

After World War One Galicia passed to Poland, but the rest of the Ukraine was designated a Soviet Socialist Republic in 1921. In a severe famine (1921–22) 2.5 million Ukrainians died. Forcible collectivisation of agriculture, ploughing of the steppes and expropriation of food by the Soviets caused a further famine (1932–33) with 7 million deaths. Galicia was returned to the

Ukraine when Russia seized eastern Poland in 1939. World War Two saw mass exterminations of Ukrainian Jews by the occupying Nazis. Parts of Romania and Czechoslovakia (Ruthenia) were added to Ukraine in 1945 and the Crimea was transferred from Russia in 1954. The country became independent in 1991 after the Soviet Union collapsed.

Government since independence has been feeble and allegedly corrupt. Although the economy is based on abundant resources, heavy industry, mining of coal, iron and manganese, and agriculture, it is still weak, though receiving much foreign aid. Reserves of oil and gas, once large, are approaching exhaustion. The world's worst nuclear accident occurred at Chernobyl in 1986; 10,000 square kilometres of land were designated as too radioactive for human occupation. As in other ex-Soviet Slav nations, a population decline began following the collapse of the Soviet Union in 1991.

unfortunate exceptions, who are infected in spite of all their carefulness or innocence, include people who receive infected blood transfusions, babies born HIV-positive, spouses with infected partners, and women who are raped. In SOUTH AFRICA, in 2000, three rapes were said to occur every minute (1.5 million in a year), and 25% of women between 20 and 29 were HIV-positive, according to a UN study.

The incidence of the disease in South Africa is mirrored in other countries of southern and eastern Africa (e.g. BOTSWANA). Probably, in a few years' time, a majority of each national population will be infected. By 2010 the populations of these countries could be falling significantly, as deaths greatly exceed births.

The disease is less advanced in the vast populations of East and South-east Asian countries, but it appears to be developing along the same lines. India was said to have 4 million cases in 2000. Several ex-Soviet nations, both Slav and Muslim, are thought to be seriously affected because of drug addiction and unsafe sex.

It remains to be seen whether the potentially limiting factor: people deciding voluntarily to abstain from unsafe sex and unsafe drug-taking, will reverse the trend and control the disease. Possibly, of course, resistance will develop, or a cure or vaccine will be found. On balance, it seems likely that the pandemic will get much worse before it gets better. Charles Darwin might interpret it as natural selection in action. The fittest, who will survive, are those who look to their futures and safeguard them by behaving with caution. *Homo* truly *sapiens*, perhaps.

4.8   Tragedy of the Commons

This expression was coined by Garrett Hardin (*Living within Limits*, 1993). "The tragedy of common ownership" is perhaps more easily understood.

UNITED KINGDOM Area 244,000 km²

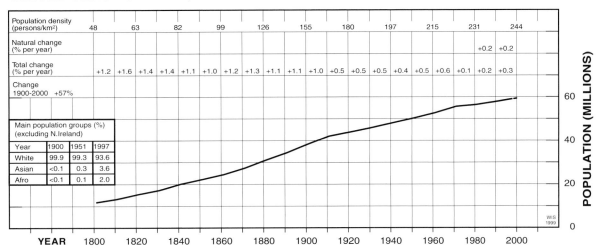

| Population density (persons/km²) | 48 | 63 | 82 | 99 | 126 | 155 | 180 | 197 | 215 | 231 | 244 |

| Natural change (% per year) | | | | | | | | | | +0.2 | +0.2 |

| Total change (% per year) | +1.2 | +1.6 | +1.4 | +1.4 | +1.1 | +1.0 | +1.2 | +1.3 | +1.1 | +1.1 | +1.0 | +0.5 | +0.5 | +0.5 | +0.4 | +0.5 | +0.6 | +0.1 | +0.2 | +0.3 |

Change 1900-2000 +57%

Main population groups (%) (excluding N.Ireland)

Year	1900	1951	1997
White	99.9	99.3	93.6
Asian	<0.1	0.3	3.6
Afro	<0.1	0.1	2.0

WIS 1999

YEAR 1800 1820 1840 1860 1880 1900 1920 1940 1960 1980 2000

POPULATION (MILLIONS) — 60, 40, 20, 0

The early history of the United Kingdom is described in the entries for the component countries: ENGLAND, IRELAND (Northern), SCOTLAND and WALES. The UK graph, based on censuses every tenth year from 1801 (except 1941), shows the steady but decelerating growth over more than 2 centuries typical of Western nations that participated in the 18th century Industrial Revolution and are still experiencing their DC population surges. Its shape closely follows that of England, which contains 81% of UK population. The first kink in the graph, before 1920, is linked to the heavy loss of young men (some 850,000 killed and 2 million wounded) in World War One, and economic depression in the 1920s and 30s; the second (1970–80) has been explained as reflecting the introduction of the contraceptive pill and the legalisation of abortion.

By the end of the 20th century immigrants to the UK greatly outnumbered emigrants. In the last years of the 1990s net immigration exceeded natural change (births minus deaths) as the cause of population growth. 8000 illegal immigrants arrived hidden in trucks in 1998. In 2002 a Government report predicted that ethnic minorities would account for half the increase in the working population to 2010.

18 – 180 million

Fish in the ocean belong to nobody (except in 'territorial waters' claimed by coastal nations). They are a resource in common ownership. If and when the oceanic fish catch is so great that the total resource begins to shrink, no one nation will try to conserve what is left by unilaterally reducing its catch. If it did, there would simply be a little more for less scrupulous nations to take for themselves.

On the other hand, nations controlling extensive territorial waters can, in theory, restrict fishing in them in order to conserve fish stocks for the future. Some coastal nations claim exclusive rights to fishing within a 200 mile wide 'economic zone'. In practice, short-term exigencies often lead to agreements with other fishing nations that erode national fish stocks. A few clear-sighted governments have avoided overdoing this risky procedure; Iceland's care for its stocks of cod is an outstanding example. The UK lost its chance to protect its own fish stocks when it made them a 'common' by joining the European Common Market.

People used to say "there are more fish in the sea than ever came out of it", but the concept became unlikely many years ago and now it is certainly untrue. It was quoted to refute accusations of imprudent or wasteful consumption of a resource. Equally spurious claims could be made for other resources such as oil in the ground, or trees in the forest.

The real tragedy of the commons is the inability of nations to co-operate in sustainably managing a common resource. As Hardin remarks, the 'dream of a global village' in which responsible villagers take only their fair share of global resources, has dissolved into a Darwinian 'nightmare of global pillage'.

Our most vital common is the world's atmosphere. At the Hague Conference of November 2000, nations failed to agree on measures to reduce atmospheric pollution, largely because (especially in the case of the USA, which produces a quarter of the world's greenhouse gas emissions) governments fear that restricting their peoples' use and enjoyment of the motor car would make them unpopular. If such straightforward planning to escape

UZBEKISTON   Uzbekistan   Area 447,400 km²

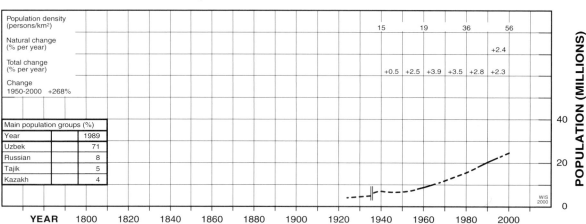

| | | 15 | 19 | 36 | 56 |
Population density (persons/km²): 15, 19, 36, 56

Natural change (% per year): +2.4

Total change (% per year): +0.5 +2.5 +3.9 +3.5 +2.8 +2.3

Change 1950-2000 +268%

Main population groups (%)		
Year		1989
Uzbek		71
Russian		8
Tajik		5
Kazakh		4

POPULATION (MILLIONS): 40, 20, 0

WIS 2000

YEAR 1800 1820 1840 1860 1880 1900 1920 1940 1960 1980 2000

The Kizyl Kum desert occupies west and central Uzbekiston and includes half the Aral Sea, a saline lake bigger than Switzerland but now shrinking dramatically as its feeder rivers are diverted into canals irrigating cotton plantations. The Silk Route of trade between the Chinese and Roman empires traversed the mountains, oases and fertile valleys of east Uzbekiston, which from antiquity were conquered by invaders from Persia (Cyrus the Great), Macedonia (Alexander the Great) and by Arabs and Turks spreading Islam in the 8th century AD. The Mongols under Genghis Khan sacked Bukhara and Samarkand and butchered their half-million citizens in the 1220s.

Another Mongol, Tamerlane, made Samarkand his magnificent capital a century later. The region was annexed to the Russian Empire in the 1860s. Uzbek peasants rose several times against Russian plans for giant cotton plantations and in 1924 the Soviets created the Uzbek Soviet Socialist Republic. The present borders were finalised in 1936. During World War Two Stalin relocated 160,000 Georgians to Uzbekiston. From 1957 huge Soviet irrigation projects began the desiccation of the Aral Sea and the poisoning of its surroundings by wind-blown salt and agrochemicals: one of the world's greatest ecological disasters.

Uzbekiston became independent, and changed its name from Uzbekistan, when the Soviet Union collapsed in 1991. Most of the population is Sunni Muslim, but the government is currently secular and is harassed by Muslim fundamentalist guerrillas infiltrating from Tajikistan.

catastrophic planetary pollution (climate change) is beyond them, it must be silly to hope that agreement can be reached on the even more urgent need to reduce the numbers of polluters – us.

• • • • • • • • • • • •

"If you deny your history, you cannot learn from it"

(Tony Benn)

Chapter 5

Myth and Reality in the Population Debate

Every now and then a letter or article appears in the media discussing some aspect of human population statistics and its relevance to human society. Some are well thought out, others not. As an example of the latter, a Mr Templar wrote to the *Western Daily Press* on 9 January 1995 claiming that because the total world population of 5 billion could stand on the Isle of Wight (area 380 square kilometres) the world is not overcrowded. The irrelevance of this statement is compounded by the fact that, even assuming that 10 people can squeeze onto one square metre, it ceased to be true about 1970.

This chapter examines some of the subjects most frequently discussed.

5.1 Percentage Growth and Numerical Growth – the Tricky Difference

The growth of a population from year to year can be quoted either as the extra number of people (the numerical increase) or as the percentage increase. For example, the population of BRAZIL rose by 19 million in the decade 1950–1960, an average increase of 3.5% per year. From 1990 to 2000 the numerical growth was greater, 25 million, but the percentage increase was only 1.7% per year. The reason is a simple mathematical one: Brazil's population in 1950 was 52 million, whereas in 1990 it was 144 million. Obviously, 19 as a percentage of 52 is greater than 25 as a percentage of 144.

A similar calculation is sometimes expressed in terms of fertility rates. The *total fertility rate* (TFR) of a nation is the total number of children born to an average woman of that nation. It varies year on year. In Gaza Strip and Congo Democratic Republic it is now about 7; in Pakistan about 6; in India about 3; in China 1.8 and in Europe 1.4 (Population Reference Bureau statistics, 2001). TFR figures have declined in most countries in recent decades, but numerical growth has not necessarily followed suit, usually (net migration can confuse the issue) for the same reason as applies to percentage growth. The average woman may have fewer children, but there are more average women. Even when TFR drops to the 'threshold' of 2.1 that would, if maintained, create a population that neither rises nor falls, stability would not be reached for many years as the preceding 'bulge' worked through several generations.

Why is the distinction between numerical growth on the one hand, and TFR and percentage growth on the other, important? It is because journalists, charity workers and others, from Fred Pearce (e.g. in *New Scientist*, 11 July 1998), to Matt Ridley (e.g. in *Sunday Telegraph*, 9 November 1997), to Bernadine Prat of Oxfam (*in litt*, 14 November 1995), have interpreted declining TFR and percentage growth figures to mean that population growth also is declining. Malthus was misguided, they deduce, and the real world demographic crisis is likely to be too few people, not too many.

The graphs in this book represent numerical growth. Data rows alongside them allow direct comparison with percentage growth. Draw your own conclusions!

VENEZUELA Area 916,500 km²

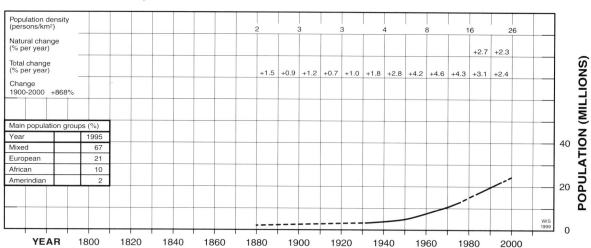

Population density (persons/km²)			2	3	3	4	8	16	26
Natural change (% per year)								+2.7	+2.3
Total change (% per year)			+1.5	+0.9	+1.2	+0.7	+1.0	+1.8	+2.8	+4.2	+4.6	+4.3	+3.1	+2.4
Change 1900-2000	+868%

Main population groups (%)		
Year		1995
Mixed		67
European		21
African		10
Amerindian		2

Located by Columbus in 1498 and named 'Little Venice' because some of the native Amerindian coastal villages were built on the water, Venezuela belonged to the Spanish viceroyalties of Peru and New Granada until the early 19th century. A long war for independence was successfully completed in 1830. After a series of military dictatorships, democratic government was established in the 1950s.

Huge oil resources were discovered in 1915. Oil production on a large scale began in the 1920s, but population did not begin to boom until modern medicines and agrochemicals revolutionised death control (the DC population surge) in the 1940s. Although forests cover about 30% of the country, and agriculture, tourism and manufacturing are important to the economy, oil and oil products provide nearly 90% of Venezuela's export revenue. They have given the country the highest average income *per capita* in Latin America, but due to the uneven distribution of wealth, 80% of the population was living below the poverty line in 2000.

5.2 The Demographic Transition Theory

For six decades the existence of this theory has enabled politicians, business interests, charities and others to assume that human population growth is not a serious problem. As explained by Adolphe Landry in 1934, the theory proposes that as people become more prosperous they will, for a variety of reasons, have fewer children. On this basis the population of a poor country can be stabilised by providing such financial and technical aid as is necessary to end its poverty (and the peasant's traditional need for many children to care for him in his old age). The proposition is attractive because the delicate subject of population control is avoided by deliberate generosity, which may, as a bonus, induce feelings of righteous pleasure in the donor. And when all humankind is lifted out of poverty, world population growth will naturally cease, according to the theory.

Unfortunately for the world, neither the theory nor its practical application have had much success, as the graphs in this book demonstrate. The most positive results have been in prosperous European nations, where the theory is persuasive *at the personal level*. In Italy the TFR (total fertility rate) has dropped to 1.2, compared to the 2.1 necessary for population stability, and in spite of much immigration the population shrank very slightly in the 1990s. Spain's population growth almost ceased in the 1990s. In the United Kingdom the TFR fell below replacement level in 1972 but the population has continued to increase at 0.2% to 0.3% per year, mainly due to the surge in immigration from –30,000 in 1981 to +185,000 in 1998 (Shaw, 2000). The French population appeared to stabilise in the 1930s and 1940s, but then resumed growth. In spite of its manifest affluence the USA's population increases by nearly 3 million (1.1%) per year, due largely to immigration and the high birth rate among recent immigrants. In Japan, where immigration is low, TFR has fallen to 1.3, but the population was still just rising in 2001.

Poverty in most developing nations is getting worse, no matter how much aid they have received. That is inevitable when population growth, stimulated by aid, outpaces economic growth (section 6.4, the *Micawberish Rule*). Donor nations may not realise it, but in practice,

VIETNAM Area 330,500 km²

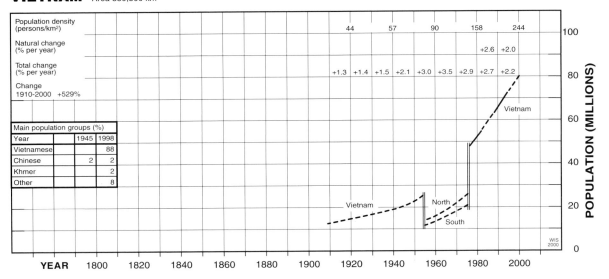

Population density (persons/km²)			44	57	90	158	244
Natural change (% per year)						+2.6	+2.0
Total change (% per year)			+1.3 +1.4 +1.5 +2.1 +3.0 +3.5 +2.9 +2.7 +2.2

Change 1910-2000 +529%

Main population groups (%)

Year	1945	1998
Vietnamese		88
Chinese	2	2
Khmer		2
Other		8

Northern Vietnam was ruled by China for a thousand years before the 10th century, when one of the repeated Vietnamese rebellions finally succeeded. The Vietnamese then pushed south into the Champa kingdom which they subjugated in the 15th century. By 1800 they had conquered Cochin-China in the extreme south, creating the Vietnamese Empire of Annam. In the mid-1800s France attacked the empire and negotiated protectorates over it. With Laos and Cambodia, it became French Indo-China in 1887. The French developed a colonial economy specialising in rice and rubber.

After Japanese occupation in World War Two France fought and lost an eight-year war against the communist Vietminh guerrillas, conceding independence in 1954 when the country split into a communist North Vietnam and a non-communist US-backed South Vietnam. The Vietnam War began in 1959 with a revolt by communist Viet Cong guerrillas in the South, backed from the early 1960s by Northern troops. It ended in 1975 when, in spite of massive US support, the South was defeated and some 200,000 refugees fled the country. The new communist Vietnam, united from

1976, began an invasion of Cambodia in 1978 and fought a border war with China in 1979 which caused the exodus by sea of up to 700,000 ethnic Chinese and other 'boat people', seeking asylum in Hong Kong and elsewhere. Many were sent back to Vietnam, and 40,000 were still detained in camps in Hong Kong in 1995. China deported them when it took over Hong Kong in 1997.

Since the Soviet Union's collapse in 1991 the regime has become more liberal, but the country remains very poor. Rice is the main export, but there are large offshore oil deposits (claimed by China). The densely populated Mekong Delta in the south is subject to disastrous floods (especially in 1999 and 2000), accentuated by deforestation further upstream. Like Bangladesh, it will be one of the first world regions to be damaged by rising sea levels. Peasant farmers, including some 50 ethnic minority forest tribes, comprise 70% of the work force. Taoism and Buddhism are the dominant religions. The high fast-rising population density is devastating wildlife habitats. Officially, families of more than 2 children are discouraged.

population control must *precede* poverty reduction, not the other way round. The other essential is peace: poverty is vastly increased by destruction of private and public assets, that war causes.

High birth rates and high TFRs are the norm in oil-rich Muslim nations such as SAUDI ARABIA. Influxes of foreign workers lift total growth rates to astronomical levels (e.g. 33% per year in the 1970s in the United Arab Emirates). In traditionally patriarchal Muslim societies, whether rich or desperately poor like AFGHANISTAN or the GAZA STRIP, women bear many children. Islam is a particularly ambitious faith, and Landry's theory is irrelevant to Muslims who sense that more believers equate to more power in their struggle to advance Islam.

The theory received a severe blow in the 1990s when the populations of a sharply-defined group of nations really did begin to decline. They were not prosperous nations – quite the opposite. They were independent mainly Slav nations, 15 in all, freed by the collapse of the Soviet Union in 1991. Some had been part of the USSR, some its East European satellites (Table 3.1). Their economies were struggling, their people poverty-stricken and bewildered by the sudden change from an all-powerful communist regime into a sort of vacuum with capitalist connections (see RUSSIA). YUGOSLAVIA, which was mainly Slav but independently communist and more prosperous until break-up in 1991–92, has developed differently; the populations of well-off VOJVODINA (part of Serbia) and breakaway SLOVENIA have begun to decline, but the other component populations, especially Muslim

YUGOSLAVIA   Area 255,800 km² to 1992, then 102,200 km²

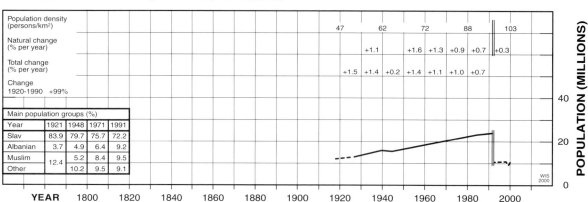

Population density (persons/km²)											47	62	72	88	103	
Natural change (% per year)											+1.1		+1.6	+1.3	+0.9	+0.7	+0.3
Total change (% per year)										+1.5	+1.4	+0.2	+1.4	+1.1	+1.0	+0.7
Change 1920-1990 +99%																

Main population groups (%)

Year	1921	1948	1971	1991
Slav	83.9	79.7	75.7	72.2
Albanian	3.7	4.9	6.4	9.2
Muslim	12.4	5.2	8.4	9.5
Other		10.2	9.5	9.1

YEAR   1800   1820   1840   1860   1880   1900   1920   1940   1960   1980   2000

POPULATION (MILLIONS)   40   20   0

This was a valiant but eventually (and predictably) unsuccessful attempt to unite eight proudly distinct ethnic and cultural groups into one 'multicultural' nation. For two millennia the inhabitants of the western Balkans (mostly Slavs with widely differing tribal loyalties) had suffered from marauding armies, including the Romans, Turks, Habsburgs and Nazis. Each occupying power favoured some groups over others. Centuries of inter-group rivalry, involving dreadful cruelty and uncounted savage massacres, had established a climate of deep-seated ethnic mistrust and hatred.

The Balkan War of 1913 ended 500 years of Ottoman Turkish rule, during which important Bosnian and Albanian populations adopted the Muslim faith. Christianity was dominant in the other groups. All groups inhabited tribal homelands recognisably similar to those they occupy today. The Kingdom of Serbs, Croats and Slovenes, re-named Yugoslavia (Land of the Southern Slavs) in 1929, was proclaimed immediately after World War One. Narrowly ethnic political parties were banned, and new administrative provinces were set up which fragmented the old tribal areas (and devalued the 1931 census, population-wise). But tribal prejudices persisted, and were exploited in World War Two by the Nazis who created a Croatian puppet state in which there took place deeds of "appalling atrocity and bestiality" (Singleton, 1985), including the slaughter of some 350,000 Serbs.

When the war ended in 1945 the old tribal provinces were reinstated, and in spite of its internal tensions the new communist Yugoslavia was held together by the iron grip of its President, the partisan war hero Marshal Tito. He died in 1980, aged 88, and the leaders who followed him were unable to deal with the economic rivalries and intensifying nationalisms within multicultural Yugoslavia (sharpened in particular by the growing proportion of Muslims/ Albanians in the population, at the expense of the Slavs), which culminated in brutal civil war, 1991–92, and secession of the tribal provinces of SLOVENIA, CROATIA, BOSNIA and MACEDONIA as independent nations. Post-1992 Yugoslavia consists only of SERBIA, including VOJVODINA and KOSOVO, and MONTENEGRO. The lesson of Yugoslavia is that multiculturalism can work, but only when the population is contented, or else controlled with an iron hand.

BOSNIA and KOSOVO, have continued to rise.

So Landry's theory has proved *correct* in that the populations of many prosperous developed nations would be stable or declining, with huge benefits to their national environments and to global sustainability, were it not for immigrants flooding in from the Third World, hoping to copy the wasteful and polluting Western lifestyle (section 6.4). The theory has proved *incorrect* with regard to the poor ex-communist Slav nations of Eastern Europe whose populations are declining. It has proved incorrect as regards some rich Muslim states whose populations are rocketing. It has proved *worse than useless on the global scale* because it has persuaded governments and aid charities to concentrate on economic development and saving lives, while ignoring birth control and family planning. The consequences to the world's poorer nations, over the six decades that the theory has been fashionable, have been devastating.

If the ecosystems of Planet Earth are failing to cope with the destructive and polluting activities of humankind, it is in large part because Landry's convenient theory has persuaded decision takers and their advisors that population growth is not a serious problem because financial aid will sort it out.

5.3 Sustainable Development

This is another concept, or catch-phrase, that induces complacency by frequent repetition. If examined more than superficially, it is found to be an oxymoron, a self-contradiction, in today's world.

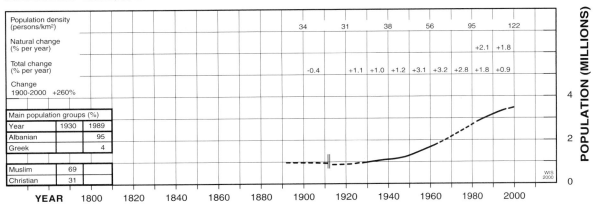

ALBANIA Area 28,700 km²

Population density (persons/km²)								34	31	38	56	95	122	
Natural change (% per year)												+2.1	+1.8	
Total change (% per year)								-0.4	+1.1	+1.0	+1.2	+3.1	+3.2	+2.8	+1.8	+0.9

Change 1900-2000 +260%

Main population groups (%)		
Year	1930	1989
Albanian		95
Greek		4
Muslim	69	
Christian	31	

YEAR 1800 1820 1840 1860 1880 1900 1920 1940 1960 1980 2000

POPULATION (MILLIONS)

WIS 2000

Part of the classical Illyria, Albania was incorporated into the Roman Empire and remained under Byzantine rule until it crumbled into feuding tribal kingdoms in the 6th century. From 1479 it belonged to the Ottoman empire. When the Turks were driven out in 1912 their provinces of Scutari and Yanina, with parts of Kossovo and Monastir, came together in an anarchic regime which after World War One became Albania, under King Zog. After World War Two Albania was resurrected as a People's Republic, linked first to the Soviet Union and then, uniquely in Europe, to China. In the late 1980s, with Albania the poorest country in Europe, relaxation of the strict communist regime began. Subsequent history has been chaotic. Unemployment, riots and harassment of the Greek minority have intensified emigration. In 1997 widespread financial frauds caused a minor civil war that was ended by UN intervention. Nearly 500,000 refugees fled the 1999 Kosovo war into Albania and returned a few months later.

Albania is the only Muslim nation in Europe. Since the 1940s its natural population growth rate has been two to three times faster than any other European country, worsening its chronic poverty. The lower gradient of the graph after 1980 is largely due to emigration. Albania's neighbours, especially Greece, blame Albanian immigrants for banditry and systematic crime, and for smuggling and piracy in the Adriatic Sea. They increased ethnic tension in Kosovo and Macedonia. In 1997 the government claimed that exporting people, to send home part of their earnings, was vital to the economy. In 2001 about one third of national income was money sent home by 700,000 emigrants.

Sustainable development "enables the present generation to satisfy its needs while ensuring that future generations can also satisfy theirs". So nations that practice it successfully must never run short of basic resources such as *food, fresh water, energy, climate stability, and adequate living space (land per person)*. On a higher plane they must ensure maintenance of a benign environment, no loss of biodiversity, no worsening of personal security, pollution, congestion, corruption, etc. Unfortunately, one has only to contemplate the current state of the world to see that vast numbers of the *present generation* cannot satisfy their needs *now*. Future generations will be far worse off, as population grows and the basic resources have to be shared among more and more people.

From North Korea to sub-Saharan Africa, *food* shortages and famines are estimated to kill between 5 and 20 million people every year. In 1990 the charity WaterAid calculated that nearly 2 billion people lacked "reasonable access to *safe water*". 700 million people had acquired it, worldwide, in the 1980s, but the achievement was more than offset by world population growth of 800 million over the same period. As regards *energy*, the demand for oil is expected to exceed supply in the present decade (Campbell, 1997, 2003), after which it will rapidly become a rich person's commodity. Future generations will be much less mobile than ours (commercial aircraft driven by nuclear, steam, electrical or wind power have yet to be invented).

By 1995 most scientists had accepted the reality of *climate change* (see WORLD) induced by human pollution of the atmosphere, and by 2002 the media were blaming it for the severe climatic events, especially floods and droughts, that were becoming unusually frequent around the world. According to OPT criteria, the world has about 10 billion hectares of 'ecologically productive land' (after deducting unproductive land such as deserts, mountains and icecaps). So in 2000 the *productive land per person*, worldwide, was 1.7 hectares (4.2 acres), whereas in 1987, with a smaller population, it had been 2 hectares.

ANGOLA   Portuguese West Africa   Area 1,247,000 km²

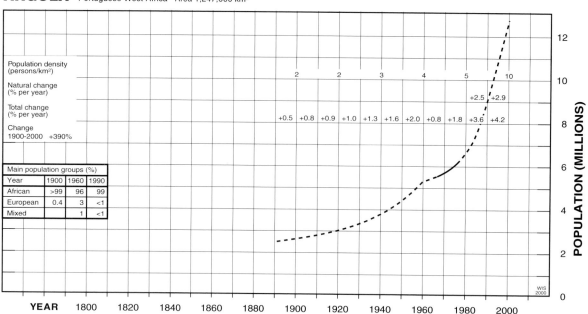

Angola was the home of warring Bantu tribes when Portuguese navigators reached it in the 1470s. The first Portuguese traders and missionaries penetrated far inland, but the main settlements were established along the coast in the 16th century. They traded with inland tribes for goods and slaves, hundreds of thousands of whom were exported to Brazil before the slave trade was officially abolished in 1836. Similar numbers were traded illegally in the later 1800s. From the 1870s Portugal gradually overcame native resistance and occupied the Angolan interior. Modern borders were defined in 1926. There followed a phase of accelerated European settlement. Coffee, diamonds, and later oil, were major exports. The country was designated an overseas province of Portugal in 1951.

Half a century of peace ended in 1961 when a revolt, organised in the ex-Belgian Congo, broke out in North Angola. The death toll in the initial massacres and reprisals was around 50,000. Up to a million refugees fled to the Congo. The revolt had been totally unexpected and Portugal was unable to prevent it spreading slowly to the whole country, as other groups backed by external powers sought to gain financial or political advantage. Portugal withdrew from all her African territories in 1975 and most Europeans departed. Since 1975 Angola has been racked by civil war, with c. 1 million deaths, financed by oil revenues (the government) and illegal diamond sales (UNITA rebels). Population estimates since an attempted census in 1970 must be highly unreliable. In the 1990s the people of this potentially rich nation were impoverished, harassed by lawless soldiers and militias, and threatened by millions of carelessly-laid land mines. There were 8000 UN 'peacekeepers' in the country in 1995. Sleeping sickness has returned. Life expectancy has fallen from 47 in 1970 to 42 in 2000. To those who remember it, the colonial period must seem like a golden age.

Clearly, the availability of basic resources to even the present generation is inadequate, and usually on a marked downward trend. The same applies to the more aesthetic resources listed above, and to many specific resources such as tropical rainforests, tigers in Asia, fish in the oceans, or limestone hills in Somerset, England. Half the world's wetlands and natural forests disappeared in the 20th century, according to the World Resources Institute in 2000. In 1900 there were 315,000 orang-utans in the world; a century later there were 21,000.

Given that 6 billion humans today are failing to conserve a single planetary resource of any significance, it is pretty obvious that the present generation is helping itself lavishly to resources which, if it was genuinely practising sustainability, it should be protecting for the use and enjoyment of its predicted 8 billion children, 10 billion grandchildren, etcetera. Not forgetting that a third at least of the present generation, in the developing world, has so few resources at its disposal that it lives in abject poverty.

Who are we kidding, we in the West, when we "do our bit for sustainability" by sending off bundles of old newspapers, or boxes of aluminium bottle tops, for recycling? Landfill, the cheap but unfashionable alternative to recycling, can when thoughtfully used be of great sustainability value, raising farmland to above the reach of floods (Stanton, 1990) and creating carbon sinks within which most plastics endure indefinitely and even biodegradable paper and food survive almost unchanged for decades – as was proved in the 1980s by sampling New York's vast heaps of garbage.

ARMENIA Area 29,800 km²

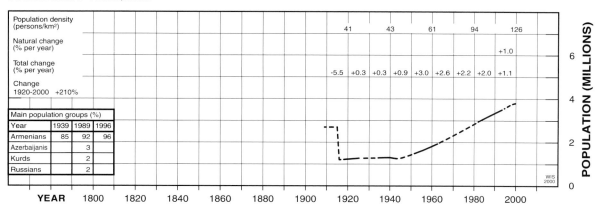

Population density (persons/km²)			41	43	61	94	126
Natural change (% per year)						+1.0
Total change (% per year)			-5.5 +0.3 +0.3 +0.9 +3.0 +2.6 +2.2 +2.0 +1.1
Change 1920-2000 +210%			

Main population groups (%)			
Year	1939	1989	1996
Armenians	85	92	96
Azerbaijanis		3	
Kurds		2	
Russians		2	

POPULATION (MILLIONS)

YEAR   1800   1820   1840   1860   1880   1900   1920   1940   1960   1980   2000

Modern Armenia, on the southern slopes of the Caucasus mountains, is smaller than the ancient kingdom of Armenia which included parts of Turkey and Persia (Iran). Under the Romans, when Armenia adopted Christianity, its territory extended to the Mediterranean. Partition between the Byzantine and Persian empires was followed by independence, 886–1045. Mongol hordes overran Armenia in the 1380s and massacred more than half the population. Four centuries of Ottoman Turkish rule followed during which the Armenians suffered religious persecution but retained their Christianity. Russia annexed the eastern region in 1828 but Turkish harassment continued and culminated, in 1915, in the horrific expulsion of some 1.5 million people. About half of them died en route to Syria and Palestine and most of the rest emigrated, adding to the worldwide Armenian diaspora.

After World War One Armenia acquired its present boundaries and joined Georgia and Azerbaijan in a Transcaucasian Soviet Socialist Republic. In 1936 it became an SSR in its own right, the smallest in the Soviet Union. In 1988 the population of Nagorno-Karabakh, an enclave of Christian Armenians in adjacent Muslim Azerbaijan, petitioned to join Armenia; this led to intermittent war with Azerbaijan, 1989–94, but no agreement has been reached and Armenia still occupies some Azeri territory. When the Soviet Union collapsed in 1991 Armenia declared independence and the rate of population growth slowed. In 1999 the economy was still suffering the effects of a major earthquake in 1989.

Sustainable development is not just a farce, for most of the world's people it is a cruel farce.

5.4 Maximum Population

In the mid-1990s United Nations demographers predicted possible futures for world population growth. Their graph shows smooth accelerating growth from half a billion people in 1800 to 5.5 billion in 1995, when the line splits into 3 alternative scenarios. In the "High" scenario, population reaches 11 billion in 2100 and continues to rise. In the "Low" scenario it peaks at about 7 billion in 2050 and falls steeply thereafter. Between the two is a "Medium" scenario in which population reaches 10 billion about 2080 and stays just above that level well into the 22nd century.

The graph from 1800 to the present is a curve based on existing data. From 2000 to 2120 the 3 alternative scenarios are best guesses. The fact that their graphs are gentle curves smoothly extending the past into the future strongly suggests that they are simple extrapolations of past trends, drawn assuming that there will be no sudden large disruptions to 'business as usual' for at least 120 years. Much has happened in the last 5 years to challenge such a view.

The world's DC population surge (section 4.3) of the 19th and 20th centuries adapted itself with great precision to the familiar global geography of our times. Deserts support few people, except where water is provided artificially. High mountains and icecaps are almost uninhabited. The densest populations are found where land is fertile, the climate is benign, and there are well-established trade routes, especially along coasts, for the movement of people and materials. Like traffic in a busy city, the world DC surge is a complex system that proceeds smoothly until, at some point, something goes wrong. A monumental snarl-up follows. In our complex global society, inflexible because it is fine-tuned to the nth degree, snarl-ups when they come will be drastic.

1.8 – 18 million

AUSTRIA Area 83,850 km²

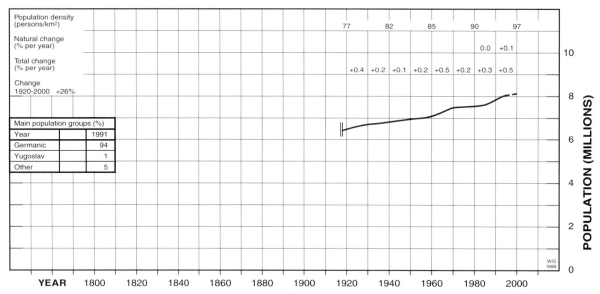

The region which is now Austria was occupied by Romans, Germanic tribes, Franks and Holy Romans before the 13th century when it came under Habsburg rule. Vienna was threatened by Ottoman Turks in the 16th and 17th centuries, but the Habsburg dynasty survived and dominated wide regions of Europe until the defeat and disintegration of Austria-Hungary in World War One. Modern Austria, reduced in size, was created in 1918 and was forcibly united with Nazi Germany in 1938, but following post-war occupation by

the World War Two Allies it regained independence in 1955.

The graph shows slow irregular growth typical of developed European nations, until the 1980s. The accelerated growth of the 1990s was not caused by natural change, which was negligible, but by influxes of foreign workers and refugees. When a government including the anti-immigration Freedom Party was formed in 2000 it was condemned and shunned by its European Union partners. Censuses: roughly every 10 years from 1923.

Suppose, for example, that global heating / climate change is real and already operating, a view to which most scientists working in the field now subscribe. On the continents, climate zones will tend to shift away from the Equator towards the Poles. Other things being equal, the loss of agricultural land by overheating and desertification might be balanced by the expansion of agriculture into regions that are currently too cold, such as the tundra of Siberia. But other things are not equal. Global heating means sea level rise. In 2000 some climate change modellers were thinking that much of the Greenland icecap could disappear by 2100, causing a sea level rise approaching 5 metres. If this were to happen, huge areas of the world's best agricultural land in deltas and coastal flatlands would be lost. A billion people, at least, would be forced out of these areas. It is not possible to forecast how they would be received elsewhere, but one thing is certain. 'Business as usual' would have ended, and in all probability there would be immense loss of human life.

Other factors could be almost as potent. The spread of HIV/AIDS, if unchecked, is capable of reversing population growth early in the present century (section 4.7). Considering that at about the same time oil will be priced out of many nations' reach (section 7.3), the foreign aid ethic will fade away, and VCL wars will become ever more common (section 4.2), a population turndown is inevitable. On this basis my best guess is that world population will reach a maximum of 7 to 8 billion before 2020, and a steep decline will follow.

5.5 Beneficial Growth: 'More is Better'

Everyone loves economic growth. Governments and businessmen depend on it because it increases wealth and power, at national and personal levels. Increasing productivity in factories and on farms, thanks to scientific and engineering discoveries and advances in efficiency, allows more goods to be manufactured at less cost, and more foodstuffs to be harvested per hectare. The people benefit.

AZERBAIJAN   Area 86,600 km²

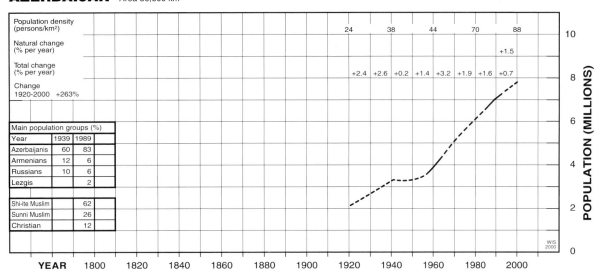

	1939	1989
Main population groups (%)		
Year	1939	1989
Azerbaijanis	60	83
Armenians	12	6
Russians	10	6
Lezgis		2
Shi-ite Muslim		62
Sunni Muslim		26
Christian		12

Population density (persons/km²): 24, 38, 44, 70, 88 — 10

Natural change (% per year): +1.5

Total change (% per year): +2.4 +2.6 +0.2 +1.4 +3.2 +1.9 +1.6 +0.7

Change 1920-2000 +263%

WIS 2000

POPULATION (MILLIONS)

YEAR   1800   1820   1840   1860   1880   1900   1920   1940   1960   1980   2000

Azerbaijan covers most of the south-eastern half of Transcaucasia, west of the Caspian Sea. It was an independent state long before it adopted Islam in the 7th century. The Mongols occupied it in the 13th and 14th centuries, followed by the Ottoman Turks and then the Persians (Iranians). Russian influence spread through Azerbaijan late in the 18th century and it became part of the Russian Empire in the 1820s. There was large-scale immigration of Russian oil workers, 1880–1900, to develop the Baku oilfields.

A chaotic period following the Bolshevik revolution of 1917, including brief independence, ended in 1920 with Soviet invasion and merging with Georgia and Armenia in the Transcaucasian Soviet Socialist Republic. In 1936 it de-merged to become the Azerbaijan SSR. The oilfields were the focus of much industrial development during the Soviet period and population continued its rapid growth after the Soviet collapse in 1991, in common with other ex-Soviet Muslim nations. Large offshore oilfields are being developed, but future prosperity is threatened by ongoing ethnic conflict with Armenia over the Christian Armenian-populated enclave of Nagorno-Karabakh. Formal warfare, begun in 1989, ended in 1994, but Armenian forces still occupy some Azeri territory.

Moreover, a thriving economy is stimulated by a growing population. More people means more demand for the produce offered for sale. Money circulates and the people prosper. Worldwide, if population didn't grow, surpluses wouldn't be absorbed; they would have to be dumped, causing recession and unemployment. So more is better, as regards numbers of people.

Well, no. Not any more. Not, in the short term, in most developing nations, nor, in the slightly longer term, in the remainder.

In the developed world, economic growth usually keeps pace with, or exceeds, population growth. The economic cake grows as fast as, or faster than, the number of people who want to share it, so the slices are big and there are some to spare that can be sold to foreigners.

In most developing nations, on the other hand, the national economic cake is growing more slowly than the fast-growing population, so the slices get ever smaller as more and more people have to share the cake. Result: worsening poverty (the *Micawberish Rule*, section 6.4), which in due course is exacerbated as discontented citizens, fighting VCL-type factional battles amongst themselves for the remaining crumbs, spend the nation's cash reserves and destroy its capital resources: buildings, precious infrastructure such as bridges or power stations, and profitable businesses. Young men migrate to the West and claim asylum. Finally, having ruined itself, the nation has to survive on charitable aid, with a foreign peacekeeping force holding the factions apart (see AFGHANISTAN). In the short term, more has become worse.

In the longer term, the developed world also must avoid letting population growth overtake economic growth. If the latter falters, a nation risks embarking on the same slippery slope to VCL as described above. Many developed nations, especially in Europe, are

1.8 – 18 million

BELARUS   White Russia, Byelorussia   Area 207,600 km² (after 1939)

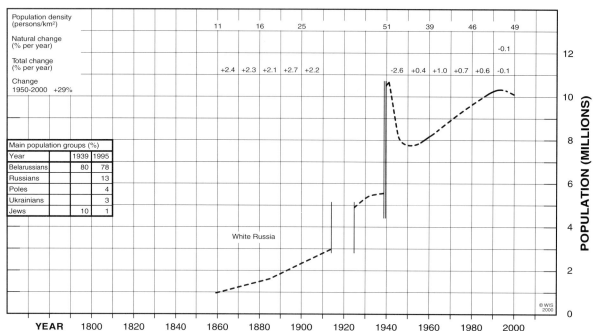

Population density (persons/km²)					11	16	25		51	39	46	49

Natural change (% per year): -0.1

Total change (% per year): +2.4 +2.3 +2.1 +2.7 +2.2   -2.6 +0.4 +1.0 +0.7 +0.6 -0.1

Change 1950-2000 +29%

Main population groups (%)		
Year	1939	1995
Belarussians	80	78
Russians		13
Poles		4
Ukrainians		3
Jews	10	1

White Russia

YEAR 1800 1820 1840 1860 1880 1900 1920 1940 1960 1980 2000

POPULATION (MILLIONS) 12 10 8 6 4 2 0

© WIS 2000

Slav tribes settled here during the first millennium AD, founding the capital, Minsk, in the 11th century. In the 13th century Belarussian and Polish armies threw back the Mongol hordes. Belarus was linked with Lithuania and Poland at times until the 17th century when Sweden temporarily occupied part of the country. The disintegration of Poland in the 1790s created White Russia as a province of Tsarist Russia. After German invasion in World War One an enlarged Belarus achieved brief independence, 1918–19, but was then partitioned between a re-born Poland and the Soviet Union.

Under Stalin the agriculture of socialist Belarus was forcibly collectivised. Some 100,000 dissidents were executed. The Soviets reoccupied western Belarus (eastern Poland) in 1939, and during World War Two more than a million Belarussians were killed, including most of the Jewish population (see group table). When the Soviet Union collapsed in 1991 Belarus became independent, but it has maintained close links with Russia. Still a largely agricultural nation, with considerable resources of fossil fuels and more than a third forest-covered, it has shared in the remarkable post-1991 population decline of the ex-Soviet Slav nations.

overcrowded and congested, their populations heavily dependent on imports of food and, critically, oil. Economic growth is balanced on a knife edge. If within a few years, as Campbell (1997, 2003) convincingly predicts, oil becomes very expensive, crowded Western nations will find economic growth hard to maintain. Alternative 'renewable' energy sources have yet to reach the market place in significant strength. Most have developmental problems (section 9.1), and all except perhaps biomass cannot easily replace oil's other uses such as the manufacture of plastics, agrochemicals, etc.

How many wind turbines per square kilometre will a Western farmer need in 2020, to energise his ploughing, sowing, fertilising, tending and harvesting the tonnage of food that he currently produces, thanks to cheap oil? If he plants biomass crops, that land is not available to grow food. And all the time, *unless policies change radically*, the populations of already congested developed countries will be swelling as refugees from developing nations arrive, seeking to escape the chaos caused by pernicious, not beneficial, growth. Western governments, struggling to preserve law, order and civilised standards as congestion worsens, will eventually succumb to internal factional violence. Their populations will have reached VCL. Darwinian conflict will run its course, creating smaller poverty-stricken populations. So in the West also, more will be worse.

BELGIUM   Area 30,500 km²

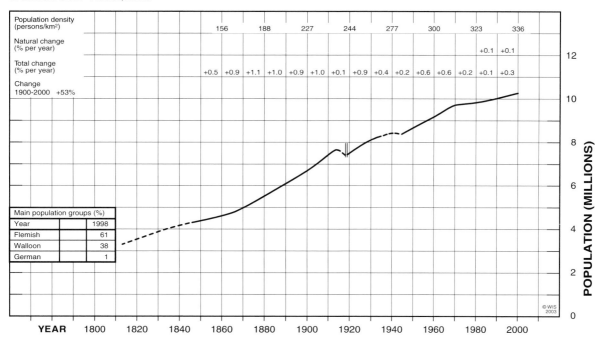

Belgium, the land of the Celtic Belgae tribes, belonged successively to Roman, Frankish, French, Habsburg, Spanish, Austrian and French nations and empires until, after the Napoleonic Wars, Belgium and Holland together formed the Low Countries or Netherlands. In 1830 the South Netherlands broke away and became Belgium. The presence of rich coal deposits fostered the development of industry, population and urbanisation earlier than in most of Europe. However, Belgium's population has grown more slowly than that of her Netherlands neighbour, which overtook Belgium in 1933 and is now 50% bigger.

In recent decades there has been persistent animosity and actual violence between Flemish-speaking Flanders in the north, and Wallonia, the French-speaking south. Political tension over the language divisions led to the collapse of several governments. Consequently, in 1993, separate linguistic communities (Flemish, French and German) and territorial regions (Flanders, Wallonia and Brussels) were created, linked in a federal system. From about 1970 influxes of immigrants, especially Turks and Moroccans, arrived to work in industry; most have since migrated to Brussels where they comprised about 40% of the city's population in 1999. They are omitted from the table, above. Albanian immigrants are said to be behind most gangland crime in Brussels.

5.6   Human Rights and the Sanctity of Human Life

In 1772 the Swedish pioneer botanist Carl Linnaeus wrote to a friend "I fear I shall not have any under-gardeners this summer… for they say they cannot work without food, and for many days they have not tasted a crust of bread… Today a wife was sent to [jail] for having cut her child's throat, having had no food to give it, that it might not pine away in hunger and tears". Linnaeus was sympathetic, but neither he nor anyone else took action. Garrett Hardin (1993) continues "in Linnaeus's day it would have done no good for the rich to donate money to a community chest because the food for a large population of needy people was not available for purchase… death from deprivation eliminated imprudently conceived infants… Few people felt that there was any community obligation to save brats whelped by the feckless".

Note the date, 1772. Hardin points out that before artificial birth control methods were available, marriage normally led to the begetting of many children, but there was no long-term population growth. In Europe during the ages when agricultural productivity was static and nations were near carrying capacity, before the mid-18th century Industrial Revolution, all but a few of the children born to poor families were doomed to die young. They died of starvation or the diseases (e.g. TB, pneumonia, dysentery) that carried away people weakened by hunger, in years when the harvest was below average. Those who survived childhood were still vulnerable to disease, or war (soldiers and their civilian victims).

Throughout the Old World, and in New World regions at carrying capacity (e.g. Central

1.8 – 18 million

BENIN   Dahomey   Area 112,600 km²

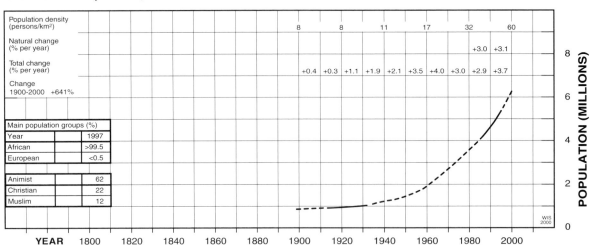

	1800	1820	1840	1860	1880	1900	1920	1940	1960	1980	2000
Population density (persons/km²)						8	8	11	17	32	60
Natural change (% per year)										+3.0	+3.1
Total change (% per year)						+0.4	+0.3	+1.1	+1.9	+2.1	+3.5	+4.0	+3.0	+2.9	+3.7
Change 1900-2000 +641%											

Main population groups (%)

Year		1997
African		>99.5
European		<0.5
Animist		62
Christian		22
Muslim		12

In the 17th century some of the Fon peoples of Dahomey lived in small coastal kingdoms that procured slaves for European traders. All-female regiments, famed for their ferocity, fought in the early 19th century, when Dahomey was celebrated for metal working and for voodoo witchcraft rituals

France declared a protectorate over the coast in 1894 and completed 'pacification' of the interior in 1914. Dahomey became part of French West Africa in 1908 and moved fairly smoothly to

independence in 1960. After 12 years of political instability involving regionally-based power struggles, a military coup in 1972 introduced a Marxist-Leninist government that achieved stability, re-named the country Benin in 1975, and impoverished the nation by disrupting trade. Multiparty elections were held in 1991, followed by more pragmatic government.

Of the nation's 42 ethnic groups the main ones are Fon, Yoruba, Adja and Bariba.

America), finite food resources meant that populations could not possibly increase. High birth rates had to be precisely matched by high death rates. Most human lives began in poverty and squalor and ended in misery or violence.

To claim in those days that poor people had 'rights', with sacred lives that must be prolonged at all costs, would have seemed absurd, because it was not possible. The extra food did not exist. If someone was saved, someone else had to die. There were, however, two exceptional circumstances. Rich and powerful people could commandeer resources as needed to ensure their own and their children's survival. Landowners joked, in medieval England, that their poorest tenants were "harvest sensitive" (Morgan, 1993). In other words, they were expected to die in lean years. The second circumstance in which lives could be saved was not in this world, but in the next. Religious leaders profited, materially and spiritually, by persuading the masses that an all-powerful deity considered the life of every individual to be important (sacred), and rebirth in Heaven, Paradise or similar was attainable, usually by humbly serving the Church on Earth.

The game of chess, popular among the aristocracy of medieval Europe, well reflected contemporary life with its powerful and mobile bishops, knights and queens protected by a screen of undistinguished and expendable pawns.

So, in the world before the Industrial Revolution, rich and powerful humans had self-awarded 'rights', but the poor had none. Their lives were cheap and unimportant. The change came when the Industrial Revolution ushered in the WROG period of Weak Restraints On Growth (section 4.5). For the first time it became possible to save lives on a large scale without causing equal numbers of deaths elsewhere. The DC surge was beginning in Britain and other industrialising nations, most of whom were Christian.

Presumably it was the Christian faith, with its basic tenets 'Love thy neighbour' and 'Thou shalt not kill', that drove human rights activists in their campaigns, especially after World War Two, to end capital punishment and to oppose proposals to legalise abortion and euthanasia. "It is our Christian duty to give criminals a second chance" said a speaker on the BBC radio programme 'Any Questions?' on 13 January 2001. Other religions have tolerated

BOLIVIA Area 1,098,500 km²

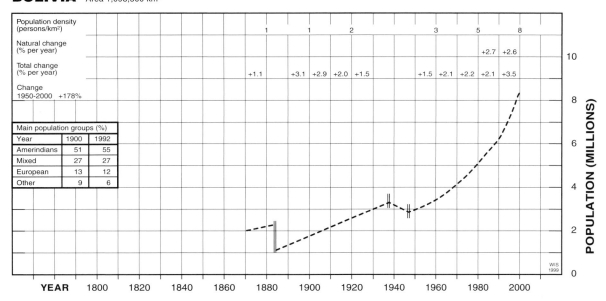

Bolivia had long been home to Amerindian civilizations, especially the Tiwanaku empire in the first millennium AD, when Spanish conquistadors arrived and occupied the country in 1538. Then for 2 centuries the Potosi mines produced huge tonnages of silver ore, followed by tin in the 19th century. Simon Bolivar fought a successful war for independence in 1825. Subsequently, Bolivia has had almost one government per year on average, many of them installed by military coups.

After a century of fast population growth in percentage terms, Bolivia is still sparsely populated. There have been large losses of territory. In the Pacific War (1879–84) Bolivia lost its Pacific coastline, with rich guano deposits on off-shore islands, to Chile. The break in the graph, 1938–47, represents the period following the Chaco War with Paraguay (1932–35) when the frontier was defined in Paraguay's favour in 1938 and "recalculated" in in 1947.

The large difference between natural and total change in the 1980s and 1990s probably reflects uncertainty in the estimates.

selective infanticide, e.g. of female babies in India and China, or the killing of opponents in Muslim holy wars, but even they tend to pay lip-service to the powerful moral authority assumed by those who claim that human life is sacred, an assumption that is a corner-stone of political correctness throughout the developed world.

In 1948 the United Nations adopted a *Universal Declaration of Human Rights*, covering "the rights and freedoms to which every human being is entitled", including among others the personal right to life, liberty, work, rest, security, privacy, property ownership, health and well-being, education and social life, and freedom of thought, religion and opinion. In 1953 the Council of Europe added freedom from slavery and torture. Campaigners have tried to extend the last-mentioned to cover corporal punishment of all kinds, even the smacking of naughty children by parents and schoolteachers to teach discipline, because physical and mental pain are inflicted.

Typically, moral zealots of whatever conviction have no doubts that their interpretation of human (or animal) rights is the *true* one and all others are *false*. If they happen to be powerful enough, thanks to their large numbers, or to a quirk of the law, or to political correctness, they have no qualms about forcing other people with equally sincere but contrary views to "do as we tell you". They are, in fact, self-righteous bullies, denying the other people their 'rights' to freedom of thought and opinion.

Just as the concept of universal human rights would have been an absurdity before the WROG period began, so it will be again when the period ends. Try to persuade participants in current VCL wars (e.g. Kosovo, Burundi, Sierra Leone) of its virtues, and they will think you crazy. Compassion is a luxury available to people enjoying peace and plenty, who are confident of their place in society. They apply it to the hungry, needy or oppressed. It makes them feel virtuous – until the needy try to take advantage of the givers (e.g. in England when it was perceived that refugees and asylum-seekers were jumping the queue for council

1.8 – 18 million

BOSNIA-HERCEGOVINA   Area 51,100 km²

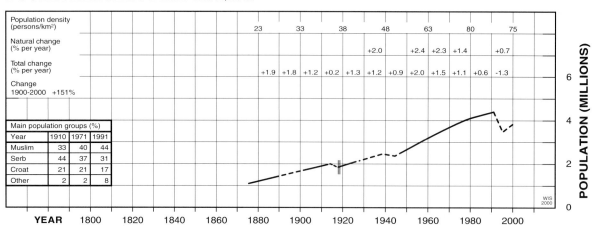

Population density (persons/km²)					23	33	38	48	63	80	75		
Natural change (% per year)							+2.0	+2.4	+2.3	+1.4	+0.7		
Total change (% per year)				+1.9	+1.8	+1.2	+0.2	+1.3	+1.2	+0.9	+2.0	+1.5	+1.1	+0.6	-1.3
Change 1900-2000 +151%													

Main population groups (%)

Year	1910	1971	1991
Muslim	33	40	44
Serb	44	37	31
Croat	21	21	17
Other	2	2	8

POPULATION (MILLIONS): 6, 4, 2, 0

YEAR: 1800 1820 1840 1860 1880 1900 1920 1940 1960 1980 2000

WIS 2000

Slav tribes settled in what is now Bosnia in the 7th century, after the disintegration of the Roman Empire. Part of the Slav population converted to Islam during the 4 centuries of Turkish rule before 1878, when, following a revolt, Bosnia came under Austrian control. The Austrian census of 1910 recorded the ethnic and religious mix that would lead to much grief. Slav nationalism was strong, and it was a Serb nationalist who assassinated the heir to the Habsburg throne at Sarajevo in 1914, precipitating World War One. After the war Bosnia joined the Kingdom of Serbs, Croats and Slovenes, which became Yugoslavia. During World War Two, as a major component of the fascist 'State of Croatia', its rugged terrain was a haven for Tito's guerrillas. Ethnic groups fought each other, and revenge massacres, so brutal that they shocked even the Nazis, transformed old rivalries into hatred.

From 1945 Bosnia was a largely autonomous republic within federal Yugoslavia, one of the more backward regions which received large subsidies, reluctantly given, from Serbia, Croatia and Slovenia. The high Muslim birth rate was denounced as a threat by the Slavs, but it was the pressure of Serb nationalism that led Muslims and Croats to combine in voting for independence in 1992.

Independent Bosnia was at once in turmoil as ethnic Serbs, assisted by Yugoslav forces, rose in revolt. All three ethnic groups committed and suffered genocidal atrocities over a three-year period. A UN peacekeeping force was ineffective, and it took NATO bombing of Serb bases to end the fighting in 1995, by which time deaths totalled c. 250,000 and 1.3 million people had fled the country. In 2001, peace was being maintained by a NATO force of 20,000. About 1 million Bosnian Muslims were refugees in their own land. Five billion dollars of foreign aid had been spent since 1995, but 60% of the population still lived in poverty. High levels of corruption were exposed in 1999.

housing, or immigrant minorities were seeking to modify the culture of the host nation to suit themselves; section 4.4).

Human 'rights' often conflict with each other. For example, if a couple insists on its 'right' to have lots of babies, the family that results may lose its 'right' to enjoy a comfortable standard of living, because its earnings are less than its expenses. Again, in Europe it is often said "Gypsies (Roma) are entitled to their own culture and lifestyle", which is fine as long as they don't clash with other peoples' 'rights' such as security and a pleasant environment. The 'right' to have lots of babies is already depriving the human race of its 'right' to enjoy the pleasure of living on an uncrowded, peaceful and bountiful planet.

If you claim 'rights' from Society, you should accept equal responsibilities to Society. This proviso is all too often ignored by zealous proponents of particular 'rights', be they human, animal, or whatever.

In 'total' war, when one country is battling with another for national survival, to escape humiliation, economic slavery and the imposition of a dictatorial alien culture, as in World War Two, the fundamentals of human nature are revealed as strictly Darwinian. Each side is striving to eliminate the other as a competitor, by superior strength and fitness. Except in defeat, as when prisoners are taken, the enemy has minimal human rights. Both sides aim to kill as many of the opposing forces as possible, and if destroying enemy infrastructure "to shorten the war" involves the random killing of enemy civilians, as in the London Blitz, the Dresden firestorm, the V1/V2 attacks on London, or the atomic wasting of Hiroshima and Nagasaki, few of the perpetrating side feel any compunction at the time. Their aim is to *win*, as quickly as possible, with minimal losses, whatever the cost to the enemy. For all they know, he may be about to deploy a new and overwhelming weapon. Popular charges that such

BULGARIA   Area 110,900 km²

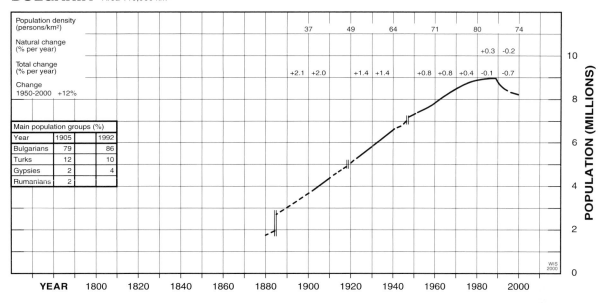

							37	49	64	71	80	74

Population density (persons/km²)

Natural change (% per year) — +0.3 -0.2

Total change (% per year) — +2.1 +2.0 +1.4 +1.4 +0.8 +0.8 +0.4 -0.1 -0.7

Change 1950-2000 +12%

Main population groups (%)		
Year	1905	1992
Bulgarians	79	86
Turks	12	10
Gypsies	2	4
Rumanians	2	

A Bulgar nomad tribe from central Asia arrived in the lands that would eventually become Bulgaria in the 7th century AD, conquering and merging with their Slav inhabitants and challenging the Byzantine Empire further south. In one campaign the Byzantine forces were slaughtered and the Bulgar chief made the Emperor's skull into a drinking cup. In the same spirit, following a Bulgarian defeat in 1014, the Byzantines put out the eyes of all the prisoners except one in every hundred, who, with one eye spared, led the others home. Mongol hordes overran Bulgaria in the 13th century, and Ottoman Turks did likewise in 1396. Ottoman rule lasted nearly 5 centuries, but the Christian religion was not suppressed. 15,000 Bulgars were massacred when they rebelled in 1876; the great powers intervened, driving out the Turks, and after a chaotic few years the basis of modern Bulgaria was laid down in 1885. The boundaries were slightly modified in 1913 after the Balkan Wars, and again after World War One. Russian domination of the region after World War Two (in which Bulgaria was again on the losing side) allowed the gain of South Dobruja from Romania in 1947.

As a Soviet satellite republic, Bulgaria stagnated economically. Late in the 1980s a programme of 'Bulgarianisation' caused an exodus of Turks, and a more gradual population decline has occurred, as in other Slav nations, following the collapse of the Soviet Union in 1991. Censuses: every 10 to 20 years from 1905.

actions were inhumane came later, when the war was nearly over and the risk of defeat was gone, from people whose personal futures were not threatened.

In a Darwinian world, a nation that inflicts great cruelties on the people of another nation in a war to steal their lands, may expect to suffer similar cruelties if the chance to retaliate arrives. "As ye sow, so shall ye reap".

The West has been involved in several wars since World War Two, but none of them were 'total'. Wars in Korea, Vietnam and the Gulf posed no threat to Western survival, so political correctness (and the existence of a second, unfriendly, superpower) demanded that the human rights of civilians be respected. So the Western forces, with one hand tied, did not achieve the 'total' victory which requires an enemy to be eliminated, or at least made completely helpless. In 2002 a Western nation, Israel, was resisting an *intifada* by Palestinian Arabs characterised by tit-for-tat revenge killings on an escalating scale. It is not yet 'total' war, but mutual hatred is exacerbated by aggressive breeding (section 4.4), and the risk is that one side, driven beyond endurance and determined to *win*, will abandon lip-service to human rights and launch a strike that will kill or expel great numbers of the enemy.

Following the September 11 2001 attack on New York, the 'war of cultures': fast-breeding Muslims with little to lose versus the slow-breeding but 'Satanic' West with much to lose, that had been under way for a decade or more, has intensified. "Islam will dominate the World" had read the placards paraded by fundamentalists marching through Luton, England, earlier in 2001. It is a Darwinian 'them against us' struggle that rejects compromise, aims at a 'total' solution, is oblivious of human 'rights', and may well lead to one side destroying or greatly diminishing the other. Politically-correct national leaders in the West ignore its significance,

1.8 – 18 million

BURKINA FASO   Upper Volta; Voltaic Republic   Area 274,100 km²

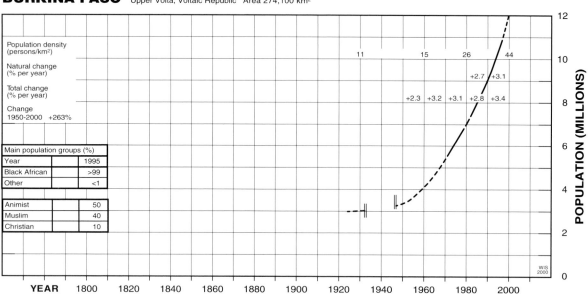

Burkina Faso is not well endowed with industry or natural resources, but its population is the highest among Sahelian nations and is fast increasing. Not surprisingly, the country is very poor. Many men migrate to work in Ghana and Côte d'Ivoire. Late in the 19th century the French began 'pacifying' powerful tribal kingdoms to establish the Upper Volta colony in 1919, mainly to provide labour for development in Côte d'Ivoire. From 1932 to 1947 the administration was divided between Mali, Côte d'Ivoire and Niger. Independence in 1960 led to a succession of weak governments dependent on foreign aid.

Frequent droughts have damaged the subsistence agriculture which supports most of the people. The failing economy has triggered occasional military coups, bloodless in most cases. There are at least 50 tribal groups. The name was changed from Upper Volta to Burkina Faso in 1984.

enabling their challengers to gain strength.

Most of the world's people are entirely unaware that they are living in the historically anomalous WROG period, which, after 250 years, is approaching its end. It has already ended for half a dozen nations whose populations have reached VCL, for by definition the restraints on growth in those nations are no longer weak. High birth rates, migration, congestion, resource shortages, cessation of foreign aid and withdrawal of peacekeeping forces will play their part in bringing many more national populations to VCL within a few decades (Chapter 9). Then, as hatred and violence between rival groups pervades society, human life will once again be cheap (see RWANDA). The average human being will be as devoid of 'rights' as the average British cow, slaughtered and burned or dumped in a mass grave, when foot and mouth disease ended her sheltered life in 2001.

5.7   Imperfections of Democracy: the Link with Racism

Democracy, "rule by the people", dates from the 5th century BC when it was introduced in Greece. In modern times it is the most usual form of government, the alternatives being oligarchy (rule by a few) and monarchy or dictatorship (rule by one). In a democratic state, political parties compete for the votes of the people in elections. In principle, the party gaining the most votes, on its own or in a coalition with others, forms the government. The representatives of the people rule the country until the next election. It is all very right and proper.

If you support the ruling party you expect to get the kind of government that suits you. So, if you are rich, you may contribute lavishly to your party's funds, to finance its election campaign. If you are devious, you may try to manipulate the voting procedures to benefit your party. If you are highly placed in the armed forces, you may impose the government you want by force. Normally though, at least in the West, the popular majority is satisfactorily represented (although there are examples of governments refusing to implement the 'will of

TURKEY Area 779,500 km² (including lakes)

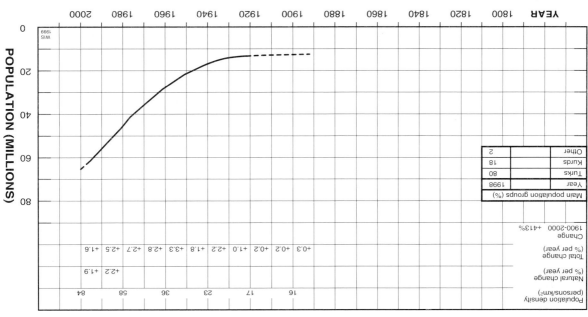

Main population groups (%)		
Year	1998	
Turks	80	
Kurds	18	
Other	2	

Change 1900-2000 +413%

Total change (% per year)	+0.3	+0.2	+0.2	+1.0	+1.8	+3.3	+2.8	+2.7	+2.5	+1.6
Natural change (% per year)								+2.2	+1.9	
Population density (persons/km²)	16	17	23		36		58		84	

The land that is now Turkey, or Asia Minor, fell to a long succession of conquerors, from Hittites in the 2nd millennium BC through Phrygians, Greeks, Lydians and Persians to the Romans, who ruled it from the 2nd century BC to the 4th century AD when it passed to the Eastern Roman or Byzantine Empire. Based on Constantinople (Istanbul), the Byzantines maintained Christianity until the 12th century when Turkish nomad tribes from Central Asia conquered the country and imposed the Muslim faith. Their leaders, the militant Ottomans, expanded their empire until, at its height in the 17th century, it stretched from Algeria through Egypt to Yemen and Iran, and through the Balkans to the borders of Austria. Thereafter its power steadily declined.

After defeat in World War One, Turkey itself was threatened with partition between Greece, Russia and Italy. Salvation came with Kemal Ataturk and his Young Turks movement, who fought the War of Independence, ending in 1923, that established modern Turkey. In 1984 the large ethnic Kurd minority in eastern Turkey began a guerrilla revolt for independence, which intensified until 30,000 were dead, and became unreal as three Kurdish factions fought each other. A truce was declared in 1999. There is large-scale smuggling in the Kurdish south-east by drug barons with private armies.

An anthropogenic (man-made) disaster occurred in 1999 when the fast-growing densely populated industrial region sited on an active seismic zone east of Istanbul was shaken by a 7.4 (Richter scale) earthquake. Thousands of recently-built sub-standard apartment blocks collapsed, killing some 40,000 people and causing damage estimated at 8 billion dollars.

Peasant farmers, whose livelihood is being lost in competition with modern intensive farming, form half the work force. They migrate to the cities, hoping for employment, but several million Turks have emigrated to find work in Western Europe, 2 million in Germany alone. Turkey has built huge dams on the Tigris and Euphrates rivers which may prejudice water supplies to Syria and Iraq. Fish stocks in the Black Sea have been severely damaged by over-fishing, river-borne pollution and an introduced jellyfish.

Frequent censuses, beginning in 1927, record the fast population growth typical of Mediterranean Muslim nations. About 98% of the people are Muslim. If Turkey joins the European Union, by 2020 it is likely to be the largest country in the Union. Economic crises in 2000 and 2001, combined with fast population growth, are worsening the national poverty (section 6.4).

TURKEY, in 1999, an earthquake on the North Anatolian Fault shattered the densely populated industrial city of Izmit, killing perhaps 40,000 people. Damage was estimated at 8 billion dollars.

The 'disaster' element of natural events such as the above is *anthropogenic*, or man-made, because the deaths and damage were the result of people choosing, or being forced by circumstance, to live in high-risk places in very large numbers.

As world population expands, violent natural events will affect more and more people and their property. The cost to any particular nation, relative to its economy, will be greatest where fast population growth is making poverty worse. Richer countries also will suffer severely from anthropogenic disasters, if the cost of insuring against events such as "extreme weather" goes on rising at its present rate. By 2065, the CGNU insurance company claims, insurance payouts "could in effect reduce the world to bankruptcy" (*The Guardian,* 24 November 2000).

So as human population density increases, we become ever more vulnerable to

4.6 Anthropogenic 'Natural' Disasters

Millions of years ago, earthquakes, floods, droughts, volcanic eruptions, hurricanes and tsunamis were just as common and powerful as they are now, but they were not disasters as we understand them. We weren't around. Many animals would have died, vast areas of vegetation would have been devastated, but in our terms it would have been just an interesting natural event. Major asteroid strikes even determined the course of evolution, and one was good for us, in a perverse way, inasmuch as by exterminating the dinosaurs it eased the rise of mammals, and eventually *Homo sapiens*. (A future dominant species might say the same if an asteroid wiped out the human race.)

A *disaster*, to us, is an event that severely harms human lives and/or property. When human populations were very low there was a good chance that, when a big natural event occurred, nobody would be affected. Even if they were, people with a nomadic or hunter-gatherer way of life would usually escape with little more than a fright. When, however, people began to congregate in permanent stone-built cities, the term disaster became appropriate. The eruption of Vesuvius in 79 AD, which overwhelmed three Roman towns, was perhaps the first well-recorded one.

The destruction of Pompeii, Herculaneum and Stabiae, towns which were small in modern terms though important in their day, had little effect on the wealth and strength of the great Roman Empire. Today, many millions of people dice with destruction by living in places like the San Andreas Fault Zone, the Central American hurricane belt, low-lying coastal regions like the Nile and Ganges deltas, and even close to Vesuvius. Where, a thousand years ago, the cost of making good the damage caused by earthquake, hurricane or tsunami in such places would have been fairly small, it is now huge. Hurricane Mitch, in 1998, killed 10,000 people in HONDURAS and caused damage estimated at 5 billion dollars, a disaster not just to the people involved but to the economy of that small, already chronically poor, nation. In

Rice was domesticated in and around north Thailand some 7000 years ago, but the first historical records are of Thai peoples moving south from China in the 8th and 9th centuries AD, invading earlier kingdoms. The first Thai state was the Buddhist Sukhotai kingdom in the 13th century. Successor kingdoms were occupied in repelling bloody invasions from Burma and in cleverly preserving Thai independence in face of European colonial pressures in the 19th century. The first king of modern Siam was Phraya Chakri in 1782 and his dynasty still heads the state, although the military have been behind most governments since 1947.

Thailand (Siam until 1939) has benefited commercially from its

independence: agricultural exports especially rice, rubber and prawns are produced by its own farmers, not by large foreign concerns which drain away the profits. Nevertheless the economy (mainly agricultural but including manufacturing and tourism) was hit hard by the Asian recession of 1997, when it supported many thousands of refugees from Cambodia and Myanmar. The ten-fold population density increase during the 20th century was accompanied by massive deforestation leading to floods, droughts, habitat loss and the extinction of many species of the particularly wide range of Thai wildlife. Nearly 2% of the population was HIV-positive in 2000. Censuses: roughly every tenth year after 1929.

18 – 180 million

THAILAND Siam Area 513,100 km²

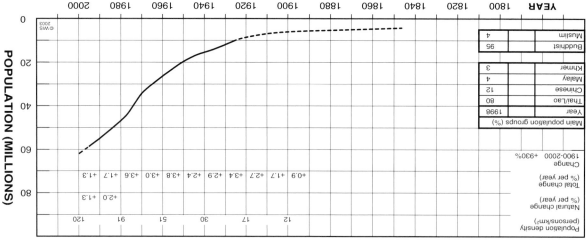

Main population groups (%)		
Year	1998	
Thai/Lao	80	
Chinese	12	
Malay	4	
Khmer	3	
Buddhist	95	
Muslim	4	

Change 1900-2000	+930%

Total change (% per year): +0.9 +1.7 +2.7 +3.4 +2.9 +3.8 +2.4 +3.0 +3.6 +1.7 +1.3

Natural change (% per year): +2.0 +1.3

Population density (persons/km²): 12 17 30 51 91 120

YEAR: 1800 1820 1840 1860 1880 1900 1920 1940 1960 1980 2000

POPULATION (MILLIONS): 0 20 40 60 80

© WIS 2003

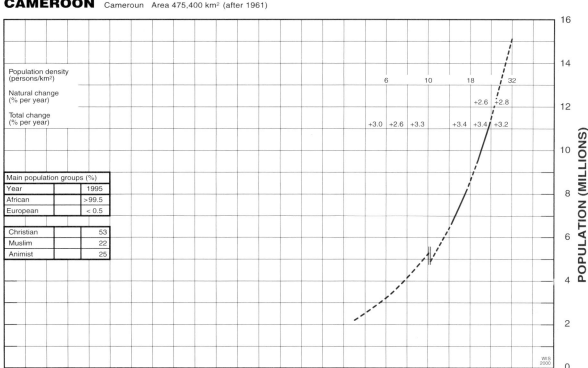

CAMEROON   Cameroun   Area 475,400 km² (after 1961)

Population density (persons/km²)				6	10	18	32
Natural change (% per year)						+2.6	+2.8
Total change (% per year)				+3.0	+2.6	+3.3	+3.4	+3.4	+3.2

Main population groups (%)		
Year		1995
African		>99.5
European		< 0.5
Christian		53
Muslim		22
Animist		25

POPULATION (MILLIONS)

YEAR   1800   1820   1840   1860   1880   1900   1920   1940   1960   1980   2000

WIS 2000

Portuguese seafarers in 1472 named this coast Camarão after its abundant crayfish. For 4 centuries the coastal peoples procured slaves from warring tribes in the interior for European slave traders. A German protectorate was established in 1884 and expanded inland from the coast. After World War One, when the Germans were expelled, the territory was divided into French (south) and British (north and west) mandates. French Cameroon became independent in 1960 and was joined in 1961 by the western British mandate. Northern British Cameroon voted to merge with Nigeria.

The first independent government was dominated by northern Fulani Muslims who crushed a rebellion by south-western Bamiléké

and maintained severe but pragmatic control until 1982 when a Christian from the southern Beti tribe became president. He began to replace the northerners, and repelled an attempted coup by them in 1984. This government's mismanagement of the potentially rich (oil, cocoa and coffee) economy led to widespread public disorder in the early 1990s but it has held on to power.

Logging activities are decimating the extensive rainforests, and corrupt officials have permitted widespread commercial hunting of elephants, great apes and forest buffalos in the Dja reserve, a world heritage site, to provide 'bushmeat'. The main ethnic groups are the Bamiléké, Equatorial Bantu, Kirdi and Fulani, but there are nearly 130 minority groups.

it as 'correct'. To criticise it is to invite accusations of racism.

The whys and wherefores of multiculturalism and its close relative, globalisation, are discussed in Chapter 7. The reason why the coincidence of *lots of babies* with *multiculturalism* is a grave threat to society is exemplified by the current British preoccupation with racism.

Fifty years ago, large-scale racism was a non-issue in Britain. There was bickering between English, Scots and Welsh, and more serious conflicts with Irish nationalists, but they were 'family' disputes over territory and dominance between members of the same, white, race. Blatant prejudice between whites and people with coloured skins was virtually absent, simply because, in 1950, only 0.4% of the population was non-white. Coloured people were not present in large enough numbers to be perceived, by the white majority, as competitors. On the contrary, there were feelings of gratitude and goodwill in Britain towards 'citizens of the Empire' who had fought alongside the British in World War Two.

What has happened since 1950 to introduce serious racism to Britain? Large-scale immigration from the Caribbean and the Indian sub-continent began in the 1950s, encouraged by governments seeking low-cost labour. It continued, embracing Asia and Africa, and by 2001 more than half of England's annual population increase of 0.2 million was net immigration and 9% of the population was non-white (Office for National Statistics

1.8 – 18 million

CENTRAL AFRICAN REPUBLIC   Ubangi Shari   Area 622,400 km²

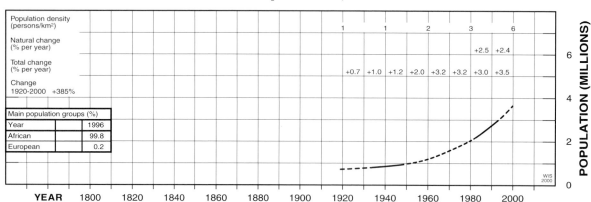

Pygmies were living in the rain forests when other tribes entered this region from east and west, a thousand years or more ago. From the 16th to the 19th centuries the area was greatly afflicted by the slave trade. Europeans on the Atlantic coast traded for slaves but Arab raiding parties from Chad and Sudan kidnapped the populations of whole villages. France acquired the country in the 'Scramble for Africa' of the 1890s. French companies used conscript labour to work cotton and coffee plantations, brutally suppressing any resistance. Political pressure achieved independence in 1960.

A military coup brought Jean-Bedel Bokassa to power. His reign (1965–79) was characterised by eccentric violence and interference with the rich diamond mining industry for his personal benefit. In 1977 he proclaimed himself Emperor, but a French-backed coup deposed him. Low export revenues, partly caused by corruption, have triggered military revolts since 1996. The main ethnic groups: Banda, Baya-Mandjia and Mbaka, comprise two thirds of the population, but there are about 80 minority groups.

website, 12.12.2002). London's population was 27% non-white in 2002. Racist incidents and accusations received non-stop high-profile attention in the media, ranging from riots in northern ghetto cities (section 5.8) and murders of blacks by whites, through perceived inequalities (e.g. disproportionately low numbers of non-white Members of Parliament) to the complicated rules governing adoption of children by adults of different races. The British police force was accused of 'institutional racism'. A 'black gun-carrying culture' had developed in the English Midlands. A Muslim Council of Britain and a Muslim Parliament of Britain had been established. Islam claimed more regular worshippers than any other British faith, and a new mosque opened every week (*Channel 4 TV*, 14.3.2002).

Pressure groups promoting the advancement of coloured people lobbied for 'positive discrimination' (e.g. where two candidates for a job were equally qualified, the coloured one should be chosen). In the USA, in December 2000, a court in Detroit upheld the University of Michigan's right to discriminate against whites in its admissions policy. In January 2001 the BBC's Director General described his organisation as 'hideously white', and called for an increase in the number of managers from minority backgrounds. Similar calls for minority quotas had been made in France and Germany in 1999.

Strong emotions were abroad in Britain, celebrating the advance of multiculturalism on one side, and deploring the loss of historical 'Britishness' on another. Chronic racial tensions had developed from nothing over 50 years. Other European nations had similar problems with large non-native groups (e.g. Algerians in France, Muslims in the Netherlands and Turks in Germany).

All the precedents indicate that as incoming minority groups, the 'new colonists', increase in size and power, the native majority reacts with resentment and fear. These emotions flower into hatred as the majority's democratic security dwindles. History records the desperation and misery of the Australian Aborigines, the North American Indians, and many other native peoples as European incomers outnumbered them and took over their homelands. Now, in many developed nations, the migratory tides have turned. As ethnic minorities advance within a Western population, the precedents point to racism intensifying until society collapses into violence – in other words, the population reaches VCL (violent

CHAD   Tchad   Area 1,284,000 km²

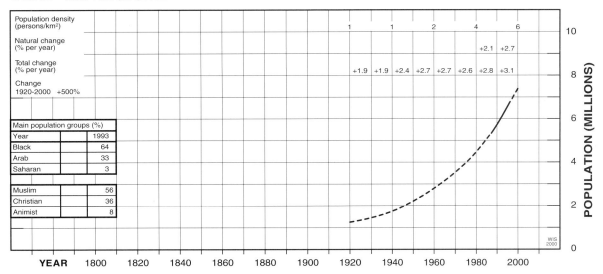

		1993
Main population groups (%)		
Year		1993
Black		64
Arab		33
Saharan		3
Muslim		56
Christian		36
Animist		8

Several millennia ago Lake Chad was a huge inland sea reaching almost to Egypt. Its north shore was the Tibesti mountains where the Saharan people made rock engravings. The lake receded with the onset of Sahelian aridity, and Berbers and Arabs from the north moved into middle Chad and established kingdoms which flourished by trading in slaves. For centuries, right up to the 20th, Arab cavalries attacked and enslaved the black tribes of south Chad, selling their 'Nubian' victims to Mediterranean states.

France claimed Chad in the 1890s, incorporating it into French Equatorial Africa in 1908 but failing to subdue the Muslim north until 1930. They ended the slave trade but forced the southerners to grow huge tonnages of cotton for export. Independence came in 1960 but the new government dominated by southerners needed French support to combat revolts by northern Muslims, assisted after 1971 by Libya which claimed the northern 'Aouzou Strip' of Chad. The civil war, which involved France confronting Libya until 1988, dragged on until 1992. North-south tension and violence continues, but the economy is beginning to recover from the 1980s low when Chad ranked as the world's poorest country. In the 1990s the climate appeared to be growing less arid.

cutback level, section 4.2). Bosnia and Kosovo are recent examples. VCL breakdowns are characterised by human behaviour of the vilest and most depraved kind, like the massacre and mutilation of up to 9000 Bosnian Muslims by Serbs in the Srebrenica 'Death March' (*Newsweek*, 31 July 1995, Brân, 2001).

One might hope that informed warnings, about current policies leading to racial violence, would deserve serious consideration. So far, however, they have only provoked knee-jerk righteous indignation and accusations of racism, such as cost Enoch Powell his political career when in 1968 he made his politically incorrect "rivers of blood" speech. It is ironic that the obsessive emotionality exhibited by both sides over issues of racial prejudice, reaching its climax during VCL scenarios, is the same force that drives the vileness and depravity mentioned above.

5.8   Religion, Race and Charles Darwin

Brainwashing a baby must be one of the easiest tasks in the world. As the infant's empty mind begins to absorb its parents' language and behaviour it also absorbs, without question, their teachings. That is why, in the most fervently religious societies, the religion passes seamlessly from parent to child. Among major religions today the Muslims are the most zealous, unselfcritical and unquestioning; thus Muslim nations remain solidly Muslim as generation follows generation.

The same overwhelming conviction is found among the most orthodox Jews and in a few breakaway Christian sects and cults. Mainstream Christianity has moved away from the obsessive dogmatism that drove the Crusades and fuelled mutual hatred between Protestants and Catholics, in Europe in the Middle Ages. True faith demands the unquestioning acceptance of (usually ancient) religious teachings, which have suffered so many debunkings at the hands of scientists that, in the science-based West, Christianity has lost most of its

CHILE   Area 510,000 km² to 1884, then 737,000 km²

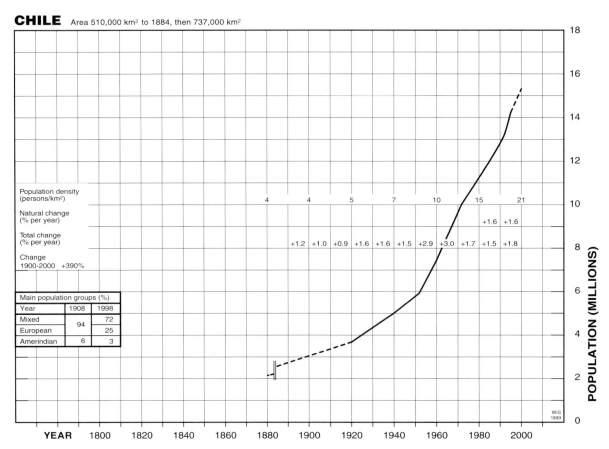

Spanish adventurers occupied Chile (named from an Indian word for snow) in the 16th century in spite of fierce opposition from Incas in the north and Araucanian Indians in the south. Two and a half centuries later the Spanish were ousted by the Army of the Andes under Bernardo O'Higgins. Independence was declared in 1818. Chile fought the War of the Pacific (1879–84) against Peru and Bolivia to control offshore islands with rich guano deposits, increasing her territory by one third and severing Bolivia's route to the sea.

Mining of the guano (natural fertiliser) and copper deposits brought prosperity, allowing final 'pacification' of the warlike Araucanians and encouraging European immigration. The mixed population of today is mainly of Spanish/Araucanian ancestry. In the 1920s the population began a steep rise that continues, in spite of typically South American swings between elected governments and military dictatorships. Many active volcanoes mark Chile's position on the Pacific 'Ring of Fire' where the Pacific crustal plate is pushing under South America. Earthquakes frequently occur; one in 1939 caused 30,000 deaths.

Chile's capital city, Santiago, has a chronic air pollution problem so severe that car traffic is sometimes legally curtailed. The Antarctic ozone hole extended over the southern city of Punta Arenas for the first time in September 2000.

credibility. (Even so, Biblical creationism is taught alongside evolution in schools in parts of the USA and, rarely, in Britain.) Religion as 'the opium of the people' is largely replaced by TV, computer games, the Internet and spectator sport, but these essentially passive pastimes cannot totally suppress the aggressive urges of rebellious youth, which, in the West, fuel outbreaks of violence at football matches or on the fringes of immigrant ghettoes. In other societies they can be harnessed in holy wars.

I was born and raised a Quaker, but my education in science, concentrating on geology, soon began to demolish in my mind the shaky religious doctrines concerning life on Earth and its history. When, in an Irish church, I picked up a leaflet detailing the exact number of years that selected saints had spent suffering the tortures of Purgatory before they were admitted to Heaven, and explaining how I, the reader, could bypass Purgatory entirely by a procedure which involved donating money, in various guises, to Mother Church, I began to think that religion was for suckers.

As human society evolved, it could hardly fail to develop mysticism and religion. In the earliest days, bright individuals would have noticed that advantage could be gained by claiming to understand mysterious natural phenomena like the transits of sun and moon, the

CONGO REPUBLIC Moyen Congo Area 341,800 km²

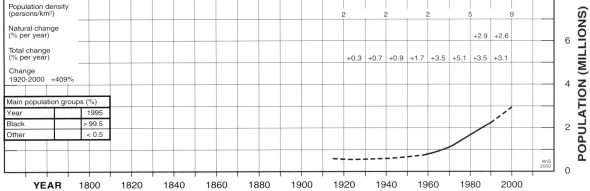

Population density (persons/km²)								2	2	2	5	9	
Natural change (% per year)											+2.9	+2.6	
Total change (% per year)								+0.3	+0.7	+0.9	+1.7	+3.5	+5.1	+3.5	+3.1
Change 1920-2000 +409%													

Main population groups (%)	
Year	1995
Black	> 99.5
Other	< 0.5

Pygmies inhabited the rain forests of northern Congo when the first Bantu farmers arrived there some 2000 years ago. When Portuguese mariners visited the Congo coast in the late 15th century they found warring tribal kingdoms who were easily persuaded to take and sell prisoners for slaves. The slave trade reached inland beyond Congo's borders and flourished for more than 3 centuries until abolition in the 19th.

The French explorer de Brazza claimed Congo for France in 1882. French exploitation of rubber and ivory, and construction of the railway from Pointe Noire to Brazzaville, involved such harsh management of forced labour that starvation and population decline was continuous into the 20th century. Rapid population growth, the DC surge, was delayed until after independence in 1960.

Post-independence Congo became a radical Marxist state in constant conflict with the army, with coups and assassinations. By 1990, with a declining economy in spite of large oil reserves, pragmatism returned and multiparty elections were held in 1992. Violence continued between territorial warlords until a ceasefire in 2000. The Bakongo are the main ethnic group, with 45% of the population, but there are 14 others, within 75 tribes.

changing seasons, death, illness, why animals migrate or why crops sometimes fail. Explanations usually involved the wills or whims of supernatural beings: spirits in the primitive world; gods and goddesses in the first agricultural civilisations. A man who understood and could communicate with the spirit world or the gods was clearly a cut above his fellows, able to prophesy, heal, condemn and, importantly for his well-being, demand. Thus arose the first witch-doctors, shamans and priests, whose teachings became the first religions. They were an early manifestation of Darwinism: the advancement of the smartest.

Religion unites and divides humankind. Devout Christians, Muslims, Jews, Hindus and the others, including large numbers of minor breakaway sects, are tightly bound together by their beliefs, which alienate them from zealots of different faiths. The break-up of the British Indian Empire, which involved the flight of millions of Muslims to PAKISTAN and millions of Hindus to INDIA, slaughtering each other on the way, caused perhaps the largest of innumerable religious conflicts in modern times. Historically, each faith has usually been convinced it is the only true faith, and has felt compelled to convert, or eliminate, the others. In today's world, Muslim fundamentalists are the most aggressive, attacking and eliminating rival believers, especially Christians, Jews, Hindus and minority Muslim groups (e.g. Sunni against Shi'ite) from Egypt through South West Asia to Indonesia and the Philippines, and avowing that their ultimate goal is a Muslim world (section 6.5). And yet, when victory is theirs and the fighting ends, they can lose their sense of purpose in a morass of puritanical trivia (see AFGHANISTAN). The struggle can be more fulfilling than victory itself (as Stevenson observed: "To travel hopefully is a better thing than to arrive").

The Jews' burning sense of identity and destiny has held them together and ensured their survival as prominent minorities over the centuries, but their self-centredness has made them a natural target for blame and persecution when the majority needs a scapegoat. Now in modern Israel/Palestine the tables are turned; the Jews are masters and the Arabs have the burning sense of oppression and injustice. The intense hatred of each religious faction for the other, driven by their rivalry to possess the land that is sacred to both, is leading to disaster as both sides 'strengthen' themselves by rapid population growth. The peaceful, impossible,

1.8 – 18 million

COSTA RICA Area 51,000 km²

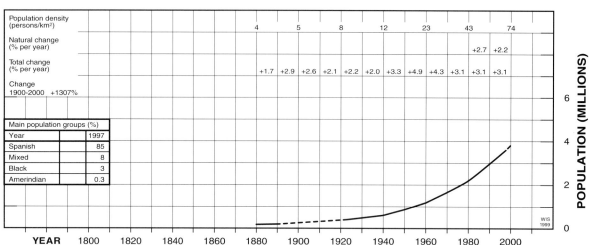

						4	5	8	12	23	43	74	
Population density (persons/km²)													
Natural change (% per year)											+2.7	+2.2	
Total change (% per year)					+1.7	+2.9	+2.6	+2.1	+2.2	+2.0	+3.3	+4.9	+4.3	+3.1	+3.1	+3.1
Change 1900-2000 +1307%													

Main population groups (%)	
Year	1997
Spanish	85
Mixed	8
Black	3
Amerindian	0.3

Columbus visited Costa Rica in 1502, noted the gold ornaments worn by the Amerindian natives, and named it 'Rich Coast'. Spanish conquest and settlement were facilitated by the accompanying European diseases which devastated the seriously hostile indigenous population. Costa Rica broke away from Spain in 1821 and was part of the Central American Federation until 1838 when it declared independence. Many Spanish immigrants arrived to grow coffee and tobacco on the fertile central plateau.

Unlike most Central American nations Costa Rica's history has been politically neutral and mainly peaceful. The army was abolished in 1948. However, population growth has outpaced the availability of agricultural produce, most of which is exported to relieve the large foreign debt. Although two thirds of the farmland rears cattle, it was claimed in 1993 that the average Costa Rican eats less beef than the average US cat.

solution would be to *reduce* populations until their densities were 50 or less, when the land could comfortably support everyone and a fair partition might be negotiated.

Humans have a profound sense of ethnic identity, which over long ages of evolution has bound people together as distinct races, tribes or clans, which defend their traditional territories, or expand into new ones, by fighting each other. Sometimes ethnic and religious groupings coincide, thus most Arabs are Muslim, especially Sunni Muslim. While religious groupings are based on faith, ethnic groupings are usually held together by historical, linguistic or blood links to a land or region which is traditionally their home and/or country of origin, as with the Irish of Ireland and North America. Ethnic supergroups, like the Slavs of Eastern Europe, are divided into subgroups living in their own regions, e.g. the Poles, Russians or Serbs. Any human can identify with one or several of a set of ethnic labels; thus a man living in London may call himself English, British or European, or he may be white, black or brown. On a slightly different basis he may think of himself as a city dweller, as opposed to a countryman, or rich, as distinct from poor.

Throughout history, and more than ever today, religious, ethnic and other groups have depended on the principle 'Unity is Strength' to defend and perpetuate their unique identities. A group of only a few members is weak, easily conquered or dispossessed. Groups with many members are, other things being equal, the strong ones. They and their identities survive because they are the 'fittest', in accordance with the laws of natural selection, elaborated by Charles Darwin.

And why are these diverse groups constantly struggling with each other to survive? Why cannot they co-exist harmoniously together, as seems so desirable and natural to politically-correct observers in the 21st century? Darwin explains (1859, Chapter 15): "The struggle for existence inevitably follows from the high geometrical ratio of increase which is common to all organic beings… More individuals are born than can possibly survive… the struggle will generally be most severe between [individuals of the same species]; it will be almost equally severe between the varieties [i.e. groups] of the same species".

In other words, adapting Darwin's analysis to the human species, the competition for

CÔTE D'IVOIRE Ivory Coast Area 320,800 km² (except 1932-47)

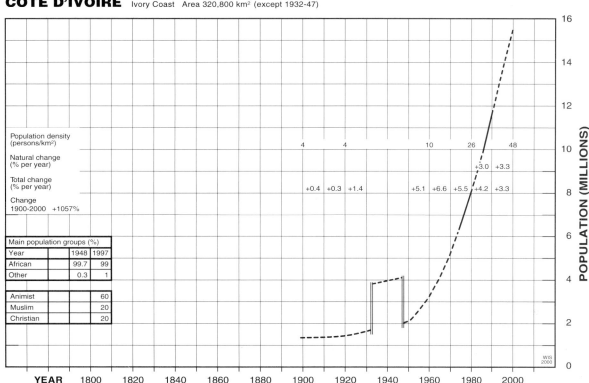

Population density (persons/km²)				4	4		10	26		48
Natural change (% per year)									+3.0	+3.3
Total change (% per year)				+0.4	+0.3	+1.4	+5.1	+6.6	+5.5	+4.2	+3.3
Change 1900-2000 +1057%										

Main population groups (%)		
Year	1948	1997
African	99.7	99
Other	0.3	1
Animist		60
Muslim		20
Christian		20

WIS 2000

YEAR 1800 1820 1840 1860 1880 1900 1920 1940 1960 1980 2000

POPULATION (MILLIONS)

Portuguese navigators reached Côte d'Ivoire in the 1460s. Dense sparsely populated rain forests in the south half of the region ensured that it was of minor interest to slavers. France established coastal trading stations around 1700, but not until the 1880s did French influence extend to the Muslim north. The country was declared a French colony in 1893; pacification of the interior continued until 1917.

In 1932 part of Upper Volta (now Burkina Faso) merged with Côte d'Ivoire, but separated again in 1947. Cash crops, especially cocoa, coffee and timber, were bringing great prosperity when the country became independent in 1960. Thereafter, President Houphouet-Boigny maintained a close relationship with France until his death in 1993, by which time corruption and unfavourable trading conditions had triggered growing social unrest.

There are some 60 different tribes, with Muslims dominant in the northern savannahs and Christians in the south, where little natural rain forest is left. About 2 million immigrants settled in the country during the period of greatest prosperity. In 1996 there were 350,000 refugees from the Liberian civil war. Muslim-Christian rivalry triggered a coup attempt in 2002 which developed into widespread revolts by several aspiring warlords.

limited resources at both individual and group levels is driven by the high fertility of *Homo sapiens*. More babies are produced than can be fed. The weaker individuals and groups are naturally selected out, and the stronger groups survive. The process is Nature's way of perpetuating the 'fitness' of our species.

Darwinian evolution ensures that most 'successful' animal species produce more offspring than are necessary to replace their parents. Surplus offspring succumb to malnutrition or predators, or have some inbuilt weakness that leads to their demise in the ruthlessly competitive Darwinian natural world. Those few that survive are likely to be the strongest and most adaptable, the 'fittest'. But *Homo sapiens* has recently found ways to evade basic Darwinian restraints. He can control death as never before. But unless he also controls birth the human population will inevitably grow to the point where Earth's resources and ecosystems can no longer support it. The anomalous WROG period has seen this happening, and the period is drawing to a close. Swarming humans are fighting for dwindling resources, and even the planet itself is reacting, Gaia-fashion (Lovelock, 1991), in ways that will limit the numbers of humans inhabiting it (see WORLD).

In today's 'informed society', ignorance of Darwinian principles is fairly normal. "I would like someone to explain why Man is so aggressive, why he fights and bombs and murders, instead of living peacefully with others", remarked a participant in a BBC Radio 4

1.8 – 18 million

CROATIA Area 56,500 km²

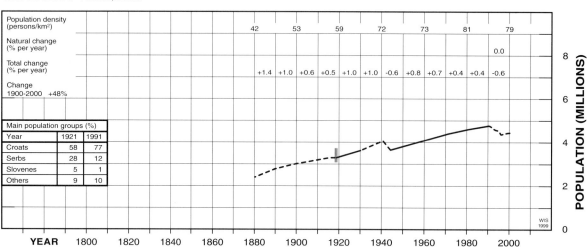

Population density (persons/km²)					42	53	59	72	73	81	79			
Natural change (% per year)											0.0			
Total change (% per year)				+1.4	+1.0	+0.6	+0.5	+1.0	+1.0	-0.6	+0.8	+0.7	+0.4	+0.4	-0.6
Change 1900–2000 +48%														

Main population groups (%)

Year	1921	1991
Croats	58	77
Serbs	28	12
Slovenes	5	1
Others	9	10

POPULATION (MILLIONS)

WIS 1999

YEAR 1800 1820 1840 1860 1880 1900 1920 1940 1960 1980 2000

In the 7th century AD Croatian Slavs settled the region, which had been a province of the Roman Empire, and adopted Christianity. From 1091 Croatia was normally part of Hungary, except for an eastern region occupied by Ottoman Turks from 1526 to 1699. It was a province of the Austro-Hungarian Empire until 1918 when, with boundary changes, it joined the new Kingdom of Serbs, Croats and Slovenes (which became Yugoslavia in 1929). Ethnic and religious rivalry between Catholic Croats and Orthodox Serbs, already long-established, intensified up to World War Two when the Nazis set up their puppet 'State of Croatia'. Within this greatly enlarged super-Croatia the Partisan guerrillas under Tito resisted fiercely. Appalling atrocities took place in the name of 'national purification', mainly against Serbs, up to 750,000 of whom were killed.

Post-war Croatia, returned to its traditional size, became largely autonomous within communist Yugoslavia. Croatia resented Serb domination of the People's Republic (though President Tito was a Croat). An ongoing grievance was the transfer of a substantial part of Croatia's rich earnings from exports and tourism on the long Adriatic coastline to aid less prosperous regions of the Republic such as Kosovo.

In 1991 Croatia voted to secede from Yugoslavia, triggering violent rebellion and mass ethnic migrations in the mainly Serb areas of Krajina and Slavonia. A UN peacekeeping force halted the war until 1995 when, immediately following partial UN withdrawal, a sudden Croatian attack expelled 200,000 Serbs from Krajina. A fragile peace survives in Slavonia. The ethnic wars have gravely damaged the economy, especially the once highly profitable tourist industry. Serb refugees who return often find their properties occupied.

discussion programme. Nobody obliged. Perhaps they were all creationists.

However racism is defined (usually involving discriminatory practices between groups or individuals with obvious differences, such as skin colour) it is only one of the many kinds of competition that mark the unceasing Darwinian struggle for advantage and resources between religious, ethnic and other groups. In the past, right up to the mid-18th century, it was life-or-death competition, because there was a ceiling on population increase. Deaths had to equal births.

These ancient certainties were overturned in the mid-18th century. From then to the present day, country after country has experienced the DC population surge, as the Industrial Revolution ushered them into the WROG period (section 4.5). Groups still compete to dominate each other, but because populations can now increase, the losers no longer have to die. Instead, their influence within society and their access to resources are curtailed. They are the so-called 'developing' nations.

WROG conditions have endured so long that to most people living today they seem normal and permanent. In consequence, some world citizens, mostly well-off inhabitants of the developed world whose access to resources is effectively unlimited, are oblivious to the fact that Darwinian competition dominates the lives of their struggling fellow humans. They scale the high ground of political correctness and moral superiority, prompted in many cases by active or vestigial Christianity, proclaiming that competing groups should forget their differences and co-exist in harmony. Unfortunately, they are living in cloud-cuckoo-land.

For the vast majority of world citizens, resources are in very short supply and Darwinian competition to possess them is instinctive. Racial discrimination, restrictive trading practices,

CUBA Area 110,900 km²

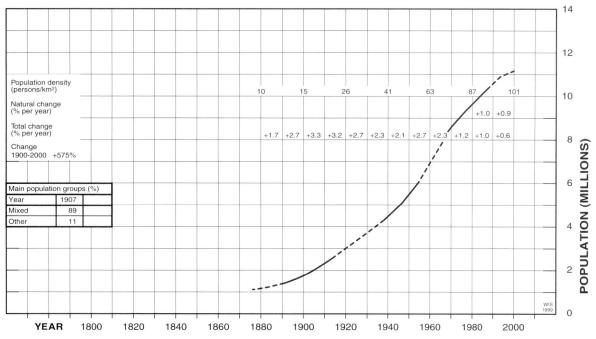

| Population density (persons/km²) | | 10 | 15 | 26 | 41 | 63 | 87 | 101 |

| Natural change (% per year) | | | | | | | +1.0 | +0.9 |

| Total change (% per year) | +1.7 | +2.7 | +3.3 | +3.2 | +2.7 | +2.3 | +2.1 | +2.7 | +2.3 | +1.2 | +1.0 | +0.6 |

Change 1900-2000 +575%

Main population groups (%)	
Year	1907
Mixed	89
Other	11

YEAR   1800   1820   1840   1860   1880   1900   1920   1940   1960   1980   2000

POPULATION (MILLIONS)

WIS 1999

Columbus visited Cuba in 1492 and claimed the island for Spain. The native Amerindians, forced to work on sugar plantations, died in large numbers from ill-treatment and European diseases and were replaced by African slaves. A mixed Afro-Spanish population developed. Three wars for independence from Spain were fought between 1868 and 1898; the last succeeded with US help and Cuba became independent in 1901. After several dictatorial pro-USA regimes Fidel Castro took power by a coup in 1959. He made Cuba the only communist state in the Americas, backed by Soviet aid and harassed politically and economically by the US. Since the Soviet collapse in 1991, Cuba's extreme socialism has gradually relaxed. The economy is based on the export of sugar (Cuba is the world's leading producer) and, more recently,

tourism. Coral reefs strictly protected from fishing are a special attraction.

As in most Latin American countries, population growth accelerated in the late 19th century as improved social conditions and agricultural productivity lowered the death rate. The percentage growth rate fell throughout the 20th century but the numerical growth rate hardly faltered until the 1990s when, in a manner recalling the ex-communist nations of Eastern Europe, the graph began to flatten. Partly this was due to a growing exodus of refugees to the USA (more than 100,000 in 1994, when the drying-up of Soviet aid had seriously damaged the economy), but the fall in the total fertility rate to 1.6 births per woman, the lowest in the Americas, is highly significant.

racist or religious killings, bullying among schoolchildren or of ordinary citizens by extremist zealots, strife between nations, bribery, post-colonial tribalism in Africa, advancing oneself or one's group by force or by trickery, aggressive breeding, chasing advantage by accusing others of racial harassment, genocide and ethnic cleansing, playing to win in sport and children's games, panic buying, betraying the trust of generous helpers – all these activities are only a few of the innumerable manifestations of normal Darwinian human behaviour: the struggle for personal or group advantage. In a world where more humans are born than their environment can comfortably sustain, the struggle is inevitable. Denying it is politically-correct folly.

True Darwinism has no moral or politically-correct (egalitarian in Western terms) dimension, as is demonstrated by fundamentalist fighters for Islam in Indonesia, Kosovo and elsewhere, who aim not just to defeat but to get rid of their cultural rivals. Politically-correct policies in many Western nations, promoting the rapid build-up of large ethnic minorities, are sowing the seeds of national disaster. In England just after World War Two racism was not a serious issue because ethnic minorities were minuscule. England had absorbed many immigrants over the centuries, but they had come in groups of a few thousand at a time, separated by long intervals during which they integrated into English society, a process made easy by the fact that they were visually almost indistinguishable from the native English. Jews and Gypsies had maintained lifestyles that proclaimed their separateness, ensuring that they

1.8 – 18 million

CZECHOSLOVAKIA   Czech Republic (Area 78,860 km²)   plus Slovakia (Area 49,030 km²)   Ruthenia excluded

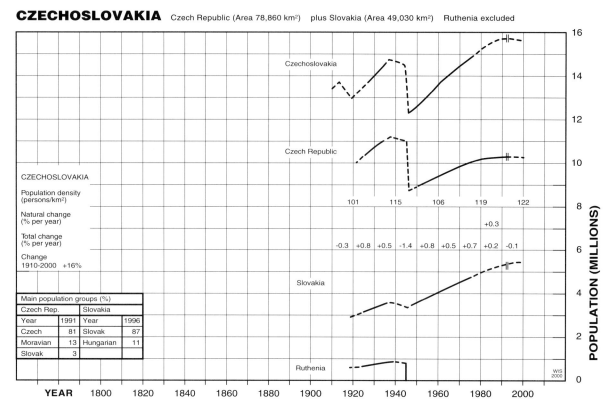

Like some of its central European neighbours, Czechoslovakia hardly existed as a nation before 1918. In the west, Slav, Magyar and Bohemian dynasties rose and fell in the millennium before Austrian Habsburgs assumed control in the 16th century. In the east, Slavs and then Magyars ruled before, in 1526, the Habsburgs took over. When the dual monarchy of Austria-Hungary was created in 1867 the western provinces were ruled by Austria, the eastern by Hungary.

After the dual monarchy was defeated in World War One, Czechs joined Slovaks to form Czechoslovakia, with the addition of sub-Carpathian Ruthenia in the extreme east. Nazi Germany annexed the largely German-speaking Sudeten regions in 1938 and the rest of the country in 1939. After World War Two

Czechoslovakia reformed and more than 2 million Germans were driven out of the Sudetenland into East Germany. Ruthenia joined the Ukraine. The Soviet Union imposed communist government on Czechoslovakia until 1989 when Soviet influence waned and a democratic government was elected. Non-violent ethnic separatism soon became irresistible (though the nation was basically Catholic) and the country divided peacefully into independent Czech west and Slovak east nations in 1992.

Following the end of communist rule the population of both countries has begun a gentle decline, in common with most east European nations that were dominated by the Soviet Union. Slovakia's Gypsy population of 0.5 million in 2000 is increasing at such a rate that it would outnumber Slovaks by 2060.

continued to suffer a degree of Darwinian discrimination.

By 2000, immigration (initially for cheap labour) mainly from the British Commonwealth had created ethnic minorities amounting to about 7.1% of the English population (Scott, Pearce & Goldblatt, 2001). Claims and counter-claims of racial harassment and exploitation were everyday occurrences.

Far from integrating with the native population, the minorities have preferred, naturally, to form ghettoes in particular towns and cities, so that they can feel comfortably 'at home' among their own kind. The Ouseley report on race riots in Bradford, Yorkshire, (July 2001) noted that the city is "in the grip of fear… communities are fragmenting along racial, cultural and faith lines… people's attitudes appear to be hardening and intolerance towards differences is growing". There is "virtual apartheid" in many schools, and the police "fear to tackle black and Asian offenders in case they are called 'racist'". The Pakistani Muslim component of the city's population had grown from 10% in 1991 to 15% in 2001. On 25 June 2000 BBC Radio 4 featured a young British Muslim recruiting fellow Muslims to fight in *jihads* (holy wars) to advance Islam. Jon Ronson (2002) visited a "Jihad training camp" near Gatwick Airport whose organiser declared he "would not rest until he saw the

DENMARK Area 43,000 km² (after 1919)

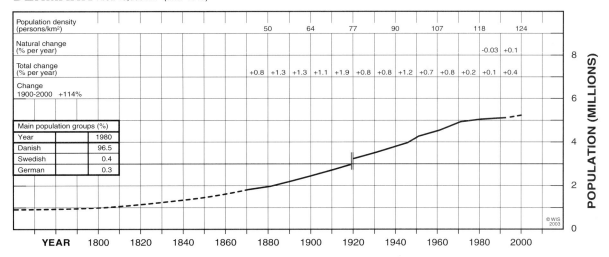

Population density (persons/km²)						50	64	77	90	107	118	124			
Natural change (% per year)											-0.03	+0.1			
Total change (% per year)					+0.8	+1.3	+1.3	+1.1	+1.9	+0.8	+0.8	+1.2	+0.7	+0.8	+0.2	+0.1	+0.4	

Change 1900-2000 +114%

Main population groups (%)		
Year		1980
Danish		96.5
Swedish		0.4
German		0.3

Denmark today is one of Europe's smaller nations, peaceful, prosperous and tolerant, but its history is of Empire and aggressive militarism. Many of the Vikings who assailed the British Isles before the 11th century AD were Danes. In the 14th century Danish rule extended from Finland and the Baltic states to Greenland. By the 19th century all was lost. The only recent boundary change has been the return of Schleswig (part) from Germany in 1919.

European-style slow population growth almost ceased in the 1970s and Denmark has often been cited as exemplifying the working-out of the demographic transition principle. However, growth resumed in the 1990s, partly from an excess of births over deaths but mainly from an excess of immigration over emigration. In 1999 ethnic minorities, mostly Asian, comprised 5% of the population. Draconian immigration controls were introduced in 2002.

Black Flag of Islam flying over Downing Street". Muslim solidarity, the *ummah*, transcends national loyalties.

"We should celebrate diversity", say lobbyists for refugees and asylum-seekers, but the lesson of modern Britain, and of countries like Northern IRELAND, ISRAEL, CYPRUS, SOUTH AFRICA and RWANDA, is 'for *diversity* read *divisiveness*'.

Precedents, set in places like the UK, FIJI, MACEDONIA, BOSNIA and KOSOVO, show that racial tension will become more acute as the growing minority groups seek political power commensurate with their size. Israeli settlements in the WEST BANK and GAZA, with all their problems and potential for disaster, are examples of a late stage in the establishment and evolution of alien ghettoes in overcrowded conditions. The end game in this Darwinian struggle for advantage is massive population reduction, when the total population, which is steadily growing, reaches violent cutback level (VCL).

Section 4.2 explains how, once a population has reached VCL, one of the means by which peace may be regained is the elimination of group rivalry by, for example, partition to create genuinely homogeneous populations (see INDIA, PAKISTAN). Throughout Western Europe, at the start of the 21st century, politically-correct governments that allow large-scale immigration are working in exactly the opposite direction. They are creating the conditions in which Darwinian competition flourishes, leading by way of burgeoning racism to national disintegration.

The widespread hatred of Muslim/Arab peoples for the USA, exemplified by terrorist attacks that peaked (so far) in the destruction of New York's World Trade Center on 11 September 2001, killing several thousand people, is a transparent expression of Darwinian rivalry. Muslim expansionism, fuelled by a combination of fast population growth with religious fervour, is frustrated by the military and economic strength of the West. Only by suicidal or covert terrorism can militarily weak Muslims hurt the developed world, within which the USA is their strongest and most arrogant competitor. Their resentment is the greater because the working of the Micawberish Rule (section 6.4) is worsening their own poverty, which is alleviated by foreign aid financed in large part by the USA itself. Many

1.8 – 18 million

DOMINICAN REPUBLIC   Santo Domingo   Area 48,450 km²

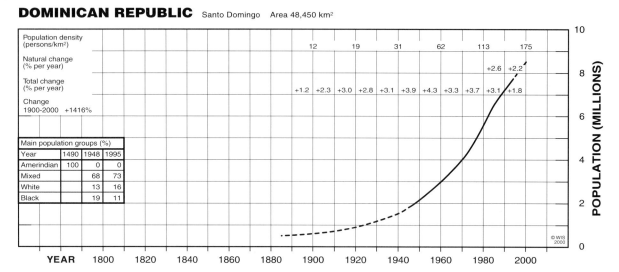

Population density (persons/km²)								12	19	31	62	113	175	
Natural change (% per year)												+2.6	+2.2	
Total change (% per year)								+1.2	+2.3	+3.0	+2.8	+3.1	+3.9	+4.3	+3.3	+3.7	+3.1	+1.8
Change 1900–2000 +1416%														

Main population groups (%)

Year	1490	1948	1995
Amerindian	100	0	0
Mixed		68	73
White		13	16
Black		19	11

© WIS 2000

POPULATION (MILLIONS)

YEAR   1800   1820   1840   1860   1880   1900   1920   1940   1960   1980   2000

When Columbus landed on and named the Caribbean island of Hispaniola (Little Spain) in 1492 the inhabitants were Arawak and Carib Indians. In 1496 Spain founded the city of Santo Domingo as the capital of its American colonies. Within a few decades the Indians were wiped out by enslavement and Old World diseases. African slaves were imported to replace them on the sugar plantations. France took the island from Spain at times in the 17th and 18th centuries. In 1844 it divided into independent nations: Haiti (west) and the Dominican Republic (east).

Political and economic troubles in the new republic brought help from, and temporary occupation (1916–24) by, the USA. Since then, military coups and dictatorships have given way to elected governments, but the economy, based mainly on agriculture and mineral extraction, plus remittances from c. 850,000 Dominicans resident in the USA, remains fragile. Tourism is growing in importance. There is ongoing large-scale illegal immigration from more densely populated Haiti; in 1937 some 20,000 immigrants were massacred. Censuses: every 5–15 years from 1950.

Muslims claim that Islam is a religion of peace and love, but their fundamentalists readily launch holy wars, like the ancient Christian *crusades*, against persons or groups who, they claim, have offended Islam.

• • • • • • • • • • • • •

"All the people like us are We, and everyone else is They"

(Kipling)

Chapter 6

Population Pressure – Some Consequences

Shortly before the invention of agriculture, 10,000 years ago, there were about 4 million humans in the world, perhaps twice as many as there had been 90,000 years earlier (McEvedy and Jones, 1978). As agriculture developed and spread, human numbers slowly increased, until by 1750 AD world population was about 600 million. Then, triggered by industrialisation, it exploded, passing 6 billion just 250 years later in 2000, when the growth rate was 80 million per year.

Already, however, the populations of a handful of nations have reached their ceilings at VCL (violent cutback level, section 4.2), others are approaching VCL, and the pressure of human population on the Earth is so high that the planet's resources and ecosystems are showing signs of failure (see WORLD). Like mould on an orange, which starts as a small spot and spreads out and round, until suddenly it has enveloped and rotted the whole fruit, humans are completing their rape of Mother Earth. This chapter looks at some of the terminal symptoms.

6.1 Failure of Natural Systems in the 21st Century

Late in the 20th century, scientists routinely monitoring the Antarctic atmosphere discovered that the high-level "ozone layer", which shields life at ground level from damaging ultraviolet radiation, was losing its natural integrity. Man-made chemicals, mainly fluorine-based (CFCs), used in aerosol sprays and refrigeration machinery, were decomposing the ozone molecules. An *OZONE HOLE* developed in spring over both Poles. Each year the holes widened, and by the late 1990s they had extended to the fringes of the inhabited continents. The danger, the cause and the remedy were undisputed, and because alternative less damaging chemicals were available many nations agreed in the 1990s to phase out their use of CFCs. Even so, some very populous developing nations, especially China, are reluctant to replace CFCs with more expensive substitutes, so it is by no means certain that the ozone holes will vanish in the foreseeable future.

Potentially far more serious than the ozone holes, more difficult to prove, and probably impossible to resolve, is the *ENHANCED GREENHOUSE EFFECT*, which is or will be the cause of *GLOBAL HEATING* and *CLIMATE CHANGE*, most scientists believe. In the late 1900s environmentalists began to wonder whether the warming world climate (the 1980s were then the warmest decade on record, world-wide, but the 1990s were even hotter) was a natural trend, or whether it was linked to the measured increase in the concentration of carbon dioxide, a known greenhouse gas, in the atmosphere. From 280 ppm (parts per million) before the Industrial Revolution the concentration had risen to 350 ppm in 1990. Other greenhouse gases including methane, CFCs and nitrous oxide, were increasing similarly, all due to human activities. By 2000, as heating continued and was accompanied by a series of unusual climatic events (such as exceptional floods and droughts, intensifying El Niño disturbances, retreating glaciers and

ECUADOR including Galápagos Islands. Area c.465,000 km² to 1941: 275,000 km² after 1941

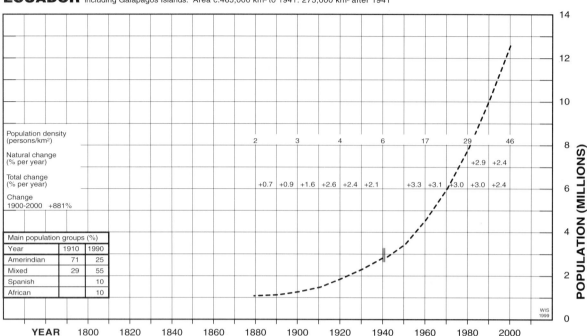

Population density (persons/km²)			2	3	4	6	17	29	46
Natural change (% per year)								+2.9	+2.4
Total change (% per year)	+0.7	+0.9	+1.6	+2.6	+2.4	+2.1	+3.3	+3.1	+3.0	+3.0	+2.4
Change 1900-2000 +881%

Main population groups (%)		
Year	1910	1990
Amerindian	71	25
Mixed	29	55
Spanish		10
African		10

YEAR   1800   1820   1840   1860   1880   1900   1920   1940   1960   1980   2000

POPULATION (MILLIONS) 14 12 10 8 6 4 2 0

WIS 1999

When Spanish conquistadores reached Ecuador in 1526 they faced and soon defeated the complex civilization of the Incas, converting the indigenous tribes to Catholicism and importing African slaves to work plantations and mines. Three centuries of Spanish rule ended with separatist rebellions and independence for Ecuador in 1830. The republic has been plagued by political in-fighting and instability. It lost a border war with Peru in 1941, forfeiting vast sparsely populated territories in Amazonia. Border fighting flared again in 1995 but the dispute was settled in 1998.

Ecuador's population growth in the 20th century was faster than either of her neighbours, Colombia and Peru. It far outpaced the competence of the weak economy to provide new basic resources such as adequate food, so the outcome has been a reducing standard of living (67% in poverty), huge foreign debts and growing environmental damage. Settlers in the GALAPAGOS ISLANDS have threatened to kill the unique wildlife that inspired Charles Darwin if their demands for development are refused. In the table, the appearance of Spanish and African categories in 1990 is largely due to redefinition of the 1910 'mixed' group.

thinning Arctic ice), the need to reduce emissions of greenhouse gases was generally agreed. Unfortunately, at several international conferences, politicians had failed to decide how this should be done.

The trouble is that greenhouse gas emissions are the product of human need to grow food, and human ambition to achieve an ever more agreeable standard of living. Methane gas is generated in rice paddy fields, in melting permafrost throughout much of Alaska, Canada and Siberia, in stagnant waters retained by great dams, and in the guts of cattle, sources that grow in importance as the population grows. CFCs and nitrous oxide (from vehicle exhausts and nitrate fertilisers) are similarly linked to increasing human numbers and well-being. Carbon dioxide, the main greenhouse gas, is created in vast quantities (22 billion tonnes annually) by burning fossil fuels (coal, oil and natural gas) to generate energy for all purposes, including powering the vehicles that modern humans consider essential to the enjoyment of life. The only large-scale alternative to fossil fuels at present is nuclear power, which most nations shun because of its perceived danger, and unless nuclear fusion is developed it is a relatively short-term solution.

The natural cycle of plant and animal death and regrowth, especially trees and oceanic plankton, has always produced huge amounts of carbon dioxide and methane, which are recycled by Earth's natural biosystems. These are able to cope when humans speed up the natural cycle by felling forests for wood and to create farmland. However, the extra greenhouse emissions from unnatural sources: fossil fuels and man-made chemicals, have clearly proved too copious for the biosystems to deal with. The polluting emissions are

EL SALVADOR Area 21,050 km²

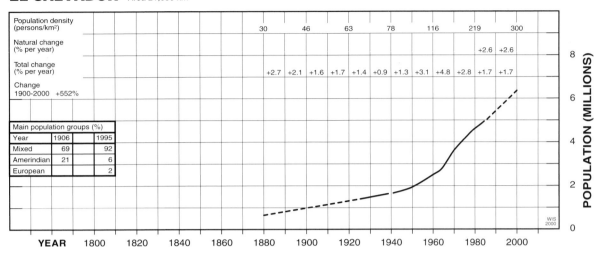

Population density (persons/km²)									30	46	63	78	116	219	300		
Natural change (% per year)														+2.6	+2.6		
Total change (% per year)									+2.7	+2.1	+1.6	+1.7	+1.4	+0.9	+1.3	+3.1	+4.8	+2.8	+1.7	+1.7
Change 1900-2000 +552%																	

Main population groups (%)		
Year	1906	1995
Mixed	69	92
Amerindian	21	6
European		2

Amerindian hunter-gatherers took part in the invention of Mesoamerican agriculture about 8000 years ago. Abundant food allowed the development of advanced cultures culminating in the Aztec empire which was destroyed by Spanish invaders. They conquered El Salvador in 1524 and it remained a Spanish colony until 1821 when it seceded to join neighbour states in the Central American Federation. This dissolved in 1839. Independent El Salvador suffered from dictatorial regimes and armed harassment by bigger neighbours. A peasant uprising against wealthy coffee planters was put down in 1932. Population began to boom in the 1950s and by 1970 overpopulation was recognised as a problem. From the 1970s left-wing FMLN guerrillas gained control of some areas, defying the government in spite of economic and military aid from the US. In 1992 the FMLN signed a peace agreement and entered the political arena. The civil war had cost 75,000 lives.

El Salvador is the smallest nation in Latin America and by far the most densely populated. Its murder rate is the highest in Central America. It suffers earthquakes and volcanic eruptions and has been mostly deforested. Earthquakes in 1986 and 2001 caused mudslides that buried suburbs sited below steep hillsides, with thousands of deaths; anthropogenic disasters directly related to the high population density. Emigration is increasing and in 1995 about a million Salvadorians were living abroad. Censuses: roughly every tenth year from 1930.

accumulating in the atmosphere, driving temperature and climate change.

In *The Carbon War* (1999) Jeremy Leggett describes the failure of conference after international conference to address the carbon dioxide problem, mainly because of powerful resistance by fossil fuel producers and by the leaders of rich energy-wasteful nations, the USA in particular, who fear the electoral unpopularity that would follow, say, the imposition of a heavy tax on carbon-based fuels. At the Hague Conference in November 2000 several developed nations backtracked on commitments made at the Kyoto Conference three years earlier to reduce their emissions to 1990 levels or less by 2010. Leggett looks to international insurance companies to force a change, when they refuse to insure people and businesses located in areas prone to anthropogenic disasters such as floods, hurricanes and tornados as these become ever more frequent and costly.

Conspicuously absent from the climate change debates has been any serious discussion of population, more particularly overpopulation. The omission is inexcusable, because overpopulation is the fundamental cause of most environmental damage. This assertion is best explained by a *reductio ad absurdum* argument, as follows:– Suppose that Earth's human population was only ten million. Those lucky people could enjoy a high standard of living, use resources wastefully, fell forests, slaughter wild animals, fight each other, be careless about pollution – and the planet's ecosystems would serenely continue their natural recycling, coping effortlessly with the little patches of human mess and destruction and maintaining an essentially pristine planet. Those humans would have a genuinely sustainable lifestyle, not because of any creditable efforts on their part, but because their damage could be no more than a pinprick to the natural world. On the other hand, they would be too few to run all the industries whose products we appreciate today, and they would undoubtedly feel that life would be more diverse and interesting if there were more people around the world, in interesting cities with different cultures.

1.8 – 18 million

ERITREA Area 100,000 km² (approx.)

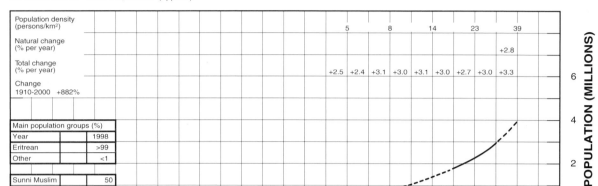

Population density (persons/km²)		5	8	14	23	39
Natural change (% per year)						+2.8
Total change (% per year)		+2.5	+2.4	+3.1	+3.0	+3.1	+3.0	+2.7	+3.0	+3.3
Change 1910-2000 +882%					

Main population groups (%)	
Year	1998
Eritrean	>99
Other	<1
Sunni Muslim	50
Christian	50

Eritrea was part of the Christian kingdom of Axum through much of the first millennium AD. Deforestation over the centuries caused massive soil loss and desertified the land. Arab immigrants brought Islam to coastal areas from the 8th century and power passed back and forth between coastal Muslims and Ethiopian Christians until the 16th century when the region became an outlying part of the Ottoman Empire. Italy established its colony of Eritrea in the 1880s and introduced thousands of Italian peasants to initiate agricultural development. Italian investment brought peace and considerable prosperity. In 1935 Italy attacked and conquered Abyssinia (Ethiopia), creating, with Somalia, Italian East Africa, which was dismantled in World War Two.

Eritrea federated with Ethiopia in 1952 but Ethiopian 'colonialism', leading to annexation in 1962, was so oppressive that Eritrean separatists launched guerrilla war. Thirty years of increasingly savage fighting between Ethiopian forces and Eritrean factions (which initially, as Muslim and Christian groups, fought each other) ended with Eritrean victory in 1991 and independence in 1993. Although the country was drought-stricken, destitute, totally dependent on foreign aid, and trying to cope with half a million returning refugees, the government seized Red Sea islands from Yemen in 1996 and allowed persisting border disputes with Ethiopia to flare into devastating conflict in 1998. This ended in defeat, and the arrival of a UN peacekeeping force of 4000, in 2000.

Now suppose that world population stabilises at 6 billion, the present figure. The great majority of people have a low standard of living, and some 2 billion exist in abject poverty. There is no prospect of any real improvement, because resources of all kinds are being depleted unsustainably (section 5.3). Human nature being Darwinian, well-off people and nations intend to maintain their own high standard of living for as long as possible. The poor are trying to better themselves, but their aspirations cannot possibly be gratified because, all the time, deserts are expanding, soils eroding, productive lands shrinking, aquifers drying, and 'natural' disasters proliferating. It is a grim overcrowded world, full of frustration and suffering, spiralling downward.

So by *reductio ad absurdum* it is demonstrated that Earth's ideal population, at which humans could live contentedly and sustainably without causing disastrous climate change, is somewhere between these absurd extremes. My best guess is that one billion would have suited *Homo sapiens* very well, had the population stabilised at that figure. Now, because we have depleted finite resources, especially oil, so recklessly, the ideal figure may be half that. I refer, of course, to a good quality of life for all humanity, not just well-off people in developed nations.

The early greenhouse effect models predicted fairly slow and small changes in global climate, but more recent ones are horrific. Climate modellers at the University of East Anglia have lately calculated that by 2100 average temperatures may have risen by 4°C to 7°C over many of the world's land masses. This is not global warming, it is *global baking*. And then what? Do temperatures go on rising after 2100? Children alive today may see sea levels 5 metres or more higher as the Greenland and Antarctic icecaps shrink. The implications in terms of human suffering are almost unimaginable.

It seems unbelievable that delegates to the Hague climate conference of November 2000, which post-dated the above predictions, could do no better than bicker over irrelevant details, securing a 'business as usual' future for greenhouse gas emissions. In a rational world,

FINLAND Area 305,000 km² (excluding lakes)

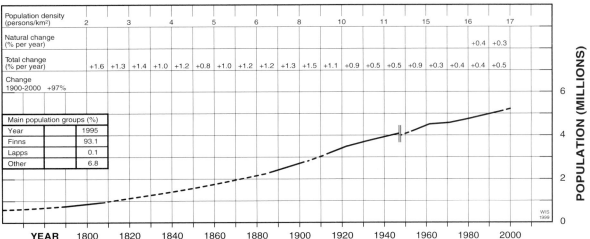

Population density (persons/km²)	2	3	4	5	6	8	10	11	15	16	17

Natural change (% per year): +0.4 +0.3

Total change (% per year): +1.6 +1.3 +1.4 +1.0 +1.2 +0.8 +1.0 +1.2 +1.2 +1.3 +1.5 +1.1 +0.9 +0.5 +0.5 +0.9 +0.3 +0.4 +0.4 +0.5

Change 1900-2000 +97%

Main population groups (%)	
Year	1995
Finns	93.1
Lapps	0.1
Other	6.8

POPULATION (MILLIONS)

YEAR 1800 1820 1840 1860 1880 1900 1920 1940 1960 1980 2000

The original Sami (Lapp) inhabitants of Finland were driven to the far north early in the first millennium AD by nomadic Finn tribes migrating from Asia. Sweden invaded and occupied Finland in the 12th and 13th centuries, introducing Christianity. Finland became a Swedish region. Climatic instability caused periodic harvest failure and famine, as in the 1696–97 winter when more than a quarter of the population starved to death. Half the remainder died in famines and epidemics during the Russo-Swedish war of 1700–21, when Finland was a battleground; devastating hostilities broke out twice more during the 18th century. Russia finally annexed Finland in 1809. Peace brought growing prosperity, but separatist movements developed and Finland achieved independence following civil war in 1918. After World War Two in which Finland twice fought the Soviet Union without success, 12% of the country was ceded to Russia. Most of the inhabitants were resettled in Finland.

A land of lakes (10%) and forests (70%), Finland has industrialised far beyond its traditional mainstay of forest products, and prospered greatly. The graph shows the long slow population increase over 2 centuries typical of a developed European nation. Censuses: irregular from 1800.

the mere possibility that humanity is entering such a catastrophic 21st century would ensure that all nations at once subscribed to the *precautionary principle*, i.e. calculate the 'worst case' scenario and take steps to prevent it happening. And yet, if the greenhouse effect cannot be reduced without massively depopulating the world, there can be no painless solution. Perhaps it is already too late to do anything but wait for events to unfold.

Climate change and human activity can both cause ***DESERTIFICATION*** and ***SOIL EROSION***. Ponting (1991) describes in detail how the Sumerian peoples of southern Mesopotamia, starting about 3500 BC, transformed cereal-growing lands of the southern 'fertile crescent' into salt-crusted desert by intensive irrigation, which was necessary to feed the growing populations of Sumerian city states. He quotes Plato, who wrote in the 4th century BC that after deforestation, parts of ancient Greece had become "like the skeleton of a sick man, all the fat and soft earth having wasted away", so that the rainfall ran off to the sea instead of being stored in the soil and reappearing as perennial springs.

Around the Mediterranean the tree-clad hills of the post-Glacial period were transformed into rocky scrublands, typical of the region today, by the soil erosion that accompanied intensive Greek and Roman cultivation. (On the other hand the Sahara Desert had been vegetated, with a rich fauna including elephants, lions and hippopotami, until the climate *perhaps naturally* (but see AUSTRALIA) became arid, several thousand years earlier.) Further south, farmers have desertified large areas of the semi-arid Sahel, from Senegal to Ethiopia, by deforestation and overgrazing to feed growing populations. Irrigated agriculture in the Indus valley of Pakistan, in Uzbekiston and Kazakhstan, and even in Australia, is turning fertile land into salt desert.

At the start of the 21st century man-made climate change, especially global heating, is threatening to expand existing deserts all round the world. In Europe, parts of Spain, Portugal, Italy, Greece and even southern France, are said to be at risk. Once desertified, the bare ground is highly vulnerable to soil erosion by wind and water. It takes centuries or millennia

1.8 – 18 million

GEORGIA   Area 69,700 km²

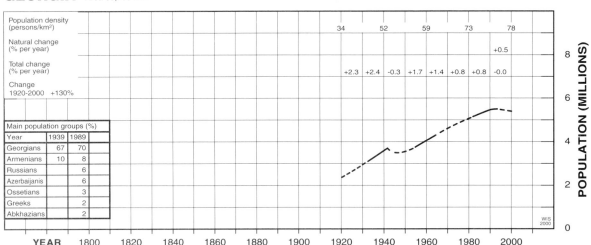

Population density (persons/km²)									34	52	59	73	78	
Natural change (% per year)													+0.5	
Total change (% per year)									+2.3	+2.4	-0.3	+1.7	+1.4	+0.8	+0.8	-0.0
Change 1920-2000 +130%														

Main population groups (%)		
Year	1939	1989
Georgians	67	70
Armenians	10	8
Russians		6
Azerbaijanis		6
Ossetians		3
Greeks		2
Abkhazians		2

The ancient state of Georgia in the western Caucasus mountains had Greek origins but was incorporated into the Roman Empire and became Christian, remaining so even under Arab and Turkish domination from the 7th to the 12th centuries. After expelling the Turks Georgia conquered much of Caucasia and enjoyed power and prosperity until it was overrun by Mongol hordes in 1240. When after nearly 300 years the Mongol grip relaxed, Turkey and Iran divided the country between them. The expanding Russian Empire forcibly absorbed Georgia late in the 18th century.

Following the Bolshevik revolution in 1917 Georgia declared independence, but the Soviets invaded and merged it into the Transcaucasian Soviet Socialist Republic. In 1936 Georgia became an SSR in its own right. 200,000 ethnic Meskhetians were deported to Central Asia during Stalinist purges and agricultural collectivisation. Growing nationalist agitation led to independence in 1991, coinciding with collapse of the Soviet Union, but civil war broke out with the internal provinces of South Ossetia and Abkhazia which wished to secede. The cease-fire arranged in 1994, with Russian peacekeepers, still held in 2002, but ethnic hostility, corruption and chronic lawlessness are ongoing, to the detriment of economic progress. Revenues from mining, manufacturing and tourism all fell in the 1990s. Deforestation to replace lost timber imports from Russia has been savage.

In common with other ex-Soviet non-Muslim nations, population growth ceased soon after the Soviet collapse in 1991.

for soil to form, so the stony wastes that remain after wind erosion are, in human terms, lost to agriculture for ever.

Five hundred years ago, when human activity had made little impression on the **MARINE FOOD CHAIN**, European fishermen discovered wonderfully rich fishing grounds off the coast of Newfoundland, North America. Particularly notable were the mature cod that cruised the Grand Banks in fabulously large numbers. They were increasingly targeted, and by the late 20th century their stocks had been so reduced that fishing for them was banned. Then, to everyone's surprise, their numbers recovered only feebly, if at all. It seems that even the legendary fecundity of cod, with each female producing a million eggs, can now hardly satisfy the appetite of marine creatures that normally prey on cod eggs and juveniles. Similarly excessive catches are depleting stocks of cod in the North Sea and many other fish species around the world.

The world's marine fish catch reached 90 Mt (million tonnes) in 1990 and has fluctuated around that figure ever since (Brown *et al*, 1997 and UNEP, 1999). A significant proportion of the marine fish catch does not feed people directly; for example, in 1999 Danish fishermen caught about 1 Mt of sand eels to process into food for fish farms. Sand eels are the main food of many wild marine fish (as well as sea birds such as the puffin), so although the output of fish farms world-wide has increased rapidly (to c. 20 Mt in 1997, roughly 20% of world fish consumption), fish farms can lessen the productivity of natural fisheries. In 2001 it was calculated that nearly one third of the wild fish catch goes to feed farm animals and farmed fish.

So far, marine fish catches have been maintained by the discovery and exploitation of new fisheries and species as old ones suffer depletion. Yields are reduced by factors such as coastal pollution and climate change as well as by overfishing, so UNEP's warning "if no

GREECE   Area 65,000 km² to 1913,   130,000 km² to 1947,   then 132,000 km²

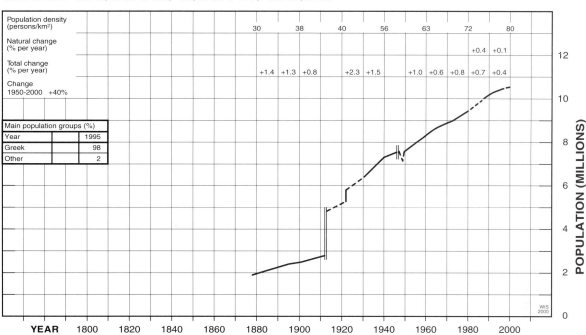

Population density (persons/km²)		30	38	40	56	63	72	80		
Natural change (% per year)								+0.4	+0.1	
Total change (% per year)		+1.4	+1.3	+0.8	+2.3	+1.5	+1.0	+0.6	+0.8	+0.7	+0.4
Change 1950-2000 +40%										

Main population groups (%)		
Year		1995
Greek		98
Other		2

The city-states of ancient Greece, including Crete, although constantly at war with each other and with Persians and others, nurtured the first great European civilisations and cultures. After more than a thousand years of intellectual glory they were conquered by Rome in the 2nd century BC. When the western Roman Empire collapsed in the 5th century AD Greece continued under Byzantine rule until the Ottoman Turkish takeover of the 15th century. Part of Greece gained independence in 1829 and in the Balkan wars of 1912–13 the Ottomans were driven out of the remainder. In 1947 Italy ceded the Dodecanese Islands to Greece.

Greece was relatively unaffected by World War One but the Nazi occupation in World War Two was devastating. Guerrilla warfare against the Germans turned into civil war between left- and right-wing Greek factions. Volatile and damaging politics have plagued post-war Greece, but the old enmity with Turkey was somewhat relaxed when the great Turkish earthquake of 1999 attracted prompt Greek aid.

The sudden population increase in 1922 had followed a Greek attack on the Turkish mainland at Smyrna (Izmir) which was repulsed. The action led to the expulsion of a million Christians from Turkey and half a million Muslims from Greece. The other kink in the graph (1947–49) marks the civil war between communists and royalists, which generated half a million refugees. Immigration has increased in recent years, initially from chaotic Albania. In 1996 Greece blamed ethnic Albanian illegal immigrants for 40% of national crime. By 2001 immigration, mostly illegal, had reached c. 250,000 per year, mainly from Asia. Censuses: about every tenth year, with gaps, from 1881.

In the 4th century BC the philosopher Plato deplored soil erosion on the Greek mainland, caused by deforestation and overgrazing to support the large grain-importing population: "what now remains… is like the skeleton of a sick man, all the fat and soft earth having wasted away". Stony littorals all around the Mediterranean bear witness to the same process, the effect of busy pre-industrial civilisations.

effective action is taken soon, production could decline" is unsurprising.

The **GULF STREAM** is one of the many natural systems that transfer energy around our planet. Solar energy is moved from low to high latitudes by atmospheric currents, the winds, and by slower oceanic currents, which often flow in different directions at the ocean surface and at depth. Energy released by radioactivity deep inside the Earth causes convection currents in the plastic mantle that, immensely slowly but most drastically of all in the long term, cause crustal plates including the continents to drift about over the surface of the globe, generating earthquakes and volcanoes as minor by-products.

Few of these energy-transfer systems are so fragile that they can be affected by human activities. The Gulf Stream may be one such. Its warm tropical surface waters flow northward along the east coast of the United States, then bathe the western coasts of Europe, ensuring, thanks to the prevalent south-westerly winds, that western Europe enjoys a very mild climate for its high latitude. When they reach the Norwegian Sea the waters have cooled and are dense enough, given their high salinity, to sink to the ocean floor where they return southwards.

1.8 – 18 million

GUATEMALA Area 109,000 km²

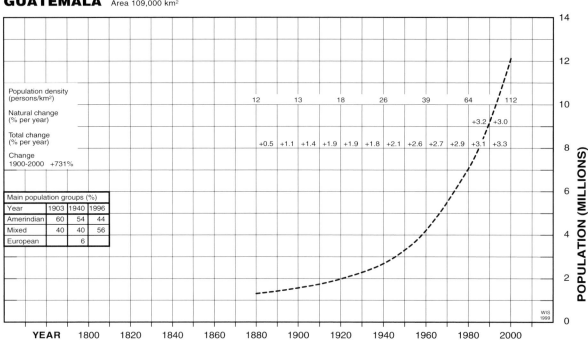

Population density (persons/km²)	12	13	18	26	39	64	112

Natural change (% per year): +3.2 +3.0

Total change (% per year): +0.5 +1.1 +1.4 +1.9 +1.9 +1.8 +2.1 +2.6 +2.7 +2.9 +3.1 +3.3

Change 1900–2000 +731%

Main population groups (%)			
Year	1903	1940	1996
Amerindian	60	54	44
Mixed	40	40	56
European		6	

Ancient Guatemala was a central area of Mesoamerica, where hunter-gatherer tribes invented agriculture some 8000 years ago, independently of and somewhat later than the Old World. Adequate food supplies nurtured the gradual development of civilisations culminating in the Maya, whose cities housed up to 100,000 people and included huge ceremonial pyramids and other religious monuments. Mayan civilisation was in decline when Spanish conquistadores arrived in 1523 and easily defeated its forces. The Mayan population had grown beyond its sustainable limit. Its farmland created by deforestation soon lost fertility and was vulnerable to drought and soil erosion. As it became less productive, the cities, with inadequate food, were depopulated.

Guatemala was a Spanish colony until 1821 when it broke away with Mexico and joined the short-lived Confederation of Central America. When this dissolved in 1839 Guatemala became independent, entering a turbulent period of military dictatorships alternating with elected governments. Left-wing guerrillas were active from the 1940s, harassing the right-wing governments (supported by the US) which responded with 'death squads'. More than 150,000 civilians were killed, many with extreme brutality, or disappeared, in the 1970s and 1980s. Fears of a communist takeover receded in the 1990s and a cease-fire was signed in 1996. Civil war had been continuous for 36 years. In 2002, hundreds of violent youth gangs wre eroding law and order. Coffee, bananas, sugar, timber and tourism are currently the mainstays of the economy. The forests, re-grown since the Mayan era, are again suffering, with 35% lost since 1954.

Censuses in 1940 and 1950 appeared to show a population decrease of half a million over the intervening decade. It is not clear why, or whether, this happened in reality.

Some climatologists believe that if anthropogenic global heating melts the ice crust of the Arctic Ocean, fresh water may dilute the saline current until it is not dense enough to sink. Then, the northern Gulf Stream circulation could cease. North-west Europe would become as bleak and cold as Newfoundland and Labrador, at the same latitude on the opposite side of the Atlantic, with mean temperatures 5°C to 10°C lower than now, and totally unsuited to supporting a dense population. The hypothesis is not proved, but the risk to Europe's future is evident. If, however, the rest of the world is overheating it is just possible that parts of Europe could benefit, if they escaped flooding by sea-level rise.

Another climate system that may be affected by global heating is the periodic El Niño disturbance of the equatorial Pacific Ocean. Every few years, a complex interaction between trade winds and oceanic currents both warm and cold brings abnormally warm and humid weather to South America, and abnormally cool and dry weather to Indonesia and its neighbours. When the El Niño event is particularly severe the knock-on events can include catastrophic floods, droughts and forest fires in America, Africa and Asia. It is thought by some investigators that El Niño events are strengthening because of global heating.

Water evaporating from the sea and elsewhere, forming clouds, falling as precipitation (rain, snow or hail) and returning to the sea in rivers or through underground aquifers, is the

GUINEA   French Guinea   Area 246,000 km²

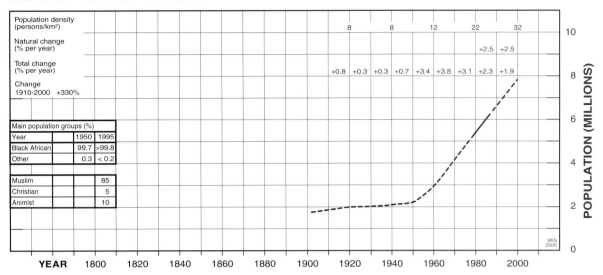

Population density (persons/km²)			8	8	12	22	32
Natural change (% per year)						+2.5	+2.5
Total change (% per year)		+0.8	+0.3	+0.3	+0.7	+3.4	+3.8	+3.1	+2.3	+1.9
Change 1910-2000 +330%							

Main population groups (%)		
Year	1950	1995
Black African	99.7	>99.8
Other	0.3	< 0.2
Muslim		85
Christian		5
Animist		10

When Portuguese seafarers reached Guinea in the 15th century they found it part of a powerful Muslim empire centred on present-day Mali. Local tribes occupied the coastal belt, where Portuguese and other Europeans settled to trade in slaves and ivory. French occupation began in the 19th century. The country was declared a French colony in 1893 and part of French West Africa in 1904. Independence was achieved in 1958. President Sekou Toure developed a repressive Marxist state, causing much emigration. The economy deteriorated in spite of the income from rich mineral deposits including the world's largest bauxite mines. After Toure's death in 1984 multiparty government was introduced but political instability and civil strife developed and the economy remains depressed. In the 1990s Guinea became involved in Sierra Leone's civil war, driving out many thousands of Sierra Leone refugees in 2000.

There are more than 20 ethnic groups.

familiar *WATER CYCLE*. It is a natural system unique to planet Earth, so far as anyone knows, and is so vital to the existence of life on land, which depends on the naturally distilled fresh water, that we neglect it at our great peril. The volume of fresh water transferred from sea to land is roughly constant year on year, and is, after loss by evaporation, the maximum amount available to sustain terrestrial life (with the trivial exception of fresh water produced by desalination).

The water cycle cannot, of course, be stopped, unless the planet becomes so cold that all water turns to ice, or so hot that it all vaporises. The former possibility is less likely than the latter, given the state of the planet Venus, where the atmospheric greenhouse effect is so magnified by 'runaway feedback' that the temperature at ground level is about 460°C. On Earth, a degree of runaway feedback could occur if global heating began to melt the permafrost of the Siberian and North American tundra, allowing frozen dead vegetation to decompose into methane and carbon dioxide, and releasing methane gas from solid methane hydrates. Methane being a potent greenhouse gas, its higher concentration in the atmosphere could raise the temperature further, melting more permafrost, releasing yet more methane, and so on: a runaway feedback.

Human activities are already affecting the water cycle's bounty in less exotic ways. Diverting water from rivers to irrigate arid lands can have disastrous consequences, as in ancient Mesopotamia (IRAQ) where over centuries the evaporation of irrigation water precipitated so much dissolved mineral salt that the soil was sterilised, and the land desertified. The same has happened in modern times in the Indus valley of Pakistan, and in river basins tributary to the Aral Sea (see UZBEKISTON).

In humid regions the groundwater pumped out of underground aquifers for water supply or agriculture is usually replaced by infiltrating rainfall, but in drier regions the aquifers may be gradually emptied. In deserts this can cause springs and wells supplying oases to dry up. The effect of pumping Libya's "Great Man-Made River" on Saharan oases has yet to be determined

1.8 – 18 million

HAITI   Area 27,750 km²

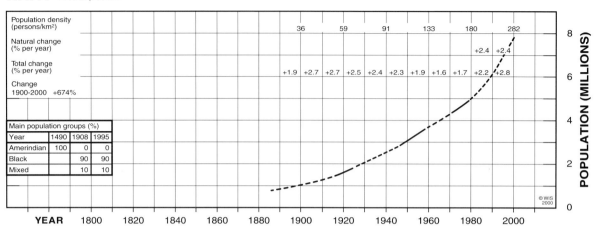

Population density (persons/km²)				36	59	91	133	180	282
Natural change (% per year)								+2.4	+2.4
Total change (% per year)			+1.9	+2.7	+2.7	+2.5	+2.4	+2.3	+1.9	+1.6	+1.7	+2.2	+2.8
Change 1900-2000 +674%			

Main population groups (%)

Year	1490	1908	1995
Amerindian	100	0	0
Black		90	90
Mixed		10	10

Haiti is the western third of Hispaniola, a large Caribbean island which supported several hundred thousand Arawak and Carib Indians when Columbus landed there in 1492. Spanish settlement in Hispaniola followed and soon caused the extermination, by enslavement and Old World diseases, of all the Indians. African slaves were imported to work mines and plantations. In 1697 Spain ceded Haiti to the French, who developed it into a prosperous colony growing sugar, coffee, cotton and indigo. A revolt by the huge slave population, beginning in 1791, led to independence in 1804, but the new nation's prosperity evaporated in a succession of coups and bloody civil wars.

The chaotic situation improved when the US intervened and occupied Haiti (1915–34), but chaos and corruption returned after

the Americans left and culminated in the Duvalier family dictatorships (Papa Doc and Baby Doc, 1957–86, who maintained power by their private militia, the fearsome Tonton Macoutes). The Americans intervened again (1994–95) to deal with renewed chaos, and a UN peacekeeping force succeeded them. In 1999 Haiti was the most poverty-stricken nation in the Americas, with about 50% of the workforce unemployed.

Ongoing chronic poverty, corruption and violence, a dense population and the loss of the best soil through centuries of deforestation and over-cultivation, has for decades provoked large-scale emigration to other Caribbean countries, the US and Canada. The high proportion of blacks in the population reflects the early expulsion of Europeans, before much intermixing had occurred.

(see LIBYA). Non-renewable groundwater in the enormous Ogallala Aquifer, which extends 1500 kilometres from South Dakota to Texas in the USA, is fast being used up as farmers and water supply engineers pump as much as 15 cubic kilometres of water from it every year. Extraction of fresh water from aquifers near the sea coast can draw in useless saline water, as happens even in such rainy places as Avonmouth, England, when the pumping is excessive.

About 97% of the world's water is saline, in the oceans, but animal and plant life on land has little use for salt water. 2% is fresh water locked up as ice in glaciers and ice-caps, and most of the remainder, less than 1% of the total, is fresh water in lakes, rivers and aquifers. This is the portion that is partly renewed every year by the water cycle. More than half of the recycled amount is used by mankind, which is why aquifers are being depleted, and why some great rivers are heavily polluted, or, like China's Yellow River and the USA's Colorado, do not always reach the sea.

At the Johannesburg Earth Summit in 2002 much was said about the growing worldwide shortage of fresh water. But there is as much water as ever. The shortage is caused by the ever-increasing numbers of people wanting to share it. The more people, the fewer litres per person.

In arid countries with fast-growing populations, such as North Africa and South-west Asia, fresh water resources, mainly in aquifers, are already being tapped at seriously unsustainable rates. In Israel/Palestine the shortage of fresh water is a major cause of friction between Israelis and Arabs, which will worsen rapidly as, due to rapid population growth on both sides, the quota of fresh water per person shrinks. Fresh water resources in arid countries are particularly susceptible to pollution, because there is no spare water to dilute or wash away concentrated pollutants. Thus the shallow aquifer beneath the Gaza Strip is being so heavily polluted by sewage, pesticides and saline intrusion that much of the drinking water has to be imported (UNEP, 1999).

HONDURAS Area 112,100 km²

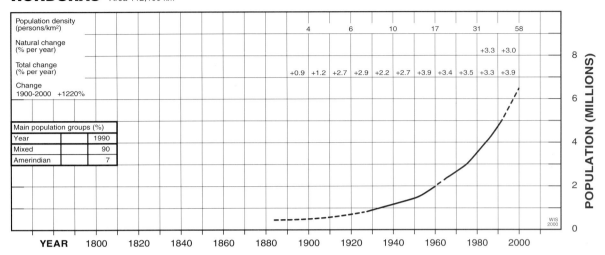

Population density (persons/km²)								4	6	10	17	31	58
Natural change (% per year)												+3.3	+3.0
Total change (% per year)						+0.9	+1.2	+2.7	+2.9	+2.2	+2.7	+3.9	+3.4	+3.5	+3.3	+3.9
Change 1900-2000 +1220%													

Main population groups (%)		
Year		1990
Mixed		90
Amerindian		7

YEAR   1800   1820   1840   1860   1880   1900   1920   1940   1960   1980   2000

Humankind developed agriculture in Mesoamerica, including Honduras, independently of and about 2000 years later than in the Old World. Maize was the most important crop, but tomatoes, peppers, papayas and avocados originated here. The Mayan civilisation was in decline (succeeding several others) when Columbus visited Honduras in 1502. Spanish settlers brought Old World diseases and overcame native resistance to occupy the mountainous interior and the country was ruled by Spain until 1821. After brief mergers with Mexico and a Central American Federation, Honduras became independent in 1838. Political instability, with military dictatorships exceeding civilian governments, has characterised its subsequent history.

The cultivation of bananas by a US company has for decades provided half the nation's export earnings. Very fast population growth in a mainly rural economy short of good farmland has ensured chronic poverty, dramatically worsened in 1998 when the winds and floods of Hurricane Mitch killed more than 10,000 people and devastated crops, communications and settlements. The damage, estimated at 5 billion dollars (greater than Honduras' international debt) is expected to set back the nation's development by a generation. Probably, given the very high birth rate, there will be no real recovery.

It is said that in the cities 'street children' are routinely culled by police to reduce petty crime. Censuses: at irregular intervals from 1930.

• • • • • • • • • • • • • •

The reason why modern civilisation is particularly vulnerable to the failure of natural systems is that the worldwide DC surge (section 4.3), building great population concentrations in what are currently the most favourable regions for human survival, was guided and determined in every detail by the natural systems *as they then operated and, for the most part, still operate.* If the systems fail, some or many of the favourable regions will become unfavourable. Huge numbers of people will find life unbearable through inundation, drought, heat, cold, hunger or disease. They will attempt to re-settle elsewhere on the planet, in places where the existing inhabitants will not welcome them. Basic elements of civilisation including tolerance, respect for others, and the rule of law, will be unlikely to survive.

6.2 Congestion: Less Freedom and More Vulnerability

As I worked on the geological survey of Angola in the 1950s and 1960s, my two or three tents, in their temporary clearing, were sometimes the only human settlement for several kilometres in any direction. The population density of my little 'empire' was less than one. As long as I got on with the job and behaved in a civilised fashion, my team could go anywhere and do anything. There was no private property and no neighbours to upset us or be upset by us. We enjoyed the same freedoms as the hunter-gatherers of old, but we missed the social opportunities that go with living in an organised society.

The opposite extreme, in a *reductio ad absurdum* appraisal of congestion, could be the state of England today. England's population density of 382 in 2000 meant that if the country

1.8 – 18 million

HONG KONG Area 1,090 km²

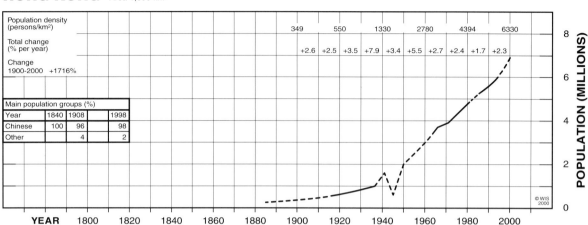

Population density (persons/km²)						349	550	1330	2780	4394	6330	
Total change (% per year)						+2.6	+2.5	+3.5	+7.9	+3.4	+5.5	+2.7	+2.4	+1.7	+2.3
Change 1900-2000 +1716%												

Main population groups (%)				
Year	1840	1908		1998
Chinese	100	96		98
Other		4		2

Hong Kong was an unremarkable small island at the mouth of the Pearl River in 1842 when it was ceded to Britain by China. Kowloon and the New Territories were added in 1860 and 1898 respectively. Anglo-Chinese trade focussed on the excellent natural harbour and the colony grew into a great commercial and industrial centre. A steady flow of economic migrants from China was augmented by 700,000 refugees when Japan attacked China in 1937.

During the Japanese occupation (1941–45) half the population was deported, but most of them returned after the war. There was a huge influx of refugees in the 1950s following the 1949 communist revolution in China. Thereafter, Hong Kong's booming laissez-faire economy continued to attract floods of immigrants (e.g. there were 120,000 resident Filipinos in 1999) despite the imposition of strict border controls in the 1980s. Reversion of the territory to China in 1997 coincided with a significant drop in revenue from tourism, but the economy remains very strong and population growth rapid. China expelled the last 35,000 'boat people' who had been held in detention camps after fleeing Vietnam in the early 1980s.

Hong Kong's high population density is acceptable to the people because they are rich enough to import all the resources, especially food, that they need.

were equally divided between its inhabitants, the living space allotted to each one would be about one quarter of a hectare. If the English were all self-sufficient peasants, a family of three would have to win its basic survival needs: food and drink, water, building materials for a house, waste disposal facilities, clothing, and energy for travel, heating, lighting, appliances and cooking, from an area about the size of a football pitch, all while remaining on good terms with their equally challenged neighbours.

Luckily for the English they are no longer peasants, and their social and survival needs are supplied by their complex organisation of specialists. The most important of these, today, are the farmers who produce two-thirds of England's food, and the specialists who by their expertise in manufacturing or services, earn the foreign currency that buys from less densely populated overseas lands the food and raw materials, including oil, that maintain England's Western standard of living. The Optimum Population Trust has calculated that the area of 'ecologically productive land' necessary to provide each English citizen with a satisfactory standard of living (a "modest ecological footprint" for each citizen) is 3 hectares at average worldwide productivity levels (Ferguson, 1998). On this basis, referring to the previous paragraph, the English are squeezed into an area at least 12 times smaller than could support, on its own, their actual standard of living.

The vulnerability of a congested country is obvious. If the western counties of IRELAND had been sparsely populated when a fungal blight destroyed the potato crop in 1845–46 there would have been no Great Famine. In fact, excessive population growth had rendered them so congested that only one management strategy kept the malnourished peasants alive: grow more and more potatoes. When the strategy failed, anthropogenic disaster inevitably followed.

The shortage of space in a congested country frustrates people. In England, which is more densely populated than any other Western nation of its size (and lacks precedents, therefore, to guide its journey into hyper-overpopulation) proposed developments such as

HUNGARY   Area 93,000 km²

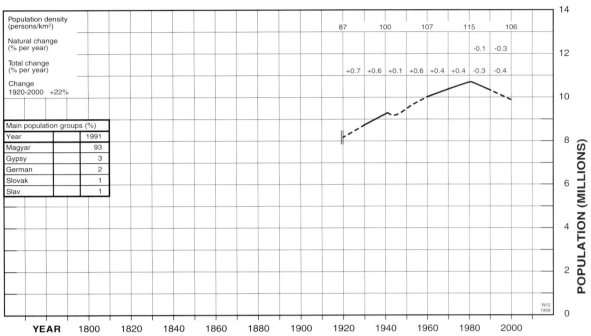

Population density (persons/km²)		87	100	107	115	106
Natural change (% per year)					-0.1	-0.3
Total change (% per year)		+0.7	+0.6	+0.1	+0.6	+0.4	+0.4	-0.3	-0.4
Change 1920-2000 +22%						

Main population groups (%)		
Year		1991
Magyar		93
Gypsy		3
German		2
Slovak		1
Slav		1

YEAR   1800   1820   1840   1860   1880   1900   1920   1940   1960   1980   2000

POPULATION (MILLIONS)

WIS 1999

Modern Hungary was created in 1920 by a treaty which awarded three quarters of its pre-war territory to Romania, Czechoslovakia and Yugoslavia. The region lies between Germanic and Slav homelands and has been fought over since Roman times, when it was part of the provinces of Dacia and Pannonia. Magyar tribes arrived in the 9th century. The next 500 years saw great but not abnormal brutality. In 1241 Mongol hordes swept through, killing one third of the population. In 1514 70,000 rebel peasants were tortured and killed and their leader burned alive on a red-hot iron throne. In 1526 at the Battle of Mohacs the Turks wiped out four-fifths of the Hungarian army, conquering most of the area. Austrian Habsburgs drove out the Turks in 1699 and contained a growing Hungarian nationalism by devising the dual monarchy of Austria-Hungary in 1867. This powerful state was defeated and dismantled in World War One. The new small Hungary sided with Germany in World War Two and was liberated by the Soviet army. The post-war Soviet-dominated communist dictatorship brutally quelled a revolt in 1956, when 200,000 people fled the country, but gave way to democratic government in 1989.

The slow increase of Hungary's population has been normal for a developed European nation, but the post-1980 decline is remarkable. Similar declines characterise the ex-communist non-Muslim states of eastern Europe and the old Soviet Union. The Hungarian decline began about a decade before most of the others. There are severe restrictions on immigration into Hungary, but the borders are naturally porous, given the large Hungarian minorities in adjacent states. In 2000 the Roma (Gypsy) minority had increased to about 600,000 (6% of the population) due to a birth rate six times as high as that of native Hungarians. Most were living off state benefits.

Censuses: about every 10 years.

new roads, housing, factories, quarries or garbage disposal sites generate hosts of protesters who claim that local and/or national amenities would be permanently damaged. They are branded 'nimby' (not in my back yard) and selfish by opponents who argue that the development will benefit society in general. Vastly expensive public inquiries are held, delaying decisions for as much as 7 years (in the case of a proposed new terminal for London's Heathrow Airport).

Congestion is the scourge of individuality. In a small village, few people are total strangers to each other, but in a big city individual anonymity is the rule. People in crowds become depersonalised, rather like telephone numbers: my number 35 years ago was my village name followed by 3 figures, but now it is a mere sequence of 11 figures.

An overcrowded Western nation may, by clever management, be able to cope with some or all of the problems commonly associated with congestion, such as overstretched public services, lack of waste disposal sites, urban squalor including shortage of safe play areas for children, traffic slowdown, air and water pollution, rapid spread of contagious diseases, expensive housing and land, administrative complexity, racial tension, creeping deterioration of countryside and wildlife habitat, etc. But when something happens that disrupts the clever management, like a strike, extreme weather, a terrorist threat, or foot and mouth disease, the

1.8 – 18 million

IRELAND (EIRE) and N. IRELAND   Area 84,400 km² (total): Ireland 70,300 km²: Northern Ireland 14,100 km²

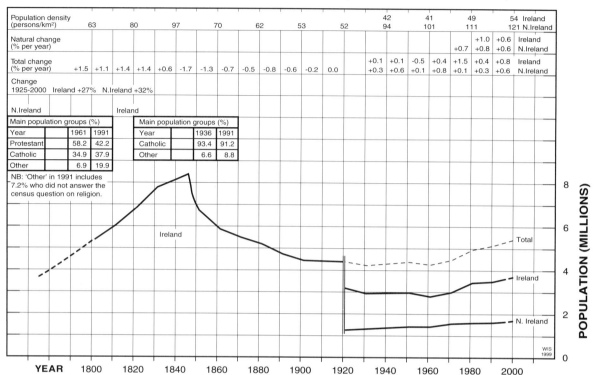

Population density (persons/km²)		63		80		97		70		62		53		52		42 / 94		41 / 101		49 / 111		54 Ireland / 121 N.Ireland

Natural change (% per year)															+0.7	+1.0 / +0.8	+0.6 Ireland / +0.6 N.Ireland

Total change (% per year): +1.5 | +1.1 | +1.4 | +1.4 | +0.6 | -1.7 | -1.3 | -0.7 | -0.5 | -0.8 | -0.6 | -0.2 | 0.0 | +0.1 / +0.3 | +0.1 / +0.6 | -0.5 / +0.1 | +0.4 / +0.8 | +1.5 / +0.1 | +0.4 / +0.3 | +0.8 Ireland / +0.6 N.Ireland

Change 1925–2000   Ireland +27%   N.Ireland +32%

Main population groups (%) N.Ireland		
Year	1961	1991
Protestant	58.2	42.2
Catholic	34.9	37.9
Other	6.9	19.9

Main population groups (%) Ireland		
Year	1936	1991
Catholic	93.4	91.2
Other	6.6	8.8

NB: 'Other' in 1991 includes 7.2% who did not answer the census question on religion.

Graph labels: Ireland; Total; Ireland; N. Ireland. Y-axis: POPULATION (MILLIONS) 0, 2, 4, 6, 8. X-axis: YEAR 1800, 1820, 1840, 1860, 1880, 1900, 1920, 1940, 1960, 1980, 2000.

WIS 1999

The island of Ireland was settled by Iron Age Celts from Britain and continental Europe late in the first millennium BC. Rome never conquered it, in fact raiders from Ireland harassed Roman Britain in the 4th century AD. Saint Patrick converted the Irish to Christianity in the 5th century. Vikings pillaged monasteries and towns in the 8th century, establishing their own coastal settlements. Anglo-Norman (English) invasions began in the 12th century and powerful barons carved out great personal estates. Many of them gradually merged with the native Irish. English rule was forcibly re-established, often with much bloodshed on both sides, every now and then from the 16th to the 19th centuries. In the 17th century, part of Ulster was 'planted' with imported Scottish and English settlers; a political initiative which by mixing Catholic Irish with Protestant foreigners to create multicultural Northern Ireland has caused more than 3 centuries of factional strife.

Ireland's 'Great Famine' (1845–51) was the world's first well-documented case of Malthusian cutback, caused by a sudden collapse of regional carrying capacity. In the western counties the climate was unsuited to growing most cereals, but the potato, introduced about 1600, thrived. The prolific and nutritious tuber could support a large family on a tiny acreage, enabling the population to double, and double again, abnormally fast, in the 18th and early 19th centuries. By 1840 millions of people were living on a knife-edge with little to eat but potatoes. Periodic poor harvests had already caused severe hunger and local starvation; in the 1739–41 famine one third of the (much smaller) population had died. The vulnerability of a population that depended on a single species of plant for survival was remarked by many writers. Finally in 1846 an imported fungal blight wiped out the whole crop. The people had no money to buy other food. More than a million died of starvation and diseases of malnutrition, and the ensuing decades saw mass emigration, mainly to America.

The disaster exacerbated the ancient hatred of the Irish for their British rulers. Political and terrorist campaigning led to partition in 1921 when the largely Catholic South separated from the mainly Protestant North, which remained British. Continuous emigration from both countries, more or less balancing their persistently high birth rates, has limited the rivalry between them to terrorist activity by both sides, mainly in Northern Ireland where each side has sub-factions within which intermittent violence continues in defiance of a 'peace process'. In the North the Catholics appear to be overtaking the Protestants in terms of numbers. The Irish ambition to unite south and north as one nation may seem surprising. Controlling a rebellious Protestant faction would be proportionately more difficult for Dublin than the present confrontation is for London.

severity of the consequences is linked, as in anthropogenic disasters, to the degree of national congestion. The Netherlands, in 1997, planned to reduce its pig-rearing industry by 25%, because the pig population was so dense that an outbreak of swine fever could not be contained until 8 million pigs had been destroyed. Reducing a human population by 25% would be more controversial.

Clever management implies efficiency, which often means inventing clever new ways of dealing with problems caused by congestion. Some ways were not so clever, as in Belgium and France where waste oils and sewage respectively were added to animal feed. When the

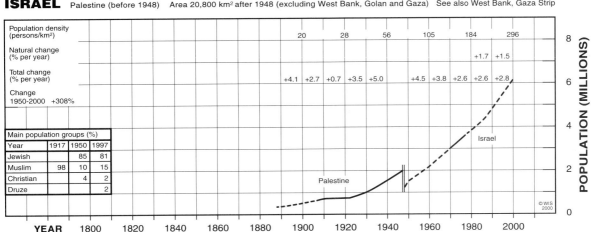

ISRAEL   Palestine (before 1948)   Area 20,800 km² after 1948 (excluding West Bank, Golan and Gaza)   See also West Bank, Gaza Strip

Israel/Palestine is a Holy Land to three religious groups: the Jews, who believe it was promised to them by God; the Christians, because Jesus Christ lived and taught there; and the Muslims, because Mohammed ascended to Heaven from Jerusalem. Religious zealotry and intolerance has brought hatred and strife to the Holy Land during the last several millennia. The ancient history of the area figures largely in the Old Testament books of the Christian Bible. It was part of the 'fertile crescent' which saw the invention of agriculture more than 10,000 years ago. Jericho, which was a settlement in 8000 BC, must be one of the oldest cities on Earth.

After the Roman and Byzantine occupations Palestine was invaded and conquered by Muslim Arabs in 637 AD. Christian nations backed by the Pope launched Crusades between 1096 and 1272 to recover the Holy Land from the 'infidels', with short-lived success. Palestine was incorporated into the Ottoman Empire by conquest in 1517. In World War One it was occupied by British forces and in response to Zionist pressure Britain expressed support for a Jewish homeland in Palestine, which became a British mandate in 1920. Increased Jewish immigration, which had begun in the late 19th century, provoked both Arab unrest and Jewish terrorism aimed at ending the British administration.

In 1948 Britain relinquished control over Palestine, which became Israel as the Jews declared independence, promising "complete equality of social and political rights to all its inhabitants irrespective of religion, race or sex". Israel was at once attacked, unsuccessfully, by Syria, Lebanon, Jordan, Iraq and Egypt. Arabs fled in great numbers from Israel to Jordan (which had seized the West

Bank), Lebanon and Gaza, and there was massive Jewish immigration. In 1967, in a pre-emptive strike against massed Arab forces, Israel seized Gaza, the West Bank, East Jerusalem, the Golan Heights and Sinai. The last-named was returned to Egypt in 1979. Israel invaded Lebanon in 1978 and 1982 to subdue Palestinian guerrillas, with indifferent success, evacuating an occupied frontier strip in 2000.

In Israel proper, Jews outnumber Arabs five to one. If Israel allowed the five million Palestinian refugees to return, the Jews would be a minority. Israel dominates its Arab neighbours by virtue of its advanced technology, military strength, and aid (several billion dollars annually) from the US. Israel's Arab neighbours are gaining strength by rapid population growth ('aggressive breeding'), which is imitated by some Orthodox Jewish minorities and offset also by ongoing Jewish immigration. Mutual race hatred is intense, there are frequent acts of terrorism, and as resources per person diminish (in 2001 over-abstraction of fresh water from the river Jordan catchment area was causing the level of the Dead Sea to fall by more than a metre per year) a showdown leading to Malthusian cutback will not be long delayed. Given the powerful backers of both sides, and the region's position on the edge of the world's greatest oil pool, such a showdown would have world-wide consequences.

A low-key Arab uprising or *intifada* began in October 2000. US attempts to end it were frustrated by Palestinian insistence on the right of their refugees to return to Israel proper. It escalated through 2001 until Palestinian suicide bombers and Israeli military reprisals were causing up to 50 deaths in a day.

British agricultural industry began pulverising its mountains of slaughterhouse wastes (animal bones, skins and offal) to produce 'meat and bone meal' which was then fed to cattle in place of imported vegetable protein, the process was hailed as a marvel of efficiency. Unfortunately, the unnatural diet led to the outbreak of Bovine Spongiform Encephalopathy (BSE) which severely damaged Britain's cattle industry.

A few years later, in 2001, inefficient Customs procedures (the underfunded service was unable to control the vast import-export traffic through congested ports and airports) allowed the foot and mouth disease virus to enter Britain, probably in contaminated meat. Several million farm animals had to be killed and buried or burned, huge areas of countryside were closed, and the cost to the nation, in particular through loss of tourism revenues, totalled many billions of pounds.

England during World War Two had a PD (population density) of about 300, and the English would have starved in 1940 had not convoys of merchant ships imported between 0.7 and 1.2 million tonnes of food and raw materials, including fertilisers but not counting oil, into

1.8 – 18 million

JAMAICA Area 11,420 km²

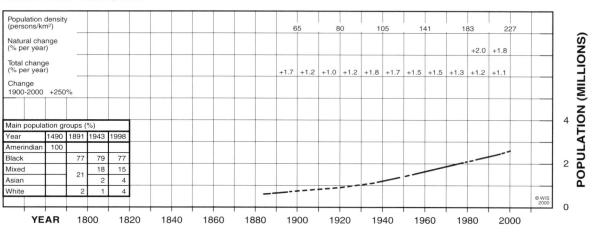

Population density (persons/km²)							65	80	105	141	183	227
Natural change (% per year)											+2.0	+1.8
Total change (% per year)						+1.7	+1.2	+1.0	+1.2	+1.8	+1.7	+1.5	+1.5	+1.3	+1.2	+1.1
Change 1900-2000 +250%												

Main population groups (%)

Year	1490	1891	1943	1998
Amerindian	100			
Black		77	79	77
Mixed		21	18	15
Asian			2	4
White		2	1	4

POPULATION (MILLIONS)

YEAR   1800   1820   1840   1860   1880   1900   1920   1940   1960   1980   2000

© WIS 2000

Late in the first millennium AD Jamaica was settled by Arawak Indians, who numbered about 100,000 when Columbus visited the island in 1494. He was marooned there for a year when his ship sank in 1503. Spain claimed Jamaica ('Island of Springs') in 1509, and developed sugar plantations, enslaving the Indians who all perished from ill-treatment and Old World diseases. African slaves began to replace them from 1517 onwards. Britain captured the island in 1655 and expanded sugar production, importing huge numbers of slaves. British buccaneers based in Jamaica attacked Spanish ports and shipping throughout the Caribbean. Slave revolts broke out at intervals between 1760 and 1831 and were mercilessly suppressed.

Slavery ended in 1838 and the economy was devastated. Bananas grown on vast plantations joined sugar as the main exports. The black population, growing fast but mostly unemployed, began violent agitation for independence in the 1930s and it was declared in 1962 after 3 centuries of colonialism. In subsequent decades the high birth rate has caused poverty, violence, drug crime, deforestation, loss of wildlife and constant emigration, but revenue from bauxite exports and a growing tourism industry are bringing some economic recovery. Censuses: at irregular intervals from 1891.

the UK *every week* (Churchill, 1951). Today, in a national emergency, England could not adequately feed itself. Its population is 10 million greater than in 1940, its farmland is much reduced by built development, and its agricultural efficiency depends on agrochemicals and farm machinery which themselves depend on (offshore or imported) oil. The UK food trade deficit (the difference in value between food imports and exports) was £9 billion in 2000 (*Farmers Weekly*, 8 June 2001).

Congested England's well-being, which the English take for granted, is as fragile as a house of cards. An economic setback reducing foreign earnings, war or the threat of war (anywhere) cutting imports of food and oil, or political turmoil or civil strife disrupting the smooth functioning of government or public services; any one of these crises could paralyse the country. In October 2000, a brief strike by petrol and oil suppliers caused panic buying and threatened breakdown of the transport system.

England's predicament is shared by most other European nations, but only the Netherlands (PD 468 in 2000) is more congested. Of the world's other major nations, Bangladesh (865), Taiwan (605) and South Korea (472) have higher PDs than England (382), while Belgium (336), Japan (335) and India (318) have slightly lower ones. It is no coincidence that pressure to ease congestion by emigration has been especially high in two of these nations (Bangladesh and India) in which, in recent decades, poverty has worsened because population growth has outstripped economic growth (the Micawberish Rule, section 6.4). Congested India's vulnerability was illustrated by the anthropogenic disaster of January 2001 when an earthquake (7.9 on the Richter scale) in Gujarat state killed 35,000 people, injured 67,000 and made 600,000 homeless.

Micawberish poverty and congestion in developing nations is weakening medical services in their struggle to cope with many chronic diseases, especially TB. In 1998 TB was responsible for one in every seven adult deaths, world-wide, and the disease was worsening in step with malnutrition, drug resistance and HIV/AIDS.

Population density is not always a reliable measure of congestion. Egypt has a fairly low

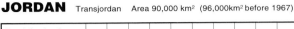

JORDAN Transjordan   Area 90,000 km² (96,000km² before 1967)

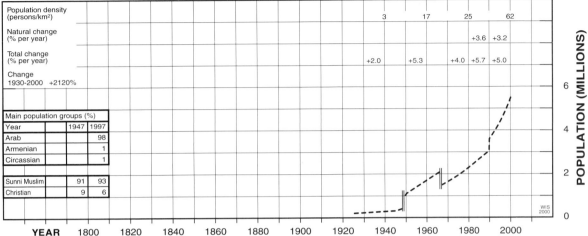

Population density (persons/km²)			3	17	25	62	
Natural change (% per year)					+3.6	+3.2	
Total change (% per year)			+2.0	+5.3	+4.0	+5.7	+5.0
Change 1930-2000 +2120%							

Main population groups (%)		
Year	1947	1997
Arab		98
Armenian		1
Circassian		1
Sunni Muslim	91	93
Christian	9	6

POPULATION (MILLIONS)

YEAR   1800   1820   1840   1860   1880   1900   1920   1940   1960   1980   2000

Jordan did not exist as a state before 1920, when it came into being as Transjordan. In Old Testament times Ammonites had lived in the region, and it became part of the Israel of King Solomon at the start of the first millennium BC. Rome conquered it in 64 BC. Rome's successors the Byzantines lost it to Arab armies spreading Islam in the 7th century AD. Briefly in the 11th and 12th centuries Christian Crusaders partly evicted the Arabs, who returned only to lose the whole region to the Ottoman Turks in 1517. When the Ottoman Empire collapsed after defeat in World War One the League of Nations placed Palestine and Transjordan under British mandates. Transjordan, a sparsely populated semi-desert country ruled by King Abdullah, separated from Palestine in 1923.

Transjordan became independent in 1946 and attempted, with other Arab states, to crush the newly created Israel in 1948. Israel won the war, but Transjordan annexed part of Palestine, the West Bank, and renamed itself Jordan, in 1949. Some 400,000 refugees left Palestine for Jordan. In a pre-emptive strike by Israel, the Six Day War of 1967, Jordan lost the West Bank but acquired more floods of Palestinian refugees. King Hussein quelled uprisings by Palestinian militant organisations, but Jordan was forced to accept vast influxes of refugees from Kuwait and Iraq during the Gulf War, 1990–91. In 1998 there were 900,000 registered refugees in Jordan.

With few natural resources, and fresh water and agricultural land in short supply, Jordan is far from self-sufficient and depends on foreign aid, though there is valuable income from tourism. In 2001 overpumping of fresh water in the river Jordan catchment was dropping the level of the Dead Sea more than a metre per year. The very fast population growth is bound to accentuate resource shortages, poverty and political instability.

Population estimates for Jordan after 1967 vary widely. For example, the UN estimate for 2000 of 6.33 million does not match the Population Reference Bureau estimate of 4.7 million for 1999. The causes appear to be partly that Jordan maintained a claim to the West Bank until 1988, and partly that some sources include refugees (also a variable figure) whereas others do not. The graph assumes that Jordan lost the West Bank in 1967 and that the population includes all refugees.

national PD of 68, but is mostly uninhabited desert. PD in the fertile valley and delta of the Nile is a dangerously high 1900. The same is true of almost all the Arab countries of North Africa and South-west Asia: although their overall PDs are low, their aridity coupled with the high Muslim birth rate of their peoples has resulted in excessive congestion of their habitable regions, causing poverty, civil unrest and pressure to emigrate, even in some of the oil-rich states.

Constantly increasing congestion *must* in due course end, usually with population reduction through genocide and/or ethnic cleansing in a VCL event (e.g. KOSOVO). Arguably therefore, the moral opprobrium including the epithet 'world criminal' directed at the political and military leaders who happen to be in power when the event occurs, is not wholly appropriate. Some of the opprobrium should attach to those people who, for self-centred politically correct, commercial or other reasons, let the congestion increase by suppressing discussion and analysis of its cause: population growth.

6.3 The Fragility of Large International Associations

History is littered with brave attempts by groups of nations to merge, forming a single nation that will, they assume, enjoy the enhanced power and influence that its greater size and population deserves. This section is concerned with *voluntary* mergers, entered into for calculating or idealistic reasons, in modern times (the 20th century). Enforced unification

1.8 – 18 million

KAZAKHSTAN   Area 2,717,000 km²

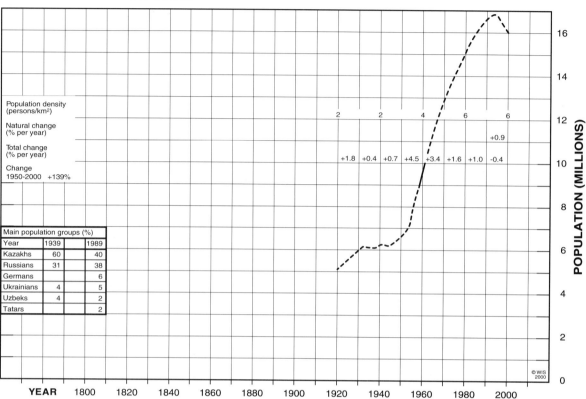

Population density (persons/km²)	2	2	4	6	6
Natural change (% per year)					+0.9
Total change (% per year)	+1.8	+0.4	+0.7	+4.5	+3.4	+1.6	+1.0	-0.4
Change 1950-2000   +139%

Main population groups (%)		
Year	1939	1989
Kazakhs	60	40
Russians	31	38
Germans		6
Ukrainians	4	5
Uzbeks	4	2
Tatars		2

© WIS 2000

Kazakh and Turkic nomadic tribes drifted west over the endless grass steppes and semi-deserts of Kazakhstan (a country the size of Western Europe) from their Mongolian homelands during the first millennium AD. They converted to Islam in the 8th century and joined with Genghis Khan's Mongol hordes that devastated cities and nations as far west as the river Danube around 1220. Mongol rule in the 13th and 14th centuries was followed by successive Kazakh khanates (chiefdoms) that were annexed into the Russian Empire in the 18th and 19th centuries. Russians and Ukrainians immigrated to farm the steppes, triggering an anti-Russian revolt in 1916 that was put down with 150,000 killed.

The Soviet Socialist Republic of Kazakhstan was created on an ethnic basis (i.e. the boundaries were drawn to include as many Kazakhs as practicable) in 1925, and the Soviets forced the 'collectivisation' of farming so brutally that a million peasants died of starvation in the early 1930s. During World War Two long-term German residents in Russia were forcibly resettled in Kazakhstan. After the war a vast influx of Russian settlers made the Kazakhs a minority in their own land. They ploughed the steppes as part of the Virgin Lands campaign, a spectacular failure that, through excessive irrigation, began the desiccation and poisoning of the Aral Sea and its surroundings. Huge oil and gas reserves were discovered.

When the Soviet Union collapsed in 1991 Kazakhstan declared independence. As in other ex-Soviet nations with a large Slav element, population growth went into reverse. If the big difference between total and natural change in the 1990s is real, it must result from migration.

by conquest, as in the creation of empires, normally lasts only as long as the conqueror's military strength endures.

It has to be said that few of the mergers last very long. Much more commonly the opposite occurs, when large nations with expanding populations break up into their component, usually ethnic or religious, parts. International associations are vulnerable to the same divisive Darwinian forces as multicultural societies (section 7.2).

YUGOSLAVIA is the prime example of a failed merger. For many centuries the mainly Slav inhabitants had fought each other, establishing well-defined tribal homelands (SLOVENIA, CROATIA, BOSNIA, SERBIA, MACEDONIA and MONTENEGRO). After World War One, hoping to end their ruinous confrontations, the tribes joined together as the 'Kingdom of Serbs, Croats and Slovenes', later renamed Yugoslavia (Land of the Southern Slavs). But the old loyalties persisted, flaring into savage warfare, especially between Serbs and Croats, in World War Two. After the war the ethnic rivalries were aggravated by demographic developments, Serb ambition, and growing ethnic and religious discontent. The

KOSOVO and Metohija   Area 10,900 km²

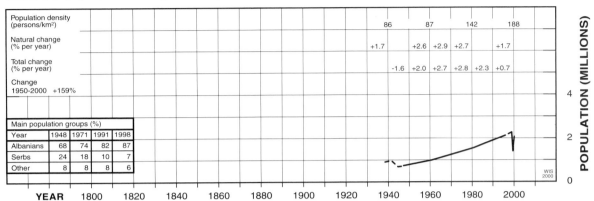

					Population density (persons/km²)	86	87	142	188
					Natural change (% per year)	+1.7	+2.6 +2.9 +2.7	+1.7	
					Total change (% per year)		-1.6 +2.0 +2.7 +2.8 +2.3 +0.7		

Change 1950-2000 +159%

Main population groups (%)				
Year	1948	1971	1991	1998
Albanians	68	74	82	87
Serbs	24	18	10	7
Other	8	8	8	6

YEAR   1800   1820   1840   1860   1880   1900   1920   1940   1960   1980   2000

POPULATION (MILLIONS)   4   2   0

WIS 2000

Kossovo was an ill-defined region of the Ottoman Empire and the Kingdom of Serbs, Croats and Slovenes. Present-day Kosovo was established in 1945 as a part of Serbia. Over the centuries there have been huge ethnic migrations into and out of the region. Rome conquered the Illyrians, ancestors of the Albanians. Slav settlers came from the north as the Roman Empire disintegrated and built a powerful Serb state centred on Kosovo, only to lose it to Ottoman Turkish invaders in the Battle of Kosovo Polje (1389). Under the Turks, Albanian settlers moved in. The Serbs returned, displacing many Albanians, after the Turks were driven out in the 1913 Balkan War. During World War Two the occupying Italians favoured the Albanians, who outnumbered the Serbs almost 3 to 1 after the war.

Since World War Two the high Albanian birth rate has greatly increased the ethnic imbalance. "Having a large number of children... was seen as ensuring for Kosovo an Albanian as opposed to a Serb future" (Vickers, 1998, page 172). By 1991 there were 8 Albanians to each Serb. Aspiring, with fellow Muslims in Macedonia, to secede and re-create the 'Greater Albania' of World War Two, they harassed the Serb minority, forcing many of them to emigrate. By 1998 there were more than 12 Albanians to each Serb. There were almost as many Roma (Gypsies) as Serbs. Kosovo was more than twice as densely populated as any other region of the former Yugoslavia except Serbia Proper, even though 600,000 ethnic Albanians had emigrated to work in Western Europe.

From about 1990 the Serbs had used increasing savagery to maintain their rule. In response the Albanians formed a guerrilla force, the Kosovo Liberation Army. By 1998 NATO was alarmed at the brutal conflict which had made more than 100,000 people homeless. After failing to achieve peace by political means, in 1999 NATO took the Albanian side and began high-altitude bombing of Serb military and economic targets throughout Kosovo and Serbia. The initial Serb response was to intensify their aggression, driving some 850,000 refugees into adjacent countries, but after 78 days the massive damage to Serbia's infrastructure forced a cease-fire. Returning Albanian refugees (plus perhaps 250,000 opportunistic immigrants from Albania) began vengeance attacks on villages and people, desecrating churches and monasteries to reduce Serb interest in the province, driving out at least 100,000 Serbs and some 80,000 Roma. Complete ethnic cleansing was and is only prevented by a NATO peacekeeping force (KFOR) 50,000 strong. In mid-2000 about 95,000 Serbs were left in Kosovo.

The cost of NATO's war is estimated at 4 billion dollars, and the cost of reconstruction around 30 billion dollars. NATO's pious hope that a peaceful multi-ethnic Kosovo can develop is unsupported by any historical precedent. By 2001 Kosovo was effectively Albanian. Kosovan fighters were behind the insurrection of Muslim Albanians in Macedonia. The ethnic cleansing of Serbs from Kosovo was the outcome of Albanian aggressive breeding (section 4.4) over half a century.

demographic changes were the population increases in every province, which were disproportionately large among the Muslim populations of Bosnia and KOSOVO. They became the backward regions of Yugoslavia, subsidised, resentfully, by the successful economies of the other provinces. In 1992 the federation disintegrated in bloody civil war, which was ended by United Nations intervention and would probably resume if the UN peacekeeping forces were withdrawn.

EGYPT and SYRIA merged in 1958 to form the socialist "United Arab Republic" (UAR), as a first stage in the pursuit of Arab unity by Egypt's President Nasser, but only North Yemen came close to joining it. The kings of JORDAN and IRAQ immediately federated their countries in a rival "Arab Union", which broke up 5 months later when the Iraqi king was killed in a coup. Syria, meanwhile, was finding Egyptian domination of the UAR too oppressive, and seceded in 1961.

In British India, Hindus and Muslims had co-existed reluctantly, and they polarised violently into INDIA and PAKISTAN when the opportunity arrived with independence in 1947. Religion was not the only divisive force; when the eastern part of Pakistan broke away in 1971 to form BANGLADESH it was to satisfy ethnic, Bengali, tribalism. Britain set up the self-governing West Indies Federation in 1958 to give its component islands some economic

1.8 – 18 million

KUWAIT Area 17,800 km²

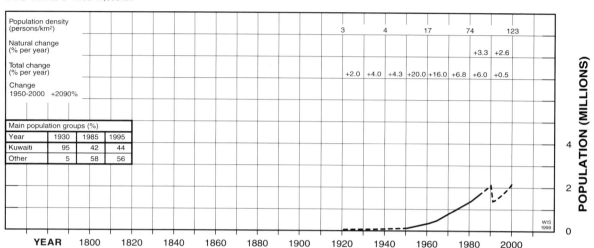

Population density (persons/km²)		3	4	17	74	123	
Natural change (% per year)					+3.3	+2.6	
Total change (% per year)	+2.0 +4.0 +4.3 +20.0 +16.0 +6.8	+6.0	+0.5				
Change 1950-2000 +2090%							

Main population groups (%)

Year	1930	1985	1995
Kuwaiti	95	42	44
Other	5	58	56

Kuwait was a small Muslim desert sheikdom at the head of the Arabian Gulf, founded in 1710, part of the Ottoman Turkish Empire until 1899 and a British protectorate from then until independence in 1961. It was totally transformed by the discovery of vast oil reserves, possibly 10% of the world total, in 1938. The need for oil industry workers caused the population to soar, and explosive growth continued as immigrants flooded in to serve the needs of newly rich Kuwaiti citizens. Oil and oil products provide almost 95% of export revenue and 98% of the nation's fresh water is obtained from the sea through desalination plants.

Iraq, which had tried and failed to annex Kuwait in 1961, occupied the country by force in 1990. More than half a million refugees fled. Western concern for the security of its oil supplies led to an international largely US force, backed by the United Nations, expelling the Iraqis in the 1991 Gulf War and restoring Kuwait's independence. The government has announced its intention to ensure a permanent Kuwaiti majority in the population by restricting foreign immigration.

The data rows illustrate how widely natural change can differ from total change (natural change plus net migration).

and political clout, but separatist forces such as the reluctance of the richer islands to subsidise the poorer ones led to its disintegration in 1962. On the other hand the Federation of MALAYSIA, created in 1963 by uniting British-ruled Malaya, Sabah, Sarawak and SINGAPORE, has survived the secession of Singapore in 1965, in spite of its ethnic mix of Malays, Chinese and Indians, largely because of its ongoing prosperity. In Africa, the Federation of Rhodesia and Nyasaland (ZAMBIA, ZIMBABWE and MALAWI) was a somewhat artificial construction by white settlers (1953–63) which was overwhelmed by the massive fast-growing black majority. Seven small sheikdoms which had been British protectorates merged into the UNITED ARAB EMIRATES when Britain withdrew in 1971. The immense oil wealth of the UAE appears to have stifled any separatist tendencies.

Elsewhere in the world, SENEGAL's brief merger with GAMBIA (Senegambia, 1982–89) failed because of political incompatibility, but the nations remained friendly. CZECHOSLOVAKIA arose as a nation in 1918 from the fragments of Austria-Hungary, and separated amicably, 74 years later, into the Czech Republic and Slovakia, because of ethnic and political ambitions only slightly more acute than those currently dividing Wales, Scotland and England in the United Kingdom. NORWAY broke away from SWEDEN in 1905, largely due to its strong and frustrated sense of national identity. Rich northern ITALY, resenting the constant diversion of national resources to the poor south, has threatened to seek independence. North and South KOREA have grown apart since they were separated in 1945.

Other nations have successfully resisted the demands of separatists, some by force, some peacefully, as in historically French Quebec, which voted by a whisker to stay part of CANADA in 1995. NIGERIA fought a vicious war, 1967–70, to prevent the secession of oil-rich, ethnically Ibo, Biafra. So did CONGO DEMOCRATIC REPUBLIC, 1961–63, to keep control over copper-rich Katanga province. Muslim fundamentalists are waging guerrilla wars to establish independent states in parts of INDONESIA, the PHILIPPINES, Chechnya and

KYRGYZSTAN Kirghizia Area 199,000 km²

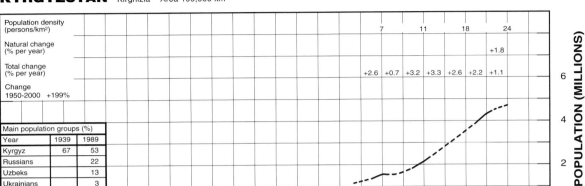

Population density (persons/km²)								7	11	18	24	
Natural change (% per year)											+1.8	
Total change (% per year)								+2.6 +0.7 +3.2 +3.3 +2.6 +2.2 +1.1				
Change 1950-2000 +199%												

Main population groups (%)

Year	1939	1989
Kyrgyz	67	53
Russians		22
Uzbeks		13
Ukrainians		3
Germans		2

© WIS 2000

POPULATION (MILLIONS)

YEAR 1800 1820 1840 1860 1880 1900 1920 1940 1960 1980 2000

The Tien Shan, or Heavenly Mountains, occupy much of this Central Asian republic. Nomadic Kyrgyz tribes came to the fertile valleys between the 10th and 16th centuries, driven from their homelands in south-central Siberia by the Mongols. After the Russian Empire incorporated Kyrgyzstan in the 1870s, Russian immigrants began settling the best land. An unsuccessful Kyrgyz revolt in 1916 was followed by brief independence as part of a Turkestan republic, but the Soviet Union created Kyrgyzstan as an ethnic (see KAZAKHSTAN) Soviet Socialist Republic in 1924. Many more Russians migrated to the scenically beautiful and climatically agreeable (but earthquake-prone) territory. Ethnic clashes marked the period of Soviet collapse (1990–91), but since the declaration of independence in 1991 the tensions have reduced, although the economy remains fragile. Islam (mainly Sunni Muslim) is the dominant religion, with Christians a sizeable minority. Muslim fundamentalists began guerrilla attacks in 1999.

AFGHANISTAN (where the Taliban briefly succeeded). Ethnic separatist wars are ongoing in SRI LANKA, SUDAN and elsewhere.

Most of this strife and rejection can be explained by *Homo sapiens'* natural preference for ethnic or religious solidarity, sticking together in mono-ethnic or mono-cultural groups, exemplified by the yearning of such groups, when divided, to re-unite, as happened in GERMANY, YEMEN and ISRAEL (the Jewish diaspora). It is the antithesis of multiculturalism. As discussed later in this chapter, it is a basic feature of Darwinian natural selection.

Distinct from sovereign nations, but swayed by the same forces, are political and economic associations such as NATO, NAFTA, the post-USSR CIS, the British Commonwealth, the old League of Nations and its successor the United Nations. The League of Nations became toothless because several major nations (e.g. the USA) never joined it, and others resigned when their nationalist ambitions could not be reconciled with its ideals. The UN has fared better, not least because almost all nations have joined it, and adequate finance has been available for most of its initiatives. One of its most important functions has been to send peacekeeping forces into civil war situations to hold the battling sides apart, but there have been major failures (see BOSNIA, RWANDA, SOMALIA) where the fighting has been particularly vicious because the population of the nation involved has been at or close to VCL.

So the destabilising forces in national and international associations are nationalist and cultural (mainly ethnic/religious) rivalry and ambition, all of which are intensified by fast population growth and a high population density. On precedent, therefore, the prospects for an enlarged European Union are not good, given the fast growth of ethnic minorities through immigration, which is likely to accelerate, the Darwinian readiness of individual members to profit from the misfortunes of others (e.g. the French ban on importing British beef long after it was officially 'safe'), and the extreme sensitivity of the existing organisation to perceived politically incorrect actions such as the incorporation into government of an anti-immigrant party (see AUSTRIA).

1.8 – 18 million

LAOS Area 236,800 km²

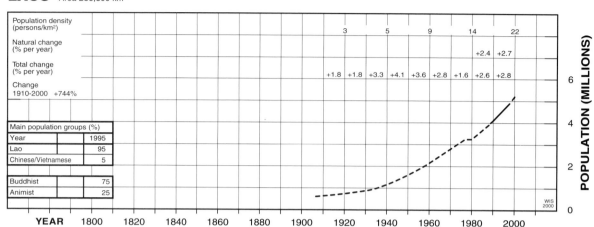

About half of mountainous Laos is covered by natural rain forest and most of the working population are peasant farmers. In this respect there has been little change since the country was part of the Khmer Empire before the 14th century AD. Many rare and unique large animal species still survive in the remote mountain forests. Lao tribes originally from China, Siam and Burma established kingdoms between the 14th century and a period of Thai rule in the 19th.

France declared a protectorate over Laos in 1893, demarcated its present borders in 1907, and ruled it as one of the three components of French Indochina until World War Two, when it was occupied by Japan. When in 1953 Laos became an independent US-backed kingdom the communist Pathet Lao, backed by North Vietnam, began open revolt which continued, devastating the economy, until 1975 when the victorious rebels set up their own republic. The next 5 years saw hunger and persecution driving some 300,000 Laotians to resettle in Thailand. Since 1989 the government has relaxed its strict left-wing regime and economic recovery has begun.

6.4 Migration, Refugees, Asylum-Seekers and the Micawberish Rule

The migration of people from place to place is nothing new. Nor is the variety of receptions that migrants encounter in lands that already have native populations. Militant migrants, invaders whose purpose is occupation and settlement, are resisted with all the force that the natives can muster. Throughout history, potential victims knew that if their resistance was overcome, many or most of them would be eliminated so that the victors (e.g. the Mongols, or empire-builders in the Balkans, or Europeans taking over the New World) could enjoy their new territory unencumbered by crowds of its previous owners. This was the normal expectation during the millennia before the mid-18th century AD, when the start of the WROG period (section 4.5) meant that a country could provide resources enough to sustain winners and losers together.

Huge numbers of refugees fled the wars, civil and international, that they were losing (see JORDAN, AFGHANISTAN, RWANDA, SIERRA LEONE, etc.). Until World War Two, economic migrants leaving European countries for the New World were numbered in hundreds of thousands annually (see IRELAND, POLAND). But since World War Two the flow of people leaving Europe has gradually reversed. Emigration became unnecessary as Europe's birth rate declined, but over the same period the population of the developing world trebled (Table 3.1). In 2000 the European Union calculated that some 500,000 *illegal* immigrants were entering it annually.

The West's economy has grown faster than its population, result: wealth. In most of the developing world, population growth has outpaced economic growth, result: poverty. In this book this simple relationship between the size of the economic cake and the number of people each wanting a slice of it (section 5.5), is called the *Micawberish Rule*, because Mr Micawber famously pinpointed a similar vital link between family income and expenditure (Dickens, 1850). So now, according to a spokesperson for the UN High Commission for Refugees, "Western Europe is seen as a haven by the rest of the world". She could have added the havens of North America and Australia, for completeness.

Developed nations faced with large-scale immigration from the developing world are in

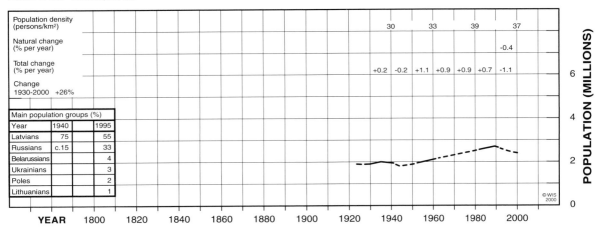

LATVIA   Area 64,500 km²

Population density (persons/km²)											30	33	39	37		
Natural change (% per year)														-0.4		
Total change (% per year)										+0.2	-0.2	+1.1	+0.9	+0.9	+0.7	-1.1
Change 1930-2000 +26%																

Main population groups (%)		
Year	1940	1995
Latvians	75	55
Russians	c.15	33
Belarussians		4
Ukrainians		3
Poles		2
Lithuanians		1

YEAR   1800   1820   1840   1860   1880   1900   1920   1940   1960   1980   2000

POPULATION (MILLIONS)   6   4   2   0

© WIS 2000

Indigenous Lett tribes in what is now Latvia suffered Viking and Slav invasions late in the first millennium AD and were conquered and occupied in the 13th century by German crusaders, the Teutonic Knights. Their state, Livonia, prospered in association with the Hanseatic League of maritime traders. From 1561 Latvia passed successively through Polish, Lithuanian and Swedish empires, ending with Tsarist Russia in 1721. After German occupation in World War One independence was achieved in 1919, but Soviet intimidation in 1939–40 forcibly created the Latvian Soviet Socialist Republic which was overrun by Nazi Germany in World War Two.

From the late 1940s agriculture in the SSR was collectivised and Latvians were deported en masse to Soviet Asia and replaced by Russian immigrants, with a view to suppressing the Latvian way of life. The nationalist reaction was strong and a Latvian Popular Front declared sovereignty in 1989. Independence came in 1991 with the Soviet collapse. Since then the population has declined, in common with all the ex-Soviet, non-Muslim nations. Latvia has large agricultural, forest and fossil fuel resources relative to its low population density, and significant industrial capacity inherited from Tsarist and Soviet times.

a *no-win* situation. If they allow in all who want to come they are quickly impoverished, according to the Micawberish Rule. The alternative, controlling the flow, is expensive. In Britain, "the cost of asylum seekers to public funds in 2000 was estimated by the Home Office to be £1235 million" (J P Duguid, *in litt*), and further expense was incurred when freight traffic through the Channel Tunnel was restricted because French police could not cope with illegal stowaways.

In 2000, the population of Western Europe was 394 million, while that of the developing world was 4900 million. With the majority of the latter living in poverty, it is a reasonable postulate that at least half would be interested in emigrating, given the opportunity, to the havens of the West. If half of that half chose Western Europe as its destination, and if all Europe's barriers to immigration were lowered (as some left-wing and humanitarian organisations say is morally desirable), Western Europe's population would rise to 1619 million. Its population density, at 485, would be higher than that of any major nation today except Bangladesh and Taiwan.

Of course, the situation could not arise, because Europe's economic growth would be so wildly outpaced by population growth that wealth would at once give way to poverty (by the Micawberish Rule, above) and the host region would cease to be an attractive haven. Possibly the left-wing and humanitarian organisations would view this as poetic justice, until they realised that their moral superiority would not protect them from the general impoverishment. North America and Australia would suffer the same fate if they played host to the other 1225 million migrants, because, in spite of their low population densities, very large parts of their total areas are too arid, cold or mountainous to support significant human populations.

When the UN High Commission for Refugees was founded in 1951, it was mainly concerned with European refugees, who were fleeing Communism or religious persecution (*Encarta Encyclopedia*, 2000). Since then, much has changed. Few if any of the politicians, pundits or organisations who deal with immigration into the developed world today appear to realise that the flow of hopeful immigrants in 2001 is a tiny trickle compared to the floods

1.8 – 18 million

LEBANON Area 10,400 km²

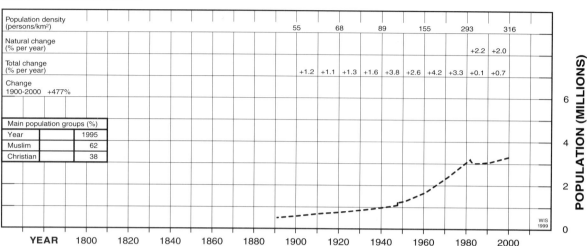

Population density (persons/km²)									55	68	89	155	293	316		
Natural change (% per year)														+2.2	+2.0	
Total change (% per year)									+1.2	+1.1	+1.3	+1.6	+3.8	+2.6	+4.2	+3.3	+0.1	+0.7
Change 1900-2000 +477%																

Main population groups (%)		
Year		1995
Muslim		62
Christian		38

POPULATION (MILLIONS)

YEAR 1800 1820 1840 1860 1880 1900 1920 1940 1960 1980 2000

Phoenicians lived on the Lebanese coast for more than 2 millennia before being conquered by Rome in 64BC. Christianity spread into Lebanon during the Roman/Byzantine occupation. Islam was introduced from the 7th century AD (including the local variant, the Druze faith, which is confined to Lebanon, Syria and Jordan) by Arab invaders, and by Ottoman Turks who ruled the area for 4 centuries from the 16th. Lebanon appeared in its present form in 1918, as a French mandated territory. It became independent in 1943.

Religious intolerance has dominated Lebanon's recent history. Thousands of Maronite Christians were massacred by the Druze in 1860. The 1943 constitution required that Christians and Muslims should share political power, but Muslims rose against Christians in 1958, and again in 1975–76 when a cease-fire was imposed by Syrian peacekeepers. The Arab-Israeli War of 1948–49 had driven more than 100,000 Palestinian refugees into Lebanon; they founded the anti-Israel PLO (Palestinian Liberation Organisation) whose guerrilla activities provoked Israeli invasions in 1978 and 1982. From 1982 there was intense civil war between Christian and Muslim militias, the city of Beirut was devastated and peacekeeping attempts were ineffective. Eventually in 1990 the Syrians enforced a fragile peace which was endangered by repeated border skirmishes between Muslim Hizbullah guerrillas and the Israeli army. In 1999 some 40,000 Syrian troops were still imposing peace. Israeli forces withdrew from southern Lebanon in 2000.

By 2001 the number of Palestinian refugees in Lebanon had multiplied to 300,000, in 12 UN-run camps. They are 'ticking time bombs', training the guerrillas whose activities are preventing economic recovery, Lebanon believes.

Rapid population growth in the prosperous 1950s, 60s and 70s abruptly ended in the chaotic 80s and 90s. When the population density reached about 300, tension between the rival factions was so intense that peacekeeping forces were brushed aside and further population growth was minimised by killings, disease and emigration. This is the scenario when a rising population reaches its Violent Cutback Level (VCL), when the death rate due to war, genocide or ethnic cleansing becomes equal to or exceeds the birth rate. A strong peacekeeping force may hold down the death rate for a while, but if population growth resumes, other options must be sought such as partition, or (much the most desirable) a fall in the birth rate.

who will want to follow them as the 21st century unfolds. A glance at the graphs of the main migrant-generating nations: INDIA, CHINA, the PHILIPPINES, Muslim nations from INDONESIA and PAKISTAN through south-west Asia to North Africa and the Balkans, black nations of sub-Saharan Africa, and Central America including MEXICO and the Caribbean, will show that there is little sign of slackening in the headlong pace of numerical population growth.

Intensified overcrowding, poverty and discontent in these demographically exploding nations cannot fail to drive emigration on an accelerating scale. Most will be economic migrants, but as civil wars proliferate, culminating in VCL situations typified by inter-group hatred, the proportion of genuine asylum seekers will rise. Victims of ethnic cleansing automatically become asylum seekers. In 1997 there were more than 15 million refugees worldwide, compared to 8 million in 1987. The rate of increase is exponential, at huge expense to receiving nations. Presently incalculable, but likely to be significant, is the extent to which sea level rise and climate change will generate more refugees, or HIV/AIDS lessen the total, or oil shortages influence the numbers one way or the other.

It is anti-social, in the modern world, for a nation to allow its population to increase beyond what it can sustain using its own national (agricultural, industrial and financial) resources. It should not rely on charitable aid, or depend on emigration to dispose of surplus

LESOTHO Basutoland Area 30,350 km²

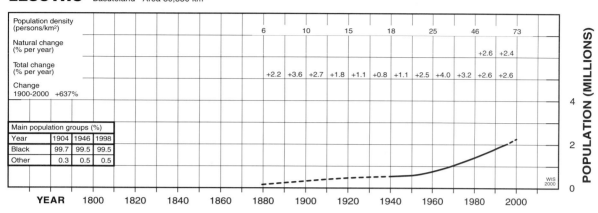

Population density (persons/km²)							6	10	15	18	25	46	73						
Natural change (% per year)												+2.6	+2.4						
Total change (% per year)						+2.2	+3.6	+2.7	+1.8	+1.1	+0.8	+1.1	+2.5	+4.0	+3.2	+2.6	+2.6		
Change 1900-2000 +637%																			

Main population groups (%)			
Year	1904	1946	1998
Black	99.7	99.5	99.5
Other	0.3	0.5	0.5

POPULATION (MILLIONS)

WIS 2000

YEAR 1800 1820 1840 1860 1880 1900 1920 1940 1960 1980 2000

Bantu tribes reached mountainous drought-prone Lesotho from the north in the 16th century and drove out the indigenous San (Bushmen and Hottentots). In the 1820s the resident Basothos were assailed by tribes fleeing from Shaka Zulu's *mfecane* (crushing) but they defeated the waves of desperate marauders and assimilated many of them to help expand and consolidate their kingdom. King Moshoeshoe I was a statesman who united the nation and countered attempted Boer takeovers by appealing to the British for protection. After decades of raids and reprisals his kingdom became the British protectorate and colony of Basutoland in 1868. Independence as the Kingdom of Lesotho came in 1966.

Worsening overpopulation has led to massive loss of soil from the steep farmed hillsides and reduced agricultural productivity, forcing many men (60% of the male labour force in 1966) to migrate to work in South Africa. In 1998 South African forces invaded to subdue political unrest, but the fighting was unexpectedly fierce, badly damaging Maseru, the capital city.

population, because that is imposing a burden on other nations that they may not wish to bear. Nor should it use moral blackmail: "If you return our emigrants, they will be tortured or killed". In 2000 the international pressure group Amnesty International argued for a declaration of "refugee rights", but receiving nations also have rights, to traditional identities and economies that should be protected from unending influxes of aliens.

The misery of Palestine, the only nation in modern times to receive so many immigrants that the indigenous peoples were outnumbered and dispossessed, becoming second class citizens in a land they had occupied for more than a thousand years, is a lesson that should not be ignored. Unlike the Palestinian Arabs, European settlers in their African and Asian colonies had 'mother countries' to retreat to, when their adopted countries achieved independence, rejected multiculturalism, and expelled them.

In 2000, European businesses and refugee lobbyists were quick to seize on a report by the UN Population Division which, perceiving Europe's population to be ageing and hardly rising, argued that the continent would need many millions of youthful immigrants to maintain a working population large enough to support the retired population. But even if there were no alternative ways to support pensioners, such as raising the compulsory retirement age, the thesis is only tenable in the very short term. In the longer term, the immigrants and their families would age and there would be *even more* old people to look after (Coleman, 2001). The UNPD director commented "It's hard to understand the fear of immigration in Europe... the US takes in 1 million immigrants a year, and yet unemployment is at a record low and the economy is booming". Astonishingly, he ignored Europe's high population density (PD). Europe is seriously overcrowded (e.g. Netherlands PD 468, England PD 382) compared to the US (PD 30). Receiving countries have physical limits, which naïve politically correct ideas do not change.

A solution to European financial problems that only works by means of constant population growth would be gravely unsustainable, given that Europeans consume a disproportionately large share of Earth's resources and generate an equally disproportionate volume of greenhouse gases (and other pollutants). This is a powerful reason for allowing Europe's population to *fall*.

1.8 – 18 million

LIBERIA Area 100,000 km²

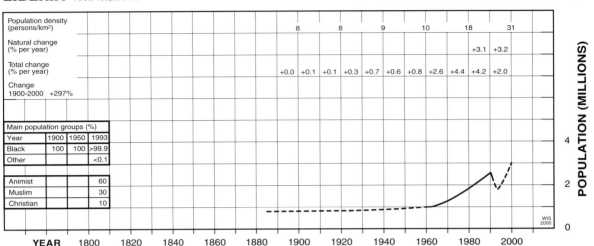

In 1822 American philanthropists established a haven for freed black slaves on the Liberian coast. It expanded and developed in much the same way as European colonies, with the settlers using indigenous tribes as forced labour to exploit timber and minerals and start plantations. One of the world's largest rubber plantations was begun in the 1920s and huge iron ore mines were opened in the 1940s.

Descendants of the settlers monopolised politics until 1980, when the native tribal population (about 95%) took power in a violent coup. President Samuel Doe ruled autocratically until 1990 when rebel forces invaded from Guinea. Doe was killed and bloody fighting involving several warlords, all seeking the presidency,

continued for 6 years. Some 200,000 people were killed and up to 2 million made homeless, many of whom fled the country. In 1996 a West African peacekeeping force, ECOMOG, persuaded the 4 warlords who controlled the country to accept a truce. Many refugees returned, plus foreign refugees from war in neighbouring Sierra Leone, but rebels from Guinea invaded in 1999 and civil war resumed.

The years of war have devastated the economy and Liberia now depends on foreign aid and, the UN says, sales of diamonds smuggled from Sierra Leone. The capital city, Monrovia, has had no mains electricity since rebels destroyed the power station in 1990, and the water supply is intermittent. Censuses: irregular from 1962.

Currently, Western Europe's population is increasing at about one million per year, almost entirely due to immigration from poor overpopulated countries. Charles Darwin's grandson, Charles Galton Darwin (1953), visualised a situation whereby, through straightforward natural selection, European national identities would be extinguished. J P Duguid (*in litt*) neatly summarises his argument: "… people who respond to social pressures to keep their families small, tend to be outbred *and replaced* (my italics) by those who for reasons of ignorance, religion or tradition, do not do so". The prospect of *replacement* by fast-breeding aliens cannot fail to arouse resentment among traditional western populations.

So the flow of immigrants, that already is seriously stretching the economies of developed nations, will increase so fast in the near future that the West will have no option but to close and defend its external borders. The sooner this is done the better, because it would force governments in the developing world to understand that their poverty is self-generated, because they do not control the growth of their populations. There is nothing immoral about Draconian birth control, Chinese fashion, as the WROG period draws to a close, because the alternative is all the human misery that overcrowding brings: famine, climate change, enforced unwanted migration, and civil strife building up to VCL disaster.

6.5 Unreal Goals

The British Chancellor of the Exchequer, addressing a meeting of the World Bank in 2000, announced "Our aim is to reduce world poverty by a half in 15 years". Rousing exhortations of this kind are often made by worthy persons in the public eye, especially politicians, high officials and newspaper editors. "Never again", they say, must there be genocides like Rwanda's, or famines in which millions die. Racism must be eradicated. The *Guardian* newspaper said of some Irish sectarian killings in December 2000: "we must leave

LIBYA   Area 1,770,000 km²

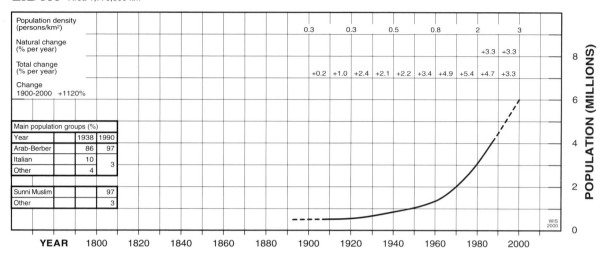

					1900	1920	1940	1960	1980	2000
Population density (persons/km²)					0.3	0.3	0.5	0.8	2	3
Natural change (% per year)									+3.3	+3.3
Total change (% per year)					+0.2 +1.0	+2.4 +2.1	+2.2 +3.4	+4.9 +5.4	+4.7 +3.3	
Change 1900-2000 +1120%										

Main population groups (%)

Year	1938	1990
Arab-Berber	86	97
Italian	10	3
Other	4	
Sunni Muslim		97
Other		3

When the Romans conquered Libya in 74 BC the one-time lush subtropical landscape, rendered semi-arid by climate change in the 2nd millennium BC, had already been colonised by Phoenicians, Carthaginians and Greeks. North Africa, especially the Libyan coastal strip, grew grain for the Roman Empire, but deforestation caused soil erosion, silting of the ports, and the apparent anomaly of ruined Roman cities surrounded by vast deserts. The Romans were succeeded by Byzantines in the 6th century AD, but the most significant conquest was by Arabs in the 7th century, who spread Islam. Ottoman Turks dominated Libya from the 16th century until 1911, when the country was invaded and occupied by Italy. Development by large numbers of Italian settlers ended when the Italians were expelled in World War Two. Independence was achieved in 1951 under King Idris. In the late 1950s large oil reserves were discovered under the eastern deserts.

A bloodless military coup in 1969 brought maverick Colonel Gaddafi to power. His Islamic-Socialist regime, financed by oil revenues, (95% of export earnings) has tried to merge with other Arab nations, without success; has supported a variety of terrorist organisations; has greatly improved the national infrastructure; and has carried out an astonishing and risky (in the long-term) project, the "Great Man-Made River", which pipes irreplaceable water to the coast from boreholes in the central deserts, in quantities sufficient to irrigate 750 square kilometres of arable land. Some 80% of the food consumed by the fast-growing population is imported, but thanks to oil, and Gaddafi's half-puritanical, half-pragmatic rule, Libya remains prosperous.

this barbarism behind us". It is "unacceptable", idealists proclaim, that eleven million people will die from infectious diseases this year because they can't afford medicines.

The trouble with such fine words is that they are meaningless while the world's population continues its explosion: the inescapable cause of poverty and conflict. At the above meeting of the World Bank the president admitted that although the percentage of people in poverty had fallen by 4% in recent years, the increase in world population meant that the actual number of people in poverty had hardly changed. A World Bank official calculated in 2001 that at least £700 billion had been disbursed in aid since 1950, without significantly improving living standards in the Third World (*The Guardian*, 22.3.2002).

Throughout history (with rare exceptions) human populations expanded when resources were available and shrank when they were not. So when the West gives aid to the developing world with no strings attached (e.g. birth control) the recipients seldom get richer; they just have more children and stay poor.

In his last months as US President, in 2000, Bill Clinton tried desperately to win lasting credit by settling the Israeli/Palestinian conflict. He might as well have tried to stop an electric kettle boiling without switching it off! He cannot have understood that the explosive population growth of both sides is driving them inexorably to boiling point, violent population cutback, VCL. If anybody told him, the most influential man on Earth, he didn't listen.

"You cannot exaggerate English sympathy for the underdog" wrote Jeremy Paxman in *The English* (1998). Save the Children, War on Want, Oxfam, WaterAid, Christian Aid – these are a few of the English or British charities which, funded by public generosity, send their workers to disaster areas all around the world where they can prolong lives by healing the

1.8 – 18 million

LITHUANIA Area 65,000 km² (after 1940)

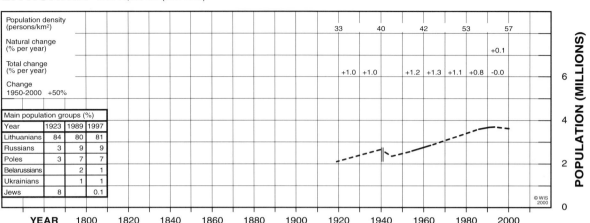

Population density (persons/km²): 33   40   42   53   57

Natural change (% per year): +0.1

Total change (% per year): +1.0 +1.0 +1.2 +1.3 +1.1 +0.8 -0.0

Change 1950-2000 +50%

Main population groups (%)			
Year	1923	1989	1997
Lithuanians	84	80	81
Russians	3	9	9
Poles	3	7	7
Belarussians		2	1
Ukrainians		1	1
Jews	8		0.1

YEAR 1800 1820 1840 1860 1880 1900 1920 1940 1960 1980 2000

POPULATION (MILLIONS) 0 2 4 6

© WIS 2000

Lithuanian tribes inhabited their wide marshy plains for a millennium before they united in the 9th to 12th centuries AD to resist the eastward expansion of Germanic peoples. In the 14th century the powerful Grand Duchy of Lithuania, in association with Poland, occupied lands through Belarus and the Ukraine as far as the Black Sea. Union with Poland in 1569 created the largest state in central Europe, but it became so weakened by wars with Russia and Sweden that, in 1795, it was annexed by Tsarist Russia. Separatist revolts in 1831 and 1863 failed, but after German occupation in World War One the independent state of Lithuania emerged in 1918.

Poland seized the Vilnius province in 1920 but the Soviets returned it to Lithuania in 1940 after the 1939 carve-up of Poland with Germany. The Lithuanian Soviet Socialist Republic was created in 1940 and was at once overrun by the Nazis who exterminated virtually all the Jewish population of more than 200,000. After the war the Soviets collectivised the peasant farms, deported several hundred thousand Lithuanians to Central Asia, and replaced them with ethnic Russians. Lithuanians reacted to this assault on their nationhood with a long campaign for autonomy that achieved independence when the Soviet Union collapsed in 1991. Population growth ceased in the 1990s, as in the other ex-Soviet non-Muslim nations. One quarter of the country is forested. The economy still has close links with Russia.

sick, feeding the hungry, providing tents and clothing for refugees, and so on. They do good – in the short term. In the long term, it can be argued, they do no good at all, apart from the personal satisfaction they and the generous public feel at being unselfish: giving their labour or money to help the underdog.

Consider ETHIOPIA. In the great famine of 1984–85, compounded of drought and war, of which so many heartrending pictures appeared on TV, more than a million lives were saved, in the short term, by medical and food aid. Ethiopia's population, which for decades had been increasing at a rate that ensured the perpetuation of national poverty, continued its growth. Like so many developing nations Ethiopia depends on charitable aid to maintain even its present low standard of living. Droughts and famines recur every few years and are combated by foreign aid; if nature were left to take its course, causing a halt in or reversal of population growth, the opportunity would arrive to reorganise the economy and improve the quality of life. That long-term benefit is made impossible by the charities' conviction that they are doing good just by saving the victims' lives. Their goal, to prevent human misery, remains unreal as long as the population continues to grow faster than the economy. Every new famine brings misery to a greater number of people, and compassionate aid will not continue for ever.

Charles Darwin was one of the first to describe this aspect of demographic reality. In *The Origin of Species* (1859) he explains: "Every being, which during its natural lifetime produces several eggs or seeds, *must suffer destruction* during some period of its life… otherwise, on the principle of geometrical increase, its numbers would quickly become *so inordinately great* that no country could support the product" (my italics). The charities ignore the fact, established so long ago, that their 'humanitarian' goals are, in practice, self-defeating.

In contrast, a very few charities, such as Population Concern and Marie Stopes International, devote their efforts to family planning in the developing world, to help those women (a very large number) who would like to limit their families, but lack the means. Their

MACEDONIA Area 25,700 km²

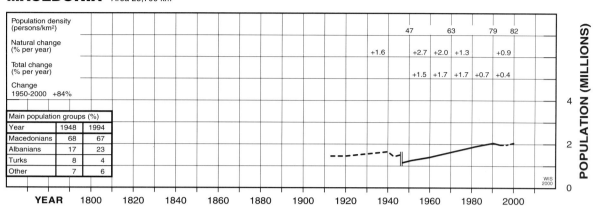

Population density (persons/km²)											47	63	79	82	
Natural change (% per year)										+1.6	+2.7	+2.0	+1.3	+0.9	
Total change (% per year)											+1.5	+1.7	+1.7	+0.7	+0.4
Change 1950-2000 +84%															

Main population groups (%)		
Year	1948	1994
Macedonians	68	67
Albanians	17	23
Turks	8	4
Other	7	6

YEAR 1800 1820 1840 1860 1880 1900 1920 1940 1960 1980 2000

POPULATION (MILLIONS)

The original Macedonia, central kingdom of Alexander the Great in the 4th century BC, extended into what are now Greece, Serbia, Bulgaria and Albania. Romans and Byzantines ruled it for a thousand years, during which period Slav settlers arrived from the north. Between the 9th and 14th centuries armies of Serbs, Bulgars and Byzantines fought over Macedonia. Some of the reciprocal atrocities are still remembered; thus when Basil 2 of Byzantium (the "Bulgar-slayer") defeated the Bulgarian army of Tsar Samuel in 1014 he blinded all his thousands of prisoners save one in each hundred, who led the others home.

Five centuries of Ottoman rule ended with the 1913 Balkan War. Between the World Wars modern Macedonia was part of South Serbia. It became a semi-autonomous Yugoslav republic, with its present boundaries, in 1946. Frustrated by Serb domination, it broke from Yugoslavia in 1992.

Recent decades have seen worsening tensions between the Christian Slav majority and the Muslim Albanian minority, which increased from 17% of the population to 23% in 46 years, due to high birth rates and in spite of much emigration to work elsewhere. The Albanians claim that they now form 40% of the population, and therefore deserve autonomy in their own state. "We will beat you in the beds", they boast to the Slavs. Macedonian Slavs claim that tens of thousands of ethnic Albanians have recently infiltrated from Kosovo and Albania to boost Albanian claims.

1999 saw the sudden arrival of 230,000 Kosovan refugees, most of whom returned home after a few weeks. An Albanian separatist uprising in 2001 was ended by international negotiation and the promise of financial aid for rebuilding. NATO forces maintain a fragile peace.

The population decrease between censuses in 1991 and 1994 is presumably due to emigration, given the positive figures for natural change. It coincided with drastic economic decline after the break-up of Yugoslavia in 1992. The economy is now very dependent on foreign aid.

goal of reducing human misery is humane, realistic and achievable, if they are given enough support.

Experienced researchers into Ethiopian famines have argued that they can be avoided by "good governance, sound growth policies, and active preparedness" but they (Webb and von Braun, 1994) concede that "the absolute number of individuals requiring assistance is becoming hard to manage".

Achieving world peace through prayer is a particularly unreal goal. Every day, in innumerable churches throughout the developed world, clerics and their congregations pray for peace, as do religious leaders on national and international occasions. Their efforts are rather obviously unsuccessful. If they *seriously* want world peace, which is constantly sabotaged by the effects of population growth, they could pray for the success of family planning charities like those mentioned above, but for Catholics and their leader this appears to be impossible.

Equally unreal is the vision, dear to many politicians, of resolving conflicts by "patient negotiation", if as the talks drag on the populations involved are increasing. For example, in Northern IRELAND, segregation and violence have become more entrenched on the local scale since the 1994 cease-fire began, according to Ulster University researchers in 2002.

Late in 2000 the British media debated *ad nauseam* the ethics of whether a pair of 'Siamese twins' should be separated. If they were (at great expense), one would die but the other could live; if not, both would die. Politically correct people were immensely concerned over this dilemma. Their unreal goal was to find a morally correct solution to one unnatural problem (in the natural world the twins and probably their mother would have died in childbirth), while contemporaneously, in the developing world, tens of thousands of children

1.8 – 18 million

MADAGASCAR   Malagasy Republic   Area 587,000 km²

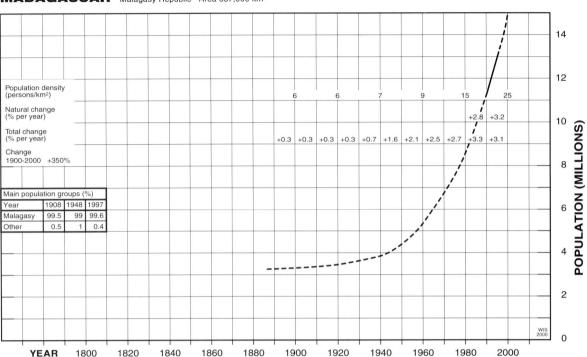

Population density (persons/km²)						6	6	7	9	15	25
Natural change (% per year)										+2.8	+3.2
Total change (% per year)						+0.3	+0.3	+0.3	+0.3	+0.7	+1.6	+2.1	+2.5	+2.7	+3.3	+3.1
Change 1900-2000 +350%											

Main population groups (%)			
Year	1908	1948	1997
Malagasy	99.5	99	99.6
Other	0.5	1	0.4

WIS 2000

YEAR   1800   1820   1840   1860   1880   1900   1920   1940   1960   1980   2000

POPULATION (MILLIONS)

Madagascar, the world's fourth largest island, was initially settled by seafarers from Indonesia perhaps 1500 years ago. Portuguese navigators were the first European visitors, in 1500. European settlement was at first resisted by the native inhabitants but it became a French colony in 1896. Independence was achieved in 1960.

Population rose slowly until the 1940s when introduced technologies and medicines triggered the DC surge in population growth that, with 86% of the labour force engaged in subsistence agriculture, and 70% below the poverty line, is causing deforestation and soil erosion on a massive scale.

A unique flora and fauna developed on Madagascar during the 70 million years since it drifted away from Africa. Gerald Durrell, pioneer in captive breeding of endangered species, transferred Madagascan lemurs and tortoises to his Jersey Zoo, but although nature reserves have been designated for some threatened species, precedent suggests that with such a fast-growing human population seeking cultivable land they are unlikely to survive for long. Population estimates before 1920 varied widely, and even in the 1990s, after a census in 1993, there is some uncertainty. In 2002, following a disputed election, the country was chaotic, with two presidents.

were dying every day of commonplace diseases.

In 1995 an Irish politician, Bertie Ahern, called on the British government to apologise for the Irish Potato Famine of 1846. In 2001 the same man, now the Irish Prime Minister, was forced to cancel his appointment to unveil a Scottish memorial to the famine victims, because of fears for his safety. The race hatred generated by an episode of Malthusian cutback that, given the context, was inevitable, was shown to have lived on for 150 years and to have itself generated equal and opposite hatred (see IRELAND). In similar vein, 'remembrance days' are sometimes designated to keep alive the memory of vast tragedies such as the Holocaust genocide of Jews (see GERMANY), but they are empty ceremonies unless the Darwinian cause of such racist actions is understood (section 5.8) and acted upon.

The goals of single-issue pressure groups, who are able to exert "the prerogative of the harlot: power without responsibility", are real to themselves but not, usually, to those they try to influence. Muslim fundamentalist guerrillas in the Philippines declared in 2000 "We dream of an entire Islamic world, and we will achieve it. Allah is with us". Woe betide the world if the fundamentalists turn it into a planetary AFGHANISTAN. Their belief that if they die fighting in a holy war their place in Paradise is assured explains why suicide bombers have been so prominent in Muslim *intifadas* and terrorism, against Israel and the USA.

Equally real to themselves is the goal of vegetarians against world hunger, who argue, credibly, that Earth could support twice as many humans if they gave up eating meat. What

MALAWI   Nyasaland   Area 94,300 km² (excluding lakes)

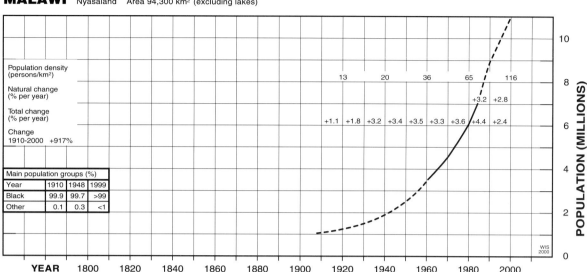

Malawi was settled from the first millennium AD by waves of Bantu migrants from north, west, east and south, usually by forcible displacement of previous inhabitants. In the 19th century the long-established slave trade, based on Arab ports along the East African coast, reached its height, with up to 100,000 negroes taken each year to Arabia and India. Slaves from Malawi were mostly prisoners taken in tribal fighting, or found guilty in trials by witchcraft; slavery being a profitable alternative to execution.

David Livingstone explored in Malawi in the 1850s and 60s. Missions were established, followed by European settlers and traders. Cecil Rhodes' British South Africa Company controlled the area by 1891 when it was declared a British protectorate,

Nyasaland. Under British rule the slave trade was abolished. Black nationalism arose in response to white settlement of tribal lands. This was small-scale compared to the Rhodesias, and the short period of federation with them (1953–63) passed without serious violence. Independence was achieved in 1964, and there followed 30 years during which the republic of Malawi was ruled, in effect, by the autocratic Dr Hastings Banda. He was peacefully deposed, aged 90, in 1994.

Faster population growth in the late 1980s was caused by the influx of nearly a million refugees from Mozambique, many of whom returned home in the 1990s. Life expectancy fell from 45 years in 1989 to 41 years in 1998, due to the rapid spread of HIV/AIDS.

is not credible is that *Homo sapiens*, who is and always has been omnivorous, would meekly agree to abstain from a foodstuff that is far more important to most humans than alcohol was to Americans in the days of Prohibition, simply because feckless or inconsiderate parents will not limit the number of babies they have. And in any case, when the surplus grain that is currently fed to cattle is no longer produced, because of oil shortage (section 9.3.a), and scarce meat comes from animals grazing grass, the unreal vision of feeding 12 billion people on grain and beans must fade away.

In 1999 an American biotech firm announced it was developing a variety of canola (British name: oil seed rape) from which plastics could be extracted, the goal being to reduce the consumption of crude oil in plastics and paint manufacture. Plants such as canola, maize and sugar cane are already processed into oil or alcohol on a large scale, and many environmentalists would like to see so much 'biomass' grown that it could replace fossil fuels in some of their functions. This is an unreal goal for two reasons: growing biofuel crops and processing them is very energy-expensive, and the farmland involved would not be available to grow food. About 100 million cubic metres of crude oil are consumed in the UK in a year (1.7 cubic metres per person); if spread on the ground it would cover two fifths of the nation's area in a layer one millimetre thick. Canola yields about one cubic metre of vegetable oil per hectare, so if it was grown on and spread over the same area the layer of vegetable oil would be a tenth of a millimetre thick. Biomass might be a useful substitute for crude oil if UK population was a few million, with farmland to spare.

In 2001, genetic researchers were speculating that before very long, by cleverly reversing evolution, they would be able to recreate dinosaurs from their descendants, birds (*New Scientist*, 21 July 2001). But if they succeeded, say in 2050, what could be done with their

1.8 – 18 million

MALI   French Sudan   Area 1,248,600 km²

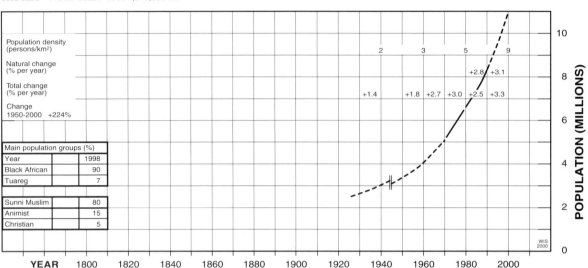

Population density (persons/km²)			
Natural change (% per year)			
Total change (% per year)			
Change 1950-2000 +224%			

Main population groups (%)	
Year	1998
Black African	90
Tuareg	7
Sunni Muslim	80
Animist	15
Christian	5

Two thirds of Mali is desert or semi-desert, but its position controlling ancient trans-Saharan trade routes from the Mediterranean to West Africa favoured the growth of powerful empires. The Ghana Empire from the 5th century AD was conquered in the 11th by Muslim Berbers from the north who founded the Mali Empire. This was superseded in the 15th century by the Songhai Empire, also Muslim, which was invaded and destroyed in 1591 by Berbers from Morocco. Their petty kingdoms were conquered by the French, who established the colony of French Sudan in 1895. There was a border adjustment with Mauritania in 1945. The country became Mali with independence in 1960.

Socialist Mali was peaceful at first, but the largely agricultural economy, damaged by mismanagement and prolonged Sahelian drought in the 1970s, steadily deteriorated. Tuareg nomads in the desert north rebelled in 1990 and fighting continued until 1996, even affecting the fertile Niger valley in the south, around Timbuktu, where the main cash crops, groundnuts and cotton, are grown. Many Malians avoid the general poverty by migrating to work in Côte d'Ivoire.

synthetic *Tyrannosaurus rex* in a world with 50% more people than today's (*World Population Data Sheet 2000*) and no wild habitat for the giant beast to live in? The animal rights people would be upset. And while the scientists had been enjoying their fascinating research, all the tigers, rhinos, orang-utans, elephants and polar bears (to name a minuscule few of our endangered relatives on Earth), the products of *natural* evolution, would have been crowded out of existence by humans struggling to survive in a Darwinian world. *Homo sapiens*, 'wise' man, would have struck again!

Huge sums of money are spent on searching for life on other planets. Our goal seems to be, at best, the discovery of primitive microbes, either underground on Mars, or in an ocean concealed beneath a thick ice cover on one of Jupiter's moons, or in the dirty ice of a comet. All the time, however, it becomes more evident that human travel into deep space will never be possible. Even the near-space International Space Station project is in financial and technical trouble and may not achieve its original goals (*New Scientist*, 18 May 2002). So one of our greatest failings, as intelligent beings, must be the tragic nonsense of searching for primitive alien life without seriously trying to preserve the wonderful variety of complex animals and plants with which Nature has endowed our own Earth.

6.6   Administrative Complexity: Enfeebling the Rule of Law

In science, the inverse square law defines many relationships. For example, when a space traveller is twice as far from the Sun as Earth is, the Sun will seem only one quarter as bright and warm as it did when the traveller was alongside Earth. Other relationships are even more skewed; thus the damage done to a road surface by a vehicle using it is related to the fourth or fifth power of axle weight.

A similarly non-linear relationship appears to govern the administration of nations and smaller population groups. If the population of a country doubles, the chances are that it is

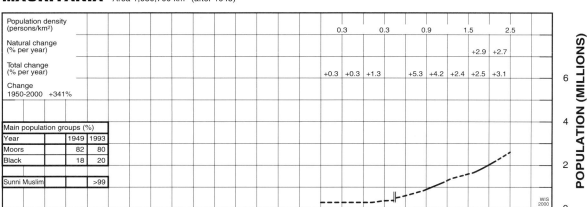

MAURITANIA Area 1,030,700 km² (after 1945)

Mauritania was a province at the western extremity of the Roman Empire, populated by Maurs or Berbers who, after the Romans departed, were conquered by empires based on present-day Mali and Morocco. It became Muslim in the 8th century. The north and centre of the country are desert (Sahara), and the south is dry (Sahel) and suffers occasional severe droughts. Slavery was officially abolished as late as 1980 but is said to persist in secret. French influence spread from Senegal in the 19th century and the country became a French colony, part of French West Africa, in 1920. Adjustment of the border with French Sudan (now Mali) in 1945 increased the population by 188,000. After independence in 1960, President Ould Daddah faced rising ethnic tension between black Sahelian farmers in the south and the dominant semi-nomadic Arab-Berber northerners. Ould Daddah began a crippling war with POLISARIO guerrillas of Spanish Sahara (now Western Sahara) when that country was divided between Mauritania and Morocco in 1975. Daddah was deposed in 1978 and Mauritania withdrew from Western Sahara in 1979, virtually bankrupt except for the income from fishing and huge iron mines in the northwest. Half the population lives in extreme poverty and foreign aid is now vital to the economy. Ongoing north-south ethnic tensions led to fighting along the Senegal border and the expulsion of 50,000 Senegalese nationals in 1989.

more than twice as complicated to administer. By the simplest of analogies, between two people only one relationship is possible, between 3 people, 3 relationships, between 4, 6; between 5, 10; and so on. For a linear increase in numbers, the increase in complexity is geometric. However, in a real society there are so many variables that a far greater range of relationships is possible, and only generalised rules can apply.

Early symptoms of excessive administrative complexity include the failure of governing bodies, national and local, to maintain standards of public service. In the UK, in 2000, long-established transport systems, especially road and rail, had become so disorganised and congested that maintenance and journey times were unreliable. Public-funded health services and school examination systems failed to cope with rapidly rising pressures. Crowds attending popular events such as the Notting Hill Carnival and the Glastonbury pop festival were so large that the police declared controlling them was "beyond our resources".

The individual becomes lost in the crowd, in which, of course, anonymity is a boon to criminals. Administrators identify people by numbers or bar-codes instead of names. Decision-takers have little time for individual representations, but are increasingly influenced by opinion polls, noisy single-issue lobbyists, and media pundits. New dimensions are introduced by the growth of ethnic minorities, sparking racist incidents, and by the dependence of government and business on computerisation, which can cope with large numbers but is vulnerable to disruption by error or incompetence, and by 'hackers' intent on vandalism, profit, ideology, exhibitionism, sabotage or whatever.

When European nations withdrew from their colonies in Africa and Asia in the 1950s and 1960s, they were influenced in no small measure by the difficulty of administering populations that were several times greater than when the colony was established, thanks to the DC surges that the colonists themselves had initiated. Law and order were failing as political opposition nurtured sabotage and revolution.

Law and order are essential to a civilised society. When the police and security services begin to lose control, civilisation is under threat. In some congested cities of Central and South

1.8 – 18 million

MOLDOVA Moldavia, Bessarabia Area 33,700 km²

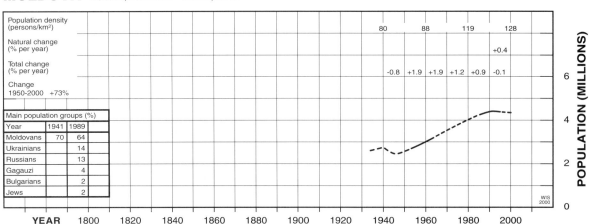

The population of this small state between Romania and the Ukraine has the multi-ethnic composition characteristic of the adjacent Balkans. Moldova's history of chronic interethnic conflict, territorial gain and loss, and conquest by larger neighbours, is typically Balkan. Originally part of Roman Dacia, Moldova (Moldavia) merged with Bukovina and Bessarabia to become a strong independent principality in the 14th century, but became subservient to Turkey in the 16th. Two to three centuries later Bukovina joined Austria, Bessarabia was ceded to Tsarist Russia, and the rump of Moldova merged with Wallachia to form an early version of Romania.

When the Soviets remodelled the region after World War One and created the Moldovan Soviet Socialist Republic in 1924 much of it was in present-day Ukraine. Modern boundaries were established in 1940 at the expense of Ukraine and Romania. After German occupation in World War Two there was famine as, as usual, the Soviets enforced industrial development, liquidation of opponents, the collectivisation of peasant farms and settlement by ethnic Russians and Ukrainians. Independence came when the Soviet Union collapsed in 1991, and in common with all the ex-Soviet Slav nations, population growth ceased. Ethnic conflict with Russian Slavs in the northeast and Gagauz in the southwest, both seeking autonomy, lasted 1991–93. Nearly half the working population, which is mainly Christian, is still in agriculture.

America, it is believed that the police themselves flout the law by culling hundreds of abandoned 'street children', who live by their wits, in the interests of 'social cleansing', without significant public protest. In the developed world, the youth of inner cities, thieving to finance their drug habits, may be risking the same fate, as the end of the WROG period approaches.

Eventually the combination of overcrowding, poverty, crime and hopelessness leads to anarchy, the collapse of law and order, as happened in ALBANIA in 1997. The next and final stage, the abandonment of civilised behaviour, is when ethnic, religious or other groups turn on each other in civil war. Population growth is ended by mass killings and ethnic cleansing. This is the VCL scenario, which in modern times first appeared in AFGHANISTAN in 1979, reached a peak of publicity in RWANDA in 1994, and is affecting more countries almost every year (see WORLD). When it strikes a major nation, such as PAKISTAN or ISRAEL/Palestine, politicians may begin to take note.

6.7 The Politically Correct Overpopulation Taboo

"The population of this country is stable, and has been stable for a long time." So said Jonathan Dimbleby, the media personality, to a telephone caller on his BBC radio programme 'Any Answers', on February 26 2000. I heard him with surprise, because my latest copy of *Population Trends*, issued 4 times a year by the UK Office for National Statistics (ONS), showed that England's population had increased by 1.94 million in the previous 10 years. UK population had increased similarly. I wrote to Mr Dimbleby, enclosing a copy of the ONS population data and suggesting that he should publicly correct the misinformation given. This seemed particularly desirable because, at the time, the BBC was interspersing its radio programmes with slogans like "The BBC – the world's reference point for news", "Trust the BBC to bring you accurate and impartial coverage of events that shape our world", and, perhaps in Freudian mode, "This is the BBC World Service, bringing you everything you need to know…".

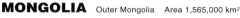

MONGOLIA   Outer Mongolia   Area 1,565,000 km²

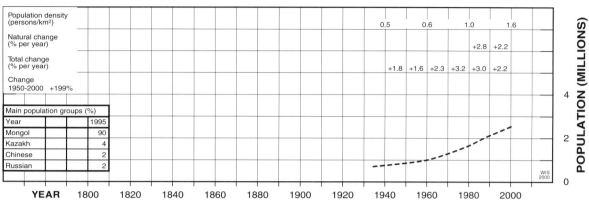

	1800	1820	1840	1860	1880	1900	1920	1940	1960	1980	2000
Population density (persons/km²)								0.5	0.6	1.0	1.6
Natural change (% per year)										+2.8	+2.2
Total change (% per year)						+1.8	+1.6	+2.3	+3.2	+3.0	+2.2
Change 1950-2000 +199%											

Main population groups (%)	
Year	1995
Mongol	90
Kazakh	4
Chinese	2
Russian	2

POPULATION (MILLIONS)

Before the 12th century AD the Mongols were scattered nomadic tribes who fought each other, reared horses and roamed the steppes and deserts of central Asia. Genghis Khan united the tribes in the 1180s and launched his hordes (armies) on an astounding campaign of world conquest on horseback. The Mongols' ferocity (usually massacring the inhabitants of cities that resisted their advance) and mobility extended their empire from Russia to the China Sea by 1250, eventually touching Austria, the Arabian Sea, India and Indonesia. After a century of splendour the Mongol Empire began to disintegrate and Mongolia itself was conquered by China in the 17th century. Three centuries of Chinese oppression ended with a bid for independence in 1911, when Inner Mongolia stayed with China, and Outer Mongolia achieved independence (with Russian help) in 1921. The new kingdom was dominated by Soviet Russia and became a communist People's Republic when the king died in 1924. A revolt against agricultural collectivization was savagely crushed, with 100,000 deaths, in 1932. Soviet domination ended when the USSR collapsed in 1991 and a non-communist government was elected in 1996.

Mongolia's dry grasslands are best suited to cattle rearing, which suffers severely from droughts and harsh winters, as in 1999 and 2000. Industry is still embryonic, so the poor and fast-growing population is very dependent on foreign aid. The DC (death control) population surge began relatively late, around 1960, and the government reversed an original pro-natal policy in favour of family planning in the 1980s. Tibetan Buddhism, headed by the Dalai Lama, is the prevailing religion.

After two follow-up letters a reply came from the BBC saying that the ONS figures "will be referred to should future programmes return to the issue". That was on May 2 2000, and although I listen to most of the Any Questions/Any Answers programmes I have not heard any such reference. A perfect opportunity on 10 March 2001, when a caller claimed "The problem with immigration is not race, but numbers", was not taken.

Mr Dimbleby's assertion was the more surprising because I had written to him on the same subject, providing ONS figures, in 1995 and 1997, when he was President of CPRE (Council for the Protection of Rural England). His polite replies proved he had received them. CPRE, an environmental charity, had consistently avoided mentioning population growth as a cause of the demand for new houses, roads, towns, etc., in the English countryside. When I put this to the CPRE Director, Fiona Reynolds, at a meeting in London, she referred me to the demographic transition theory (section 5.2) and stated that in her opinion the cause of the huge demand for new housing was the breakdown of traditional families into smaller units.

I describe these exchanges at length to illustrate my question:– why, when it is perfectly obvious that a fast-growing population will require or generate more housing, traffic, power stations, water supplies, sewage works, atmospheric pollution, environmental degradation, all the trappings, in fact, of increasing congestion, do intelligent and earnest British people and organisations duck the issue, or even deny that it exists?

If a nation's planning for the future is based on the assumption of population stability, when in fact there is significant growth, the plans are always out of date. In England, administrators and politicians often seem bewildered when public services such as local government or the police cannot cope with the demands of a growing population, or when doctors and nurses are overwhelmed by a growing number of patients. English hospitals in 2001 were so short of beds that patients sometimes had to wait on trolleys, in corridors, for as long as 2 days, but the link with population growth of 200,000 every year was seldom, if ever,

1.8 – 18 million

NETHERLANDS　Holland　Area 33,000 km² to 1920, increasing to 33,900 km² in 1999

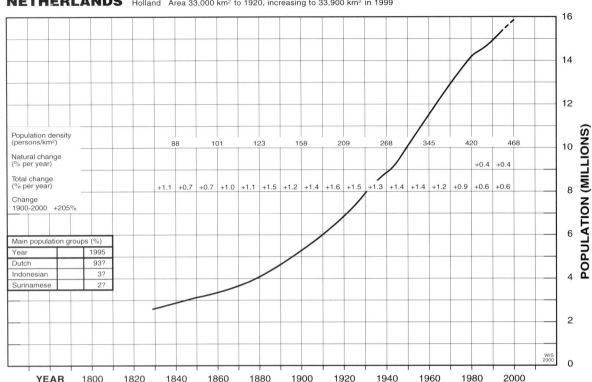

Population density (persons/km²): 88　101　123　158　209　268　345　420　468

Natural change (% per year): +0.4　+0.4

Total change (% per year): +1.1　+0.7　+0.7　+1.0　+1.1　+1.5　+1.2　+1.4　+1.6　+1.5　+1.3　+1.4　+1.4　+1.2　+0.9　+0.6　+0.6

Change 1900–2000　+205%

Main population groups (%)		
Year		1995
Dutch		93?
Indonesian		3?
Surinamese		2?

The complex early history of the "Low Countries" includes rule by Romans, Franks, Saxons, Holy Romans, Burgundians, Habsburgs and Spanish, interspersed with periods of local independence. Holland proper became independent and a great maritime power in the 16th and 17th centuries, but long rivalries and occasional wars with England and France caused exhaustion and eventual French conquest in the Napoleonic wars. In 1830 Belgium seceded from a post-1815 larger Netherlands, leaving the country in its present form. In the 20th century the land area was significantly increased by reclamation from the sea, but large areas are below sea level and vulnerable to massive flooding, as happened in 1953.

The main population increase began rather later, and has been rather steeper, than in most western European countries. Population density is much higher than any other in Europe. There has been immigration on a fairly large scale from one-time Dutch colonies and elsewhere. If any nation might be expected to have passed through the so-called 'demographic transition' to population stability it is the prosperous Netherlands, but the graph shows otherwise. In 2002 5% of the population was Muslim, and a poll showed that nearly 50% of young Dutch voters favoured zero Muslim immigration.

mentioned by politicians or the media. Nor was population growth linked to the shortage and high cost of housing.

Environmentalists claim that new roads generate extra traffic: they do, but so does population growth. In 2001 the UK government's plan to build 18 off-shore wind farms by 2010 was much praised, because they would generate electricity to power a million homes. The assumption was that fossil fuel use would be reduced accordingly, but in reality one million new homes will be needed to cope with the UK's expected population increase of 2 million over the same period.

The British charity Plantlife illustrates its appeals for money and membership by highlighting the destruction of wildflower habitats by new roads, new housing estates, and ever more intensive farming, but it was uninterested in my proffered article on overpopulation as the cause of those problems. So was *Natural World*, the journal of the British Wildlife Trusts. The editors held that population wasn't their business, but the contrary is true. Wildlife will inevitably lose out as population density goes on rising. Not all wildlife journals are so timid (Beebee, 2001).

Two centuries ago, the novel phenomenon of fast population growth in nations experiencing the Industrial Revolution was the subject of intense debate. Thomas Malthus in 1798 was only one of many authors who tried to predict the consequences of the developing British DC surge. Governments initiated censuses to measure their populations, to make

NEW ZEALAND   Area 270,000 km²

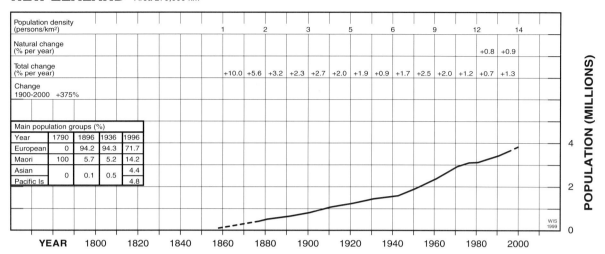

Population density (persons/km²)				1	2	3	5	6	9	12	14				
Natural change (% per year)										+0.8	+0.9				
Total change (% per year)		+10.0	+5.6	+3.2	+2.3	+2.7	+2.0	+1.9	+0.9	+1.7	+2.5	+2.0	+1.2	+0.7	+1.3
Change 1900-2000 +375%															

Main population groups (%)

Year	1790	1896	1936	1996
European	0	94.2	94.3	71.7
Maori	100	5.7	5.2	14.2
Asian	0	0.1	0.5	4.4
Pacific Is				4.8

POPULATION (MILLIONS)

YEAR   1800   1820   1840   1860   1880   1900   1920   1940   1960   1980   2000

WIS 1999

Polynesian seafarers (Maoris) reached the two large mountainous and volcanic islands of New Zealand around 1100 AD. Abundant food, in the form of seals, fish and birds including the gigantic flightless moa, that had no fear of humans, promoted a Maori population explosion. Within a few centuries most of the native animals had been killed and eaten, the large ones by humans and the small ones by the introduced Pacific rat. The Maoris had no domestic animals other than dogs, and few vegetables. Famine replaced plenty (Flannery, 1996).

The first European visitors, Tasman in 1642 and Cook in 1768, found the Maoris engaged in vicious tribal warfare in which the bodies of the slain were a prized source of food. The population was above its VCL (violent cutback level). Each village visited by Cook asked for his help to destroy the others. Tribal slaughter and cannibalism continued long after European, mainly British, settlement began in 1792. A Maori writer, Tamihana Te Rauparaha, described raids in which hundreds of the enemy were killed and many of them "cooked, as was the Maori custom" (Butler, 1980).

Immigration has been a major factor in population growth, especially after World War Two, except for the late 1970s and early 1980s when emigration dominated. From the 1960s the fast-growing Maori section of the population began claiming its historic rights of land ownership. A quarter of the land area is still forested. Censuses: every fifth year, with few exceptions, from 1878.

rational administration possible. Statistical publications such as *The Statesman's Yearbook* analysed numerical and percentage population growth rates well into the 20th century.

In fact, deliberate disinterest in population change is a very recent phenomenon. Jack Parsons (1993) points out that one of the oldest known documents, a Babylonian tablet of baked clay, "is a cri de coeur for population control". Egyptians, Greeks, Romans, Chinese and Indians all wrote of the damaging effects of population pressure on ancient societies and their environments, and the need to control exponential population growth. So, Parsons continues, did philosophers and scientists in the second millennium AD, including Thomas Aquinas, Machiavelli, Francis Bacon, Benjamin Franklin and Rousseau, all pre-dating Malthus. In Britain in 1949 a Royal Commission on Population recommended in favour of zero population growth.

Parsons describes how British politicians began to evade population issues in the 1960s and 1970s, when population increase was becoming almost synonymous with immigration and the rise of ethnic minorities, the harbingers of racism. Soon after, in the USA, "the practical issue of population control became entangled with the moral issue of abortion" (Hardin, 1993). *Political correctness* (PC), which had its origins in the Christian ethic of compassion for, and the human rights of, the hungry and underprivileged masses (section 5.6), had entered the population arena and was silencing debate on precisely those activities, population control and family planning, that could give the masses a chance of bettering themselves. Many people find PC compelling because it supposedly represents the moral high ground, but it is essentially a Western shibboleth, founded on faith and contemporary fashion (section 7.2). PC among Muslim fundamentalists, or the Hutus of Rwanda, or in Europe if Nazi Germany had won World War Two, is totally different. It is not an absolute, but a cult or fad of limited appeal.

Business interests also have been happy to observe the taboo on addressing population

1.8 – 18 million

NICARAGUA Area 131,000 km²

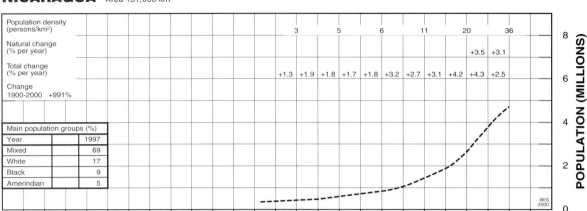

								3	5	6	11	20	36	

Population density (persons/km²): 3, 5, 6, 11, 20, 36

Natural change (% per year): +3.5 +3.1

Total change (% per year): +1.3 +1.9 +1.8 +1.7 +1.8 +3.2 +2.7 +3.1 +4.2 +4.3 +2.5

Change 1900-2000 +991%

Main population groups (%)		
Year		1997
Mixed		69
White		17
Black		9
Amerindian		5

WIS 2000

POPULATION (MILLIONS): 0, 2, 4, 6, 8

YEAR 1800 1820 1840 1860 1880 1900 1920 1940 1960 1980 2000

Mountainous Nicaragua was occupied by Amerindian farmers from around 4000 BC. Spanish colonisation began in 1522. Within a few decades, it is said, some 200,000 natives were taken away as slaves. African slaves were brought in as replacements when required. Independence was declared in 1838, except for the Caribbean coast which was ceded by Britain in 1860. US marines ousted the ruling dictatorship in 1909 and established bases from which they backed several right-wing regimes, the Somoza dynasty from 1937, against left-wing opponents, including from 1962 the Sandinista guerrillas who triumphed in 1979. The Sandinista government was opposed by right-wing Contra guerrillas backed by the US.

In 1989 after 3 decades of on-off civil war the Contras disbanded and entered politics, leaving a crippled economy and general poverty which has been exacerbated by rapid population growth, hurricanes and earthquakes. Agriculture is the mainstay of the economy, which is supported by foreign aid. A quarter of the country is still forested. Hurricane Mitch devastated much of northern Nicaragua in 1998, leading to a plague of rats in 1999.

growth. In general terms, the more people there are to sell your product to, the greater your potential profits, and you may wish not to offend potential customers in fast-breeding nations by asserting that their high fertility rates are menacing the planet. If you are a politician in the developing world you may believe that the more populous your nation, the greater is your influence in international affairs (even though your nation may be bankrupt and surviving on foreign aid).

More than any others, organisations dedicated to defending the environment and the future of the planet, such as Greenpeace and Friends of the Earth (FoE), should promote informed discussion of population issues. Both, in fact, maintain silence on the subject, but for FoE this is a fairly recent change of policy. FoE backed publication of Paul Ehrlich's *The Population Bomb* (1968) but "Since then FoE has become increasingly reticent about the effect of human population growth on the environment" (Willey, 1993). Willey notes that many people resigned from FoE "because it doesn't take population seriously". I was one of them, in 1990.

A Greenpeace spokesperson wrote to me in 1991 citing the demographic transition theory as the reason why Greenpeace saw no need to campaign on population. Earth's eco-warriors appear to be in a state of timid *denial*, prisoners of the so-called *scenario fulfilment concept*. This is jargon for self-delusion in the face of contrary evidence, or believing what you think is proper to believe – i.e. PC.

Thanks to the taboo, few people in poor fast-breeding nations understand that their high birth rate is perpetuating and worsening their national poverty. Even in the West, the taboo on discussion ensures that few people are well-informed about population statistics, population densities, the rates and histories of population increase, the burgeoning demand for and depletion of vital resources, and the sinister combination of population growth with ethnic or religious rivalry that ends in national catastrophe when the population reaches VCL. When a contributor to a BBC Radio 4 debate (February 24 2001) said that Britain should have an "open door" policy on immigration, she either knew nothing of the demographic realities that this book addresses, or she was confident that although England's population is the fifth

NIGER Area 1,186,400 km²

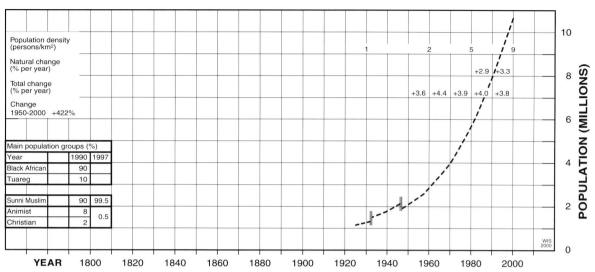

Population density (persons/km²)				1	2	5	9
Natural change (% per year)						+2.9	+3.3
Total change (% per year)				+3.6	+4.4	+3.9	+4.0	+3.8
Change 1950-2000 +422%							

Main population groups (%)		
Year	1990	1997
Black African	90	
Tuareg	10	
Sunni Muslim	90	99.5
Animist	8	0.5
Christian	2	

Niger straddles the ancient Saharan trade routes (slaves, gold, salt) crossing the desert from central West Africa to the Mediterranean. Before sea-borne trade became quicker and cheaper in the 19th century, Niger was the seat of powerful empires including the Songhai, Gobir and Kanem-Bornu, which became Islamic after the 10th century. France occupied the region late in the 19th century and declared it a colony, part of French West Africa, in 1904, but the nomadic Muslim Tuaregs of the northern deserts were not subdued until 1922.

From 1932 to 1947 part of Upper Volta (Burkina Faso) was incorporated into Niger. The country became independent in 1960 and lived by subsistence agriculture in the Sahelian south, badly affected by droughts, until uranium mining brought riches in the 1970s. In the 1980s there was unrest when uranium prices slumped. The Tuareg minority, seeking autonomy, began guerrilla activity in the 1990s.

There are at least 8 ethnic groups. In 1997 the birth rate was the highest in the world at 5.4% per year, and the total fertility rate at 7.4 children per woman was the highest in Africa.

densest in the world, it could double and double again in one or two decades without serious consequences. She represented a 'think-tank', supposedly a source of valuable ideas. I have no doubt that if her view prevailed, England's population would reach VCL, when England would cease to exist as a civilised nation, in less than 5 years.

Allowing your country, or any country, or Earth itself, to degenerate into chaos by deliberately ignoring human population growth and its consequences is unbelievably negligent and culpable. If, however, you are obsessed with PC, 'the modern cult of feeling rather than thinking', the chances are that you are contributing to such an end.

PC zealots, to whom human and other 'rights' matter more than consequences, resemble Muslim fundamentalists (or, 4 centuries ago, their Christian counterparts, as in Spain's Inquisition) inasmuch as they control the political and social behaviour of whole nations. They are intolerant of independent, objective, thinkers. In their own countries (in the developed world) they preach multiculturalism (section 7.2), which accentuates racial tensions if resources are limited and the population is growing. Abroad, in the developing world, they concentrate on healing the sick and feeding the starving, achieving moral satisfaction without reflecting (because the subject is taboo) that by saving all those millions of lives they are perpetuating the poverty of their patients, according to the Micawberish Rule (section 6.4). If they could look to the future, and make their life-saving work conditional on serious birth control, they would be doing less harm, or even much good. Unfortunately, population control is anathema to PC, which dictates that a woman should be *free* to have as many babies as she wants. But do the zealots help women to have as *few* babies as they want? If they do, we don't hear about it.

Which leads me to quote two Optimum Population Trust slogans:
"*The children people have are a gift to themselves. Those they don't have are a gift to the community*",
and
"*Fewer humans, more humanity*".

1.8 – 18 million

NORWAY Area 324,000 km²

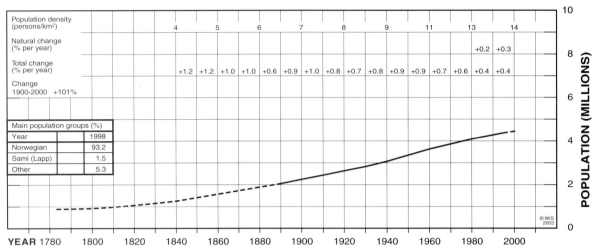

Population density (persons/km²)		4	5	6	7	8	9	11	13	14
Natural change (% per year)									+0.2	+0.3
Total change (% per year)		+1.2 +1.2 +1.0 +1.0 +0.6 +0.9 +1.0 +0.8 +0.7 +0.8 +0.9 +0.9 +0.7 +0.6 +0.4 +0.4								
Change 1900-2000 +101%										

Main population groups (%)	
Year	1998
Norwegian	93.2
Sami (Lapp)	1.5
Other	5.3

POPULATION (MILLIONS)

YEAR 1780   1800   1820   1840   1860   1880   1900   1920   1940   1960   1980   2000

© WIS 2003

Late in the first millennium AD the Germanic tribes living in Norway began to send out sea-borne raiding parties (Vikings) who plundered and settled the coastlines of north-west Europe. Some of them reached Greenland and North America. The country was a united Christian kingdom for most of the 10th to 13th centuries, but in the 14th century it merged with Sweden and Denmark. Danish kings ruled until 1814, when Norway became an unwilling junior partner with Sweden until independence in 1905.

A rugged mountainous northern land with an immensely long Atlantic coastline, Norway has specialised in fishing, especially whaling (resisting the ban on taking some species which Norwegians say are not threatened). Unlike most other oil-producing nations, Norway's population growth rate was unaffected by the start of offshore oil production in 1975, which has brought great wealth. Thanks in part to prolonged emigration, especially to America, population increase has been slow and steady, as in most Western European nations. Strong resistance has developed to non-white immigration, which is seen as causing racial problems. In the north, the indigenous Sami people had increased their reindeer herds beyond sustainable grazing limits in 2000.

• • • • • • • • • • • • •

" Facts do not cease to exist because they are ignored"

(Aldous Huxley)

Chapter 7

Globalisation and Multiculturalism

As currently used, these terms are closely linked. The first implies the spread of uniformity, in many forms, around the world; the second the spread of diversity into previously homogeneous cultures. If either process could proceed to completion, the result would be a uniform blandness around the planet, a loss of regional diversity and identity, because every nation would exhibit a similar mix of religions, economies, ethnicities, cultures and so on. This chapter looks at the effects of each process to date, and explains why, at high human population densities, neither is likely to reach completion or to benefit human society or the natural world.

7.1 Globalisation

In the context of commerce and trade, this term refers to the development and activities of multinational companies and financial organisations who can build factories to make goods anywhere in the world where costs are low. Then they transport their cars, shoes, chemicals or whatever to where the goods can be sold for maximum profits. Equally, they can buy agricultural produce such as grain, vegetables, fruit or meat cheaply, and transport it to where it is sold most profitably; thus apples may come to England from South Africa, celery from Israel and lamb from New Zealand. When, thanks to cheap labour, ideal climate, subsidies, trade agreements, or economies of scale, the combined cost of production and transport is less than the cost of production in the country of sale, producers in that country may be forced out of business.

The tendency is, therefore, for small farmers and businessmen in the developed world to go under, but for the proprietors of sweatshops and labour-intensive agribusinesses in the developing world to prosper. A limited range of goods and the most profitable crops and varieties are manufactured and grown, and offered for sale at all seasons all around the world.

In a different context, globalisation was and is the spread of a small range of dominant cultures, customs, languages, diseases, fauna and flora over all regions of the planet, dispersed by the increasing volume and speed of trade and travel. We know relatively little of the changes brought about by Roman, Arab, Polynesian and other migrants long ago, although fossil bones indicate that early human migrants exterminated many wonderful animals including mammoths, diprotodons, giant sloths, moas and aepyornes – mostly gigantic creatures that were vulnerable because they had no major predators before humans arrived.

From the 15th century AD onwards, European explorers carried diseases such as smallpox and measles to the New World and Oceania, decimating the human populations. In return, the New World gave syphilis to Europe. European domestic animals (dogs, goats, cattle, rats, etc.) diminished and often exterminated the astonishing variety of unique wildlife that had evolved on the islands of the Pacific and Indian oceans. European sailors exploited convenient sources of food such as the dodos of Mauritius and the giant tortoises of the Galápagos and Aldabra islands, annihilating the former and sparing a handful of the latter. Europe introduced coffee, wheat and

OMAN   Muscat and Oman   Area 309,500 km²

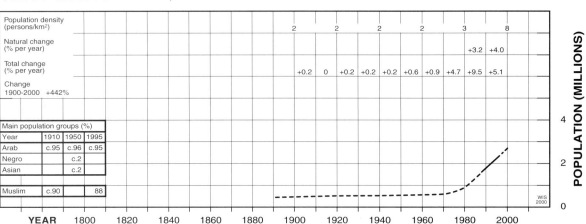

Population density (persons/km²)						2	2	2	2	3	8
Natural change (% per year)										+3.2	+4.0
Total change (% per year)					+0.2	0	+0.2	+0.2	+0.2	+0.6	+0.9	+4.7	+9.5	+5.1
Change 1900-2000 +442%											

Main population groups (%)

Year	1910	1950	1995
Arab	c.95	c.96	c.95
Negro		c.2	
Asian		c.2	
Muslim	c.90		88

YEAR   1800   1820   1840   1860   1880   1900   1920   1940   1960   1980   2000

POPULATION (MILLIONS)   0   2   4

Oman is the south-eastern coastal state of Arabia, including an isolated peninsula at the mouth of the Arabian (Persian) Gulf. From before 2000 BC Arab sailors based on this coast, under Persian control for some of the time, traded with Africa and Asia. The population converted to Islam in the 7th century AD. Arab traders established settlements on the Indian and African coasts; for many centuries they obtained slaves and ivory from the latter. The first European to reach Muscat and Oman was the Portuguese sea captain Vasco da Gama in 1498. Portugal controlled the Omani coast from 1507 to 1650 but Arab maritime trade continued and reached a peak of prosperity in the 18th and 19th centuries under the Al bu Said sultans. The territory was under British protection between 1891 and 1951.

1964 saw the discovery of significant oil reserves. From 1970 Sultan Qaboos has used the oil revenues to modernise the economy in preparation for when the oil is used up around 2020. Immigration of foreign workers is restricted to foster Omani self-sufficiency; in 1995 only a quarter of the population was foreign. The steep rise of the graph after 1970 is not so much due to immigration as to the DC surge made possible by national wealth.

barley to the New World, taking back maize, rubber, tobacco, cocoa and potatoes in exchange.

In the 21st century genetically modified plants and animals, which are seldom fully assessed as regards their capacity to change the natural ecology, are likely to spread rapidly around the globe, especially if their use is potentially profitable.

European botanists searched the world's forests, mountains and savannahs for exotic plants to enliven Europe's gardens. Some, such as rhododendrons from the Himalayan foothills and China, settled into their new environments so well that when they escaped they became pest plants, overwhelming the native flora. Water hyacinths from South America have clogged rivers and lakes in Africa so thickly as to prevent navigation. The animals brought to Europe by zoologists were mostly confined to zoos, but a few species naturalised readily and usurped the niches of unfortunate natives. In England, the American grey squirrel, signal crayfish and mink (ferocious predators released from fur farms by 'animal rights' activists) have seen off several of their British equivalents. The Nile perch, introduced to Lake Victoria in East Africa to benefit the local fishing industry, has all but wiped out the native cichlid fish.

World-wide, the architecture of cities and especially their centres has evolved into a utilitarian uniformity, within which a few old buildings, usually churches, shrines and temples, tell of the wide variety of national cultures that have been replaced. (Religious zealots sometimes destroy even these, as in Afghanistan in 2001 when Muslim fundamentalists exploded the giant statues of Buddha, carved into a cliff two millennia before. They had to be destroyed, the Taliban government said, because they were non-Islamic.) City centres in the richest countries are typically surrounded by sprawling suburbs integrated by road and rail links; elsewhere the residents occupy more compact apartment blocks or peripheral shanty towns.

"Half the world's 6800 languages are likely to vanish within two generations... we're aiming for about three or four languages dominating the world" (*New Scientist*, 12 August 2000).

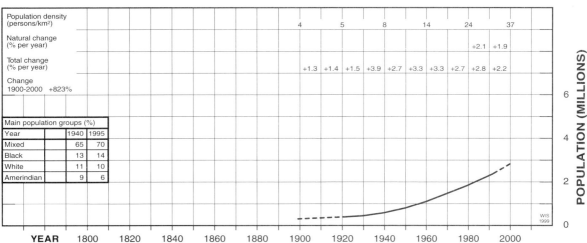

PANAMA Area 77,300 km² (including Canal Zone)

Spanish invaders took Panama from its Amerindian inhabitants at the start of the 16th century AD. Recognising its importance for Atlantic-Pacific communication they founded Panama City in 1519 and sent Andean gold and silver overland to their Atlantic-based treasure ships. The native population of some 750,000 was devastated by exploitation and Old World diseases. In 1821 Panama broke from Spain to join the Central American Federation and then, in 1830, Colombia. A revolution inspired by the US enabled Panama to declare independence in 1903. The new nation granted control of a 'Canal Zone' to the US. The Panama Canal was constructed by the US Corps of Engineers and opened in 1914.

Successive Panamanian governments have pressed for national control of the Canal, which provides 25% of national income. At intervals, anti-American rioting brought intervention by US troops, most recently in 1989. The US finally handed control of the Canal to Panama at the end of 1999. The dependable income from the Canal has enabled Panama to develop a prosperous economy, but the expansion of agriculture linked to rapid population growth is causing significant deforestation, soil erosion and loss of native fauna and flora. Censuses: every tenth year from 1930.

Globalisation enables small organisms to ride with travellers or goods into parts of the world where their arrival upsets the *status quo*. The Colorado and Chinese Longhorn beetles are introduced pests of European potato crops and timber trees. South American fire ants have invaded the USA and Australia. The zebra mussel has spread from the Caspian Sea to smother life in Europe's lakes, and Spanish slugs are swarming in Sweden. A Japanese winkle is eating its way through French oyster beds, the Australian swamp stonecrop is choking British ponds, and American jellyfish have replaced much of the Black Sea's native marine life.

At microscopic scales, a potato blight fungus from America caused a million human deaths in Ireland in 1845–46, and the foot and mouth disease virus crossed the Atlantic from Argentina in 1967 to decimate the British livestock industry. The HIV/AIDS virus, which may have jumped species from chimpanzees to humans in West Africa more than once before Europeans arrived, only to die out in the relatively static tribal societies of those times, has colonised the world in less than 5 decades, has caused 22 million deaths and may soon be influential in slowing or even reversing world population growth (section 4.7).

7.2 Multiculturalism: an Artefact of Political Correctness

This is a modern concept, idealising a 'global village' in which people of all cultures and races are supposed to live happily together. In the ancient world there was normally a direct relationship between a region and the ethnic group that occupied it, thus Israel was inhabited by Israelites, and so on. Society was essentially tribal. To the Romans, Britain was where tribes of Britons lived. Ethnic and cultural integrity was vitally important in a world of scarce resources where the Darwinian struggle for survival was won by the strongest, most coherent, groups or tribes. Even today, a gut-ambition to achieve tribal solidarity drives some nations to covet tiny alien enclaves like GIBRALTAR, Ceuta, or HONG KONG. There were a few exceptions where distinct ethnic groups exploited different ecological niches, as in

1.8 – 18 million

PAPUA NEW GUINEA Papua Area 463,000 km²

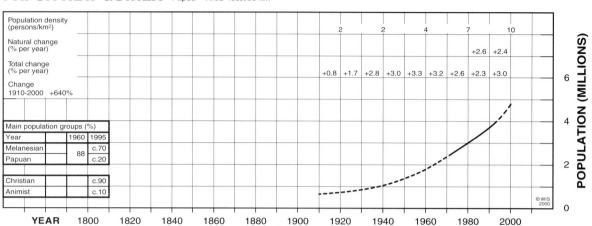

			2	2	4	7	10	
Population density (persons/km²)								
Natural change (% per year)						+2.6	+2.4	
Total change (% per year)		+0.8	+1.7	+2.8	+3.0	+3.3	+3.2	+2.6	+2.3	+3.0
Change 1910-2000 +640%								

Main population groups (%)		
Year	1960	1995
Melanesian	88	c.70
Papuan		c.20
Christian		c.90
Animist		c.10

Papua New Guinea consists of the east half of the world's second-largest island, New Guinea (the mainland), and several island archipelagos off its east coast. The mainland has been inhabited by humans for at least 30,000 years, with, periodically, waves of immigrants from the west balanced by emigrants sailing east into the Pacific. Portuguese and Spanish navigators visited in the 16th century, and Dutch, German and British merchants set up coastal trading posts in the 17th and 18th centuries. In 1884 Germany annexed the northern mainland and islands, and Britain took the southern parts.

After World War One Australia administered the whole region for the League of Nations and then, after Japanese invasion and expulsion in World War Two, for the United Nations. In 1975 Papua New Guinea became independent. Civil war broke out when copper-rich Bougainville Island tried to secede in 1990 and continued, with short-lived truces, until 1998. The country is rich in minerals, especially gold and copper, but subsistence agriculture is the main occupation. Coconut products, coffee and cocoa are important exports. About 80% of the country is still covered by tropical forest, which supports many unique animals and plants. Until the 1970s, cannibal jungle tribes preyed on the dense farming populations of the temperate highlands (Flannery, 1996).

central Africa where Pygmy hunter-gatherers traded with, and were tolerated by, Bantu farmers.

If diverse ethnic groups did live together in a region it was usually under duress, as in an empire where the conquerors lived among, and ruled, the conquered; e.g. the Romans and their subject peoples or the Spanish *conquistadores* in Central America. There might be a military or political alliance, as among the city-states of ancient Greece. Slavery was an extreme form of enforced cohabitation, universal in the ancient world, but most notorious more recently, in parts of the New World, North Africa and Arabia, where black slaves from Africa sometimes outnumbered the whites or Arabs. Failed empires often left uneasy legacies, like the coexistence of different ethnic groups in a newly independent nation (e.g. NIGERIA, SUDAN) whose colonial frontiers had amalgamated long-established tribal regions, or whose population included many settlers from the 'mother country'. Eastern Europe and the Balkans are a mélange of multicultural nations created from the remains of the Soviet, Turkish and Austro-Hungarian empires. Now that the USSR is no more, large minorities of Russian settlers in ex-Soviet republics from the Baltic states to KAZAKHSTAN and KYRGYZSTAN have suffered harassment from their new administrations.

When Europeans migrated to foreign lands with intent to settle, between the 16th and 19th centuries AD, they seldom shrank from eliminating the native inhabitants when the latter were thin on the ground and easily defeated in battle. In North America and AUSTRALIA, the numbers of 'Indians' and Aborigines were greatly depleted in the initial struggle for dominance. Africa and South Asia were more densely populated, and Europeans merely subdued the natives, by force. They imposed peace, education and health care, but remained a small ethnic minority. Eventually, after a colonial period lasting usually less than a century, they were easily persuaded or forced to leave when the thriving societies they had created, strengthened by DC surges, wanted independence.

Multiculturalism in its most modern sense began when a reverse migration started, not long after World War Two. Citizens of the newly independent nations that had been European

PARAGUAY Area 200,000 km² to 1935, then 407,000 km²

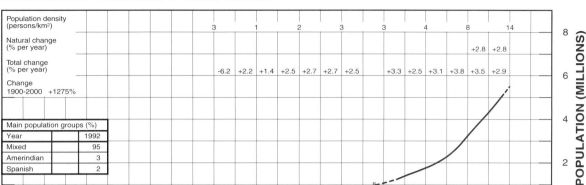

				3	1	2	3	3	4	8	14	

Population density (persons/km²): 3, 1, 2, 3, 3, 4, 8, 14

Natural change (% per year): +2.8, +2.8

Total change (% per year): -6.2, +2.2, +1.4, +2.5, +2.7, +2.7, +2.5, +3.3, +2.5, +3.1, +3.8, +3.5, +2.9

Change 1900-2000 +1275%

Main population groups (%)		
Year		1992
Mixed		95
Amerindian		3
Spanish		2

YEAR 1800 1820 1840 1860 1880 1900 1920 1940 1960 1980 2000

POPULATION (MILLIONS)

WIS 2000

Guarani and Chaco Indians were the long-established inhabitants of Paraguay when the first Europeans arrived in 1526. Spain conquered and ruled the country until 1811 when it became independent. Between 1865 and 1870 Paraguay fought a uniquely devastating war against Brazil, Uruguay and Argentina over access to the Atlantic coast; more than half the population, including 90% of the males, were killed. It took 50 years for the population to recover to its pre-war level. In the Chaco War against Bolivia (1932–35) Paraguay doubled its land area by seizing much of the vast sparsely populated plains called the Gran Chaco.

Paraguay has a history of military coups and dictatorships but their rule has been relatively benign. The largely agricultural economy has prospered and the Itaipu Dam, built with Brazil, is the world's largest hydroelectric venture. As elsewhere in South America the population growth rate has been 2% to 3% for more than a century, but the country is still thinly populated. This is a consequence of the very low population base level following the 1865–70 war.

colonies began to arrive in force in the 'mother countries', hoping for economic betterment. The ethnic migration coincided in time with, and helped to nurture, the rapid growth of political correctness (PC) in Europe (section 6.7). When a British politician, Enoch Powell, made his notorious 'rivers of blood' speech in 1968, predicting that multiculturalism would lead to racial strife in Britain, the outraged reaction showed that PC was already deeply entrenched. That Powell understood his subject has been proved by the growing numbers of race riots in British cities after only three decades (section 5.8).

In many Western nations, politically correct leaders and lobbyists felt such guilt about their colonial past (although Romans, Mongols, Turks, and countless others had preceded them in subjugating weaker countries or tribes), that the concept of national self-preservation, for which they had fought World Wars, was perceived to be obsolete and deplorable. The new concept was multiculturalism, and the growth of powerful ethnic minorities, even in urban ghettoes, was perceived to be *good*.

How are multiculturalism and PC related? PC is an emotional force. PC people have no doubt that what they do is fair, the decent thing, morally *right*. Their values are the 'eternal values'. Any opposing opinion must be *wrong*. They are as outraged as Muslim fundamentalists when their faith is questioned, as Powell discovered. He "played the race card". So in many ways PC is a modern religion, an updated version of Christianity. If you love your neighbour, protect him/her from hunger, disease and discrimination when he/she is weak or underprivileged, forgive him/her when he/she does wrong, and live your life according to the requirements of the Christian ethic, God will bless you. If you are an atheist, you will still feel the warm glow that comes with being deliberately unselfish; the certainty that you are *doing good*. Multiculturalism became appealing when foreign immigrants to Europe were perceived to be persecuted, discriminated against, exploited, made to feel second class citizens. They needed champions to *do good* on their behalf.

If PC including its offshoot, multiculturalism, is a faith, it is a blind faith, intentionally blind to the facts of population growth and its vast range of consequences. Why else would intelligent people deny facts, pervert them, or evade discussing them (section 6.7), unless they

1.8 – 18 million

PORTUGAL   including Madeira and Azores islands   Area 91,900 km²

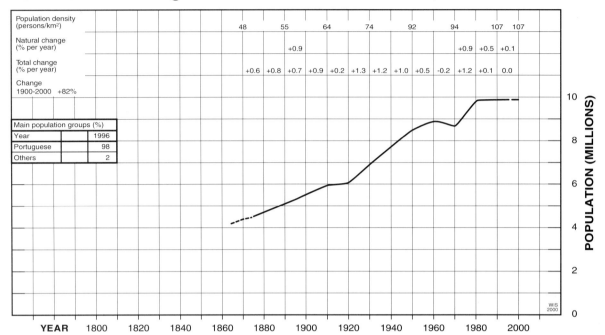

					48	55	64	74	92	94	107	107

Population density (persons/km²): 48, 55, 64, 74, 92, 94, 107, 107

Natural change (% per year): +0.9 ... +0.9 +0.5 +0.1

Total change (% per year): +0.6 +0.8 +0.7 +0.9 +0.2 +1.3 +1.2 +1.0 +0.5 -0.2 +1.2 +0.1 0.0

Change 1900-2000 +82%

Main population groups (%)		
Year		1996
Portuguese		98
Others		2

YEAR   1800   1820   1840   1860   1880   1900   1920   1940   1960   1980   2000

POPULATION (MILLIONS): 0, 2, 4, 6, 8, 10

WIS 2000

Portugal was part of Lusitania, the westernmost province of the Roman Empire, which was conquered by Muslim Moors in the 8th century AD. Three centuries of struggle against Moorish rule ended with success in the 13th century. Portugal's location at the southwest extremity of Europe led in the 15th century to a burst of maritime exploration and conquest that created a vast empire stretching from Brazil through Africa and India to China and the East Indies.

Brazil soon became independent, but the African and Asian possessions remained Portuguese until the 1970s when Portugal gave up the struggle against the nations or organisations that coveted them. Portugal's economy suffered, and dispossessed overseas residents returned home, causing large population swings. Many Portuguese left to find work abroad; in the 1990s there were 3 million such "migrant workers" (not included in the graph). Censuses: roughly every tenth year from 1878. See also MADEIRA and AZORES.

are enacting the *scenario fulfilment* charade, whereby they feel justified in bending facts to achieve an end that is, they have no doubt, the *morally correct* one? There is a similarity here to the way in which 'creationists' deny the reality of evolution.

While it is PC in Western countries to celebrate multiculturalism, the belief has few adherents in the developing world. European colonists are not praised for having created multicultural nations in Africa by demarcating frontiers that ignored tribal divisions, so that different cultures were forced to coexist in disharmony when independence came. Now civil strife is normal between tribal/cultural groups in SUDAN, NIGERIA, ZIMBABWE, CONGO DEMOCRATIC REPUBLIC and ETHIOPIA, among others. In Asia, intense tribalism in multicultural AFGHANISTAN gave that nation a particularly low VCL which was reached in 1979 (see WORLD). Multicultural British India broke apart immediately it achieved independence, and the Muslim fraction, PAKISTAN, could not hold on to its ethnic Bengali region, BANGLADESH. In FIJI, racial tension is permanently high between native Fijians and the descendants of Indians imported by the British to work on plantations. In INDONESIA, settlers exported from overcrowded Java to Borneo and Sumatra have displaced the local tribes who attack and massacre them. The divided population of CYPRUS was approaching its VCL when the UN in 1964 and then the Turkish army in 1974 intervened to prevent civil war. Even in Europe, fragments of once-multicultural YUGOSLAVIA are now themselves hotbeds of ethnic conflict which NATO and UN peacekeepers struggle to calm (see BOSNIA, SERBIA, KOSOVO, MACEDONIA).

In prosperous SWITZERLAND, on the other hand, French, German and Italian speakers coexist amicably, although in well-off BELGIUM there is strong antagonism

PUERTO RICO   Porto Rico   Area 8,880 km²

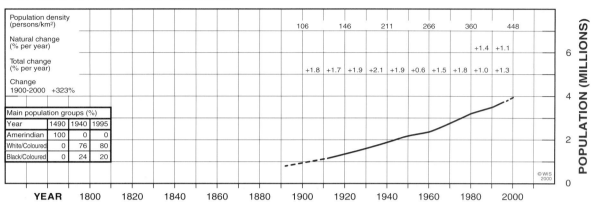

Population density (persons/km²)								106	146	211	266	360	448	
Natural change (% per year)													+1.4	+1.1
Total change (% per year)								+1.8	+1.7	+1.9	+2.1	+1.9	+0.6	+1.5	+1.8	+1.0	+1.3
Change 1900-2000 +323%														

Main population groups (%)			
Year	1490	1940	1995
Amerindian	100	0	0
White/Coloured	0	76	80
Black/Coloured	0	24	20

© WIS 2000

YEAR   1800   1820   1840   1860   1880   1900   1920   1940   1960   1980   2000

POPULATION (MILLIONS)   6   4   2   0

Puerto Rico, a large mountainous island in the eastern Caribbean, was inhabited by Arawak Indians when Columbus arrived in 1493. Spanish settlers enslaved the Indians who quickly succumbed to harsh treatment and the diseases of the Old World. The Spanish plantation economy involving sugar, coffee and tobacco was fairly easy-going, and relatively few African slaves were imported. For centuries the main occupation was sugar production, which continued following the Spanish-American War of 1989 when the island became a US colony.

After World War Two the previously stagnant economy diversified and prospered, with manufacturing industries enjoying tax advantages and tourism booming. The country became a semi-autonomous commonwealth linked to the USA in 1952. There is much emigration to the USA. Given the high population density it is not surprising that little tropical rainforest remains. Censuses: every tenth year from 1920. Ambiguity in the population groups table reflects varying interpretations of the significance of skin colour.

between French and Dutch speakers. Multicultural Northern IRELAND is no success story.

Early in 2002 a consortium of nations proposed to spend several billion dollars on "rebuilding" multiethnic AFGHANISTAN after 23 years of war, but if that country's population really is at VCL peace and economic viability are mirages unless there is unending peacekeeping and financial aid.

There were a few Jews in sparsely populated Palestine (see ISRAEL), coexisting peaceably with the Muslim majority, before the late 19th century when significant Jewish immigration began, triggering tension and violence between the two communities. In 1948 the Jewish component was strong enough to take the country from the Palestinians by force. Israel/Palestine is still multicultural, but hatred and killing dominate life in the now densely populated country.

Devout Muslim nations in particular have no use for multiculturalism; their goal is the furtherance of Islam, and their route towards it is unashamedly Darwinian. They attack and try to eliminate rival cultures. Palestinians have long claimed that they will drive the Israelis "into the sea". The rival Sunni and Shi'ite branches of Islam even massacre each other, as in IRAQ. In INDONESIA "the area [of] the Spice Islands had touted its mixed communities of Muslims and Christians as models of inter-faith neighbourliness. Then, in January 1999, it plunged into primeval war, and now thousands were dead". Dahlby (2001) goes on to describe how easily hatred is fomented by religious leaders exploiting the chaotic regime that began when President Suharto's dictatorial rule ended in 1998.

Late in 2001, following the September 11 assault on New York and Washington by Muslim terrorists, and the Western retaliation on the Taliban rulers of Afghanistan, hatred was building all around the world between Muslims and the West. Multiculturalism on the global scale was showing the same social defects as multiculturalism within a nation. Western hatred was tempered by political correctness to a mixture of fear, dislike and mistrust, but ethnic minorities in Western countries had no such inhibitions. Young British Muslims were leaving the nation in which their parents had claimed the right to a better or a safer life, to fight for Islam against British soldiers in Afghanistan (*The Guardian*, 29 October 2001). Muslim hatred worldwide was loudly proclaimed, arising as it does from the age-old resentment felt by the

1.8 – 18 million

RWANDA (part of Ruanda-Urundi to 1962)   Area 26,300 km²

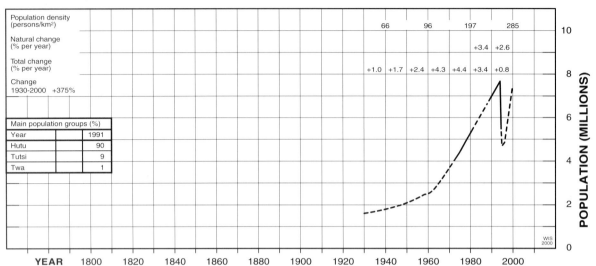

The original Rwandans, the Twa pygmies, were displaced during the first millennium AD by Hutu (Bantu) farmers coming from the west. Some 500 years later, Tutsi pastoralists (Nilotic) invaded from the north and east, dominating the majority Hutus and establishing powerful warlike kingdoms. The economy has always been based on subsistence farming, and when this could no longer feed the growing population the nation became increasingly dependent on foreign aid.

Rwanda (as Ruanda) was part of German East Africa (1890–1916), becoming the Belgian mandate of Ruanda-Urundi (1923–62) after World War One. Belgium's indirect rule through the Tutsi aristocracy fomented ethnic tension. In 1959 the Hutus rose, massacred or expelled 150,000 Tutsis, and achieved independence with a Hutu government in 1962. Ethnic warfare continued sporadically until 1990 when a strong Tutsi force invaded from refugee camps in Uganda. It was contained, with French help and much bloodshed, but when a UN force replaced the French in 1993 the Tutsis brushed it aside.

This nation of agricultural peasants (only 6% of the population is urban) now had the highest population density in Africa. Land hunger was extreme, and intensive cropping was causing massive soil erosion. Ethnic rivalry for land was behind the carefully planned genocide of Tutsis by Hutus in 1994. In 3 months, during which the Tutsi forces turned the tide and triumphed, about a million

Rwandans were slaughtered and about 2 million fled the country. Huge Hutu refugee camps at Goma in Zaire (now Congo Democratic Republic) were targeted by local Tutsi tribes and the Zairean army. In 1995–96 half a million Hutus were driven back into Rwanda. Up to 180,000 perished, mainly slaughtered, in a hopeless mass exodus westward on foot across Zaire. By 1998 1.5 million refugees were back in Rwanda, where in many cases their homes and land had been expropriated by others. In 2001, Rwandan forces were fighting and plundering in eastern Congo. 120,000 Rwandans were in prison at home awaiting trial on charges of genocide. Foreign aid supports the economy.

Ethnic killings have continued on a daily basis in Rwanda since 1994. The events of that year, and subsequently, are an archetypal model of Malthusian cutback. The population rose above its Violent Cutback Level (VCL), the point at which ethnic hatred could no longer be restrained, and the two groups set upon each other. The resident population was reduced by several million, but it is now rising again (though estimates vary widely). Peace in Rwanda is a hopeless dream and will continue so unless one side disposes of the other (as was planned in 1994), or an equitable partition is arranged, or export earnings miraculously go into surplus, allowing plentiful food to be bought from abroad, or (infinitely preferable) the birth rate falls so drastically that natural population growth is reversed and there is enough land for everyone.

poor towards the rich and privileged. Terrorism is the only kind of warfare available to the poor and weak in such circumstances, but Muslims seem not to realise that their poverty, weakness and inability to compete is largely self-generated, according to the Micawberish Rule, by their perversely high birth rate (section 6.4). Muslim nations comprised 15.7% of world population in 1970, increasing to 19.9% in 2000 (Table 3.2).

Multicultural nations are particularly vulnerable to sabotage by cultural or ethnic dissidents, when these are present in significant minorities. Their ghettoes are natural hiding-places for 'sleepers', who, when called upon to commit an act of terrorism or sabotage, don't 'stand out in the crowd' as they make their way to the scene. PC zealots who refuse to recognise that minorities are likely to include enemies who are happy to take advantage of their naïvety should have received a severe lesson on "9/11". The costs of that lesson to the West, in terms of lost security, confidence and trust in their neighbours, quite apart from business and employment, are incalculable. The actual cost to New York city was calculated to be 83 billion dollars.

SCOTLAND Area 77,200 km² (land area)

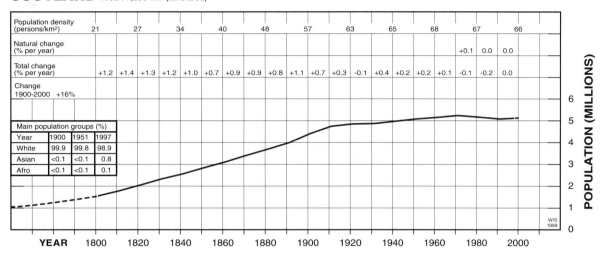

Population density (persons/km²)	21	27	34	40	48	57	63	65	68	67	66

Natural change (% per year): +0.1 0.0 0.0

| Total change (% per year) | +1.2 | +1.4 | +1.3 | +1.2 | +1.0 | +0.7 | +0.9 | +0.9 | +0.8 | +1.1 | +0.7 | +0.3 | -0.1 | +0.4 | +0.2 | +0.2 | +0.1 | -0.1 | -0.2 | 0.0 |

Change 1900-2000 +16%

Main population groups (%)

Year	1900	1951	1997
White	99.9	99.8	98.9
Asian	<0.1	<0.1	0.8
Afro	<0.1	<0.1	0.1

WIS 1999

YEAR 1800 1820 1840 1860 1880 1900 1920 1940 1960 1980 2000

POPULATION (MILLIONS)

Scotland, including the Northern Isles of Shetland and Orkney, was inhabited by farmers and fishermen 5000 years ago, but the first historical records are of the warlike Picts who successfully opposed the Roman advance into Caledonia. They merged with Celts who came from Ireland in the 6th century AD. During the Dark Ages, tribal kingdoms raided each other and the northern English tribes, and were harassed by Vikings from Scandinavia until 1018 when the Scot Duncan defeated the others and became King of all Scotland. Successive Scottish rulers, much influenced by ties with France, either co-operated with or fought against England, which claimed sovereignty after the death of the heir, a Norwegian, in 1290.

William Wallace, Robert Bruce, and their successor kings fought off English incursions until eventually in 1603 James 6 of Scotland inherited the English crown to become James 1 of England. The Scottish parliament merged with that of England in

1707, but the Scots were always resentful of domination by their more populous neighbour and in 1999 they broke away and re-elected their own parliament.

Scotland's population was at carrying capacity, about 1 million, before the Industrial Revolution in the mid-18th century. Thereafter it rose steadily at about 1% per year until World War One when, as in England and Wales, growth abruptly slowed. In the 1920s and 30s the old industrial communities of the Midland Valley experienced economic and social decline leading to high unemployment and reduced life expectancy. After World War Two economic recovery was less robust than in England. The birth rate fell, emigration to England and elsewhere increased and there was little ethnic immigration. From the 1970s the population has gradually declined. Censuses: every tenth year from 1801 except 1941.

In the West's preoccupation with PC there is a strong element of entrenched superiority, of missionary determination to save the "poor heathen" who "call us to deliver / their land from error's chain", and a complacent assumption that the superior/inferior relationship is naturally permanent. Woe betide them if it is ever reversed. PC as practised in many developing cultures sees compassion and consideration for others, loving your neighbour and turning the other cheek, as weaknesses that can be exploited.

So at the start of the 21st century the multiculturalists demand, with missionary fervour, that the lion shall lie down with the lamb, that the concept of national identity shall be abandoned, that human nature shall reverse itself so that people and groups no longer compete for advantage, or polarise into 'them and us', or obey Darwin's laws of natural selection. Discussion of the consequences of this policy shall be taboo. The multiculturalists will receive their reward in the next world, or in their egos. Somehow, they vaguely suppose, population growth will sort itself out and the planet will eventually support 9 billion or maybe 12 billion people, all living comfortably and happily in multicultural harmony.

The multiculturalists are sowing the seeds of their own destruction. They defy Charles Darwin at their (and everyone else's) peril.

7.3 Globalisation and Multiculturalism at the End of the WROG Period

The period of Weak Restraints On Growth (WROG, section 4.5), together with its spin-off cultures: globalisation, political correctness and multiculturalism, will end when powerful forces (the technologies unleashed by the Industrial Revolution) meet immovable objects. These are the depletion of finite resources, the population ceilings imposed by carrying

1.8 – 18 million

SENEGAL   Area 196,200 km²

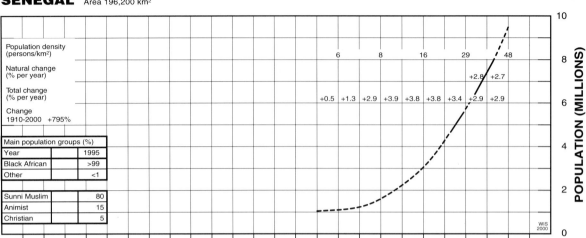

Population density (persons/km²)	6	8	16	29	48
Natural change (% per year)				+2.8	+2.7
Total change (% per year)	+0.5 +1.3 +2.9 +3.9 +3.8 +3.8 +3.4			+2.9	+2.9
Change 1910-2000 +795%					

Main population groups (%)		
Year		1995
Black African		>99
Other		<1
Sunni Muslim		80
Animist		15
Christian		5

POPULATION (MILLIONS)

YEAR   1800   1820   1840   1860   1880   1900   1920   1940   1960   1980   2000

Portuguese sailors reached Senegal in 1445 and found it part of the powerful Islamic empire of Mali, which had replaced earlier empires. The area was involved in the slave trade. France developed coastal trading posts in the 17th century and occupied the interior in the 19th, declaring Senegal part of French West Africa in 1902. Independence was achieved in 1960. President Senghor directed 20 years of slow peaceful development but after his resignation there was border warfare with Mauritania in 1989. Civil strife between internal groups, triggered by poverty, has been an intermittent problem in the 1980s and 1990s.

Northern deserts and savannahs pass southward into tropical rain forests. The weak economy is mostly agricultural, with groundnuts the main export crop. Over-fishing by foreign fleets is depleting the rich coastal waters. Unsuccessful attempts were made to federate with Mali (1959–60) and Gambia (1982–89). There are about 10 ethnic tribal groups. Censuses: 1976, 1988.

capacities and VCLs, and Darwin's laws of natural selection.

Probably the most critical factor in globalised commerce and trade is the low cost of transporting goods and people around the world at high speed, by air. This, the most energy-extravagant form of travel (one passenger flying across the Atlantic generates about as much carbon dioxide as a British motorist does in a year) has been possible up to now because crude oil, the source of aviation fuels, has been abundantly and cheaply available. For decades doomsters have been predicting the imminent exhaustion of oil reserves, but supply has always been able to meet demand, so pundits forecasting future developments tend to ignore the possibility and consequences of an oil shortage. But oil is a finite resource. Sooner or later it *will* be scarce and expensive.

According to oilman Colin Campbell (1997, 2003) that time is not far off. My own experience of assessing finite mineral resources leads me to respect Campbell's calculations. Although proved oil reserves in 1997 were very large, they would be used up by about 2030 at 1996 rates of consumption (24 billion barrels, or 3,400,000,000 tonnes, per year). New reserves have become so hard to locate that by 1997, for every four barrels of oil consumed, only two new barrels were being discovered (Figure 7.1). Campbell believes that the present decade will see demand overtaking supply, whereupon the price will skyrocket as nations compete for the vital fluid. The change will strike with little warning, because so few people have cared to think about and prepare for it. Within 2 or 3 decades world production rates will have halved. There will be insufficient time to compensate for the energy shortfall by other means: gas, coal, nuclear and the 'renewables'. Campbell concludes "The World will become a very different place with a smaller population. [It] may be a better and more sustainable place".

Campbell's work refines and updates a methodology invented in the 1950s by another oilman, M. King Hubbert. The nickname "Hubbert's Pimple" refers to the graph of world oil production over time. It starts from nothing in the late 1800s, rises to a high peak around 2000, then falls sharply and tails off to end about 2100. Hubbert used his methodology in 1956 to calculate that US oil production would peak about 1970. His peers were scornful, but he proved to be right.

Figure 7.1   The Growing Gap

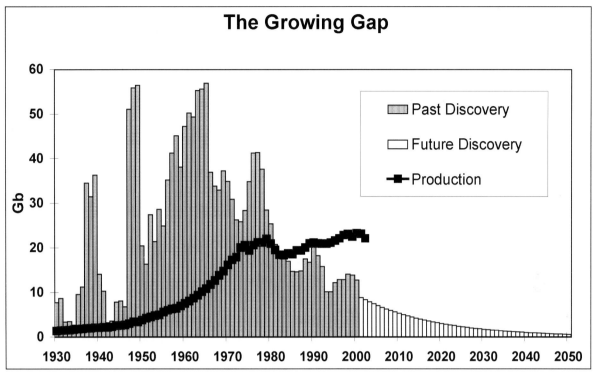

Figure 7.1 *(compiled by C J Campbell (1997, 2003)) records the history of conventional oil discovery and consumption from 1930 to 2001. Each dark vertical bar represents the amount of oil discovered in a year, measured in Gb (billion barrels); thus in 1949 new oilfields containing nearly 57 Gb were found. The pale vertical bars represent average annual discoveries from 2002 to 2050, projecting the decline that began in 1967. The line of black squares represents annual production/consumption of oil, reaching a peak of 25 Gb in 2000. When the white area below the black line equals the area of dark bars above the black line, all conventional oil will have been consumed.*

It is easy to see why, as the cost of aviation fuel increases two-, five-, or ten-fold, the era of mass travel to holidays in distant continents will grind to a halt. Cheap goods will have to be sent by sea. Desperately expensive petrol and diesel oil will revolutionise lifestyles in the West. Poor developing nations will be almost hydrocarbon-free.

Agriculture will be devastated. It was well said (Bartlett, 1986) that modern intensive farming is the conversion of petroleum and natural gas into food, via fuel and agrochemicals. (Natural gas was the feedstock, in the 1990s, for synthesising 80 million tonnes of fertilisers per year.) Unless intensive farms, which are now totally dependent on tractors and related machinery, are given special fuel quotas, they will take giant backward steps towards the days of horse or ox power. Few if any nations will have farm surpluses for export. Carrying capacities (section 4.1) will plummet, until densely populated countries experience permanent food shortage. Darwin's laws will see to it that rival groups polarise and compete (i.e. fight) for the diminishing resource. World-wide, resource shortages and national mixed cultures will cause VCLs to fall, leading to violent reduction of populations. Precedents (e.g. RWANDA) show that when the crunch of resource shortage comes, people gang up to compete for what is left, and the weakest gangs lose out.

Of course there will be a decade or two during which the world will try to adjust by crash programmes bringing in other energy-producing technologies, peacefully and legally at first. A variety of liquid fuels substituting for oil will be produced by methods ranging from

1.8 – 18 million

SERBIA   Servia   Area 48,500 km² to 1913; 88,360 km² (including Kosovo and Vojvodina) after 1918

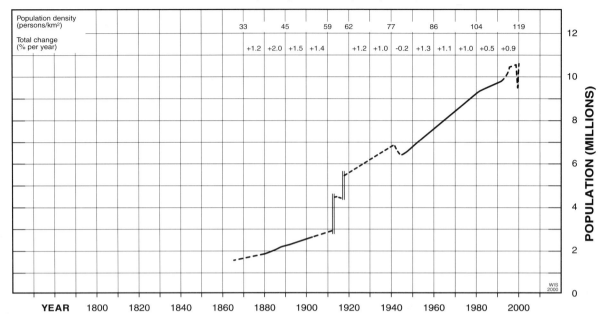

| Population density (persons/km²) | | | | | 33 | 45 | 59 | 62 | 77 | 86 | 104 | 119 |
| Total change (% per year) | | | | | +1.2 | +2.0 | +1.5 | +1.4 | | +1.2 | +1.0 | -0.2 | +1.3 | +1.1 | +1.0 | +0.5 | +0.9 |

YEAR   1800   1820   1840   1860   1880   1900   1920   1940   1960   1980   2000

POPULATION (MILLIONS)

Serbia after World War Two has consisted of 3 provinces: Serbia Proper, Kosovo and Vojvodina. Their graphs appear separately.

"We have always been victims", Serbs say. Between the 7th and 12th centuries their Slav ancestors migrated into this mountainous region (anciently Illyria and then the Roman province of Moesia) from lands that are now the Ukraine. A powerful Serbian empire developed whose armies were overcome by invading Turks at the battle of Kosovo Polje in 1389. Hundreds of thousands of Serbs fled north across the Danube, but the Turks followed and subjugated Vojvodina and Hungary. When Ottoman Turkish power weakened in the 19th century Serb guerrillas harassed the Ottomans, but the Serbs were minor players as the great powers, Russians, Austrian Habsburgs and Turks, competed to rule the Balkans. Finally a revolt in 1878 drove out the Turks and a Kingdom of Servia was proclaimed, dominated by Austria. Serb ambition achieved territorial expansion in the Balkan Wars of 1912–13.

After World War One Slav nationalism was strong enough to unite the separate Slav tribes in a new nation, the Kingdom of Serbs, Croats and Slovenes (later Yugoslavia), but as the population increased so did the other tribes' resentment of Serb domination. The Nazi occupation in World War Two provided opportunities for appalling ethnic massacres, especially of Serbs by Croats. In 1992 after vicious civil wars, Slovenia, Croatia, Bosnia and Macedonia seceded. The sharp population rises after 1992 represent influxes of Serb refugees, mainly from Bosnia and Croatia. In 1995 Croatia drove out 200,000 Serbs. Desperate Serb attempts to keep control of Kosovo province (in which the ethnic Albanians had increased by 290% from 1948 to 1998, while the Serbs had decreased by 9%) led to the civil war of 1999, in which the Serb army expelled more than 800,000 Albanians who then, as NATO bombed Serbia into impotence, returned and expelled 150,000 Serbs. The Serb view was that they had been punishing the Albanians for attempting to steal Kosovo by means of aggressive breeding – an attempt which, thanks to NATO, succeeded.

In a world mortally threatened by human overpopulation it is arguable that the ethnic Serbs, whose natural change rate is negative, had been the socially responsible element of the Kosovan population.

shale and coal distillation to electrolysis of water, but the costs will be great and the quantities inadequate for a world population of 7 billion or more. Chapter 9 examines this scenario more fully. Suffice it to say here that the global WROG period will have ended, taking multiculturalism and many aspects of globalisation with it. If Campbell's calculations are correct, the date is likely to be before 2030.

In some ways, humankind should benefit from the partial end of globalisation. Disease and pest organisms that at present are quickly carried around the globe in aeroplanes will not spread so easily. Alternative travel, by sea and possibly in airships, will be slower, expensive and less in volume. On land, the cheap fuel that in 2001 enabled trucks to routinely career hundreds of miles around Britain transporting pigs, sheep and cattle (first to market, then to the abattoir, and then to supermarkets, while distributing the foot and mouth disease virus throughout the country) will become so expensive as to enforce more rational procedures.

On the other hand, the 'global village' created by instantaneous electronic communications may not be seriously affected, though sabotage will be a constant threat in the years of conflict. Possibly the slowing and shrinking of travel and trade will encourage an

SERBIA PROPER (excluding Kosovo and Vojvodina) Area 55,970 km²

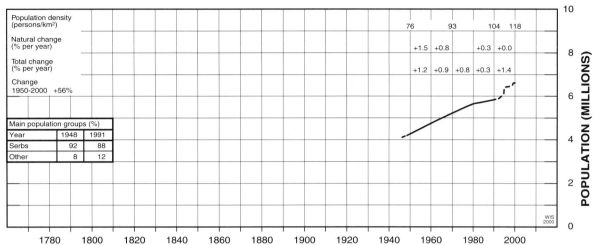

Population density (persons/km²)		76 93 104 118
Natural change (% per year)		+1.5 +0.8 +0.3 +0.0
Total change (% per year)		+1.2 +0.9 +0.8 +0.3 +1.4
Change 1950-2000 +56%		

Main population groups (%)		
Year	1948	1991
Serbs	92	88
Other	8	12

Serbia Proper, created in 1946, is the central province of the Serbian republic, with Vojvodina province to the north and Kosovo province to the south. It is roughly equivalent to the pre-1913 Kingdom of Servia. The population has always been dominantly Serb. As in other mainly Slav republics of the former Yugoslavia, the rate of natural population growth has fallen in recent decades, becoming slightly negative in the 1990s. The large irregular total increase in the 1990s was caused by influxes of refugees, mainly ethnic Serbs, from Bosnia, Croatia, Slovenia and Kosovo. Refugee numbers are now so great (c. 500,000 in late 1999) that most are living in temporary accommodation. Censuses: six between 1948 and 1991.

increase in diversity between different regions of the world.

So the end of the WROG period, the VCL scenario on a planetary scale, will affect far more aspects of life on Earth than just globalisation and multiculturalism. Ideally the reduced and chastened population will realise that to live sustainably, at peace with itself and with Nature, *growth must be outlawed.* Draconian population control will be an everyday necessity.

• • • • • • • • • • • • •

"It is rationality that makes us human"

(Lewis Wolpert)

SIERRA LEONE   Area 73,300 km²

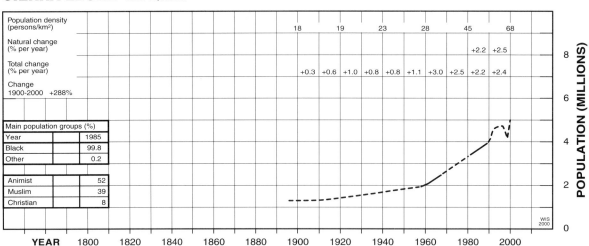

Population density (persons/km²)							18	19	23	28	45	68	
Natural change (% per year)											+2.2	+2.5	
Total change (% per year)							+0.3	+0.6	+1.0	+0.8	+0.8	+1.1	+3.0	+2.5	+2.2	+2.4
Change 1900-2000 +288%													

Main population groups (%)		
Year		1985
Black		99.8
Other		0.2
Animist		52
Muslim		39
Christian		8

In Sierra Leone, northern savannahs pass southward into rain forests. Portuguese navigators reached it in 1460, named it from the likeness of coastal mountain peaks to lions' teeth, and began a trade in slaves and ivory. Britain established trading posts in the 1600s. In 1787 English philanthropists bought land from coastal tribes and established Freetown as a haven and colony for freed slaves. About 70,000 settled there. Known as Krios, or Creoles, their dynasties became influential in the colonial administration before and after the hinterland was declared a British protectorate in 1896.

Sierra Leone moved smoothly to independence in 1961. The country continued peaceful, though suffering economic problems and rampant corruption, until 1990 when civil war spread from Liberia and rebels occupied the south-eastern provinces. Half a million refugees arrived from Liberia and anarchy became general. In 1995 war engulfed the whole country. 250,000 people fled to

Guinea and Liberia, only to flood back again when threatened by Guinea. Local warlords raised armies typically including many young children. A West African peacekeeping force, ECOMOG, has attempted to end the 'shooting and looting' which, marked by drug-induced atrocities including routine mutilation of civilians, and causing widespread starvation, has made Sierra Leone one of the world's poorest nations. Around 10% of the population is HIV-positive. In 2000 a British military force, assisting the government, was accused of neo-colonialism.

The economy, when there was one, was based on agriculture and mining, but the diamond mines soon fell to rebel forces and diamond smuggling (in which, senior UN officials alleged in September 2000, ECOMOG leaders were involved) funds their activities. Censuses: 1963, 74 and 85. There are 18 ethnic or tribal groups.

SINGAPORE   Area 648 km²

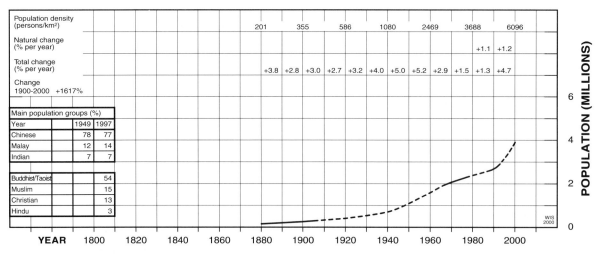

Population density (persons/km²)							201	355	586	1080	2469	3688	6096	
Natural change (% per year)												+1.1	+1.2	
Total change (% per year)							+3.8	+2.8	+3.0	+2.7	+3.2	+4.0	+5.0	+5.2	+2.9	+1.5	+1.3	+4.7
Change 1900-2000 +1617%														

Main population groups (%)		
Year	1949	1997
Chinese	78	77
Malay	12	14
Indian	7	7
Buddhist/Taoist		54
Muslim		15
Christian		13
Hindu		3

Singapore Island, at the tip of the Malay Peninsula, strategically dominating the great east-west oceanic trade routes, had been a Malay trading centre for many centuries before it was chosen for settlement by the British in 1819. With imported Chinese and Indian labour they developed it into a great port and naval base, boosted in importance after the Suez Canal opened in 1869, when it was capital of the Straits Settlement colony. After Japanese occupation in World War Two Singapore became a separate colony and founder member of the Malaysian Federation, which it left after

only two years to become an independent republic in 1965. Chinese social discipline and business acumen nurtured the city-state's booming 'Asian Tiger' economy to the point where it was not seriously affected by the Asian economic crisis of 1997.

For many years the government promoted the one- or two-child family, but this was increased to three in the late 1990s. However, the huge discrepancy between natural and total growth rates in the 1990s indicates that immigration was responsible for most of that decade's sudden population spurt.

Chapter 8

Population Sense and Sentimentality

Sentimentality can be defined as showing emotion rather than reason (Oxford Reference Dictionary, 1986) or as the elevation of feelings and image above reason and reality (Anderson & Mullen, 1998). Like political correctness it is an emotive attitude that avoids or denies the exercise of common sense. Sentimentalists prefer not to think seriously about aspects of life or the world that are not 'nice'. Candide, who accepted that everything was for the best in the best of all possible worlds (Voltaire, 1758), was a sentimentalist. Realists who warn that the smooth broad highway the sentimentalists are following leads over a cliff are dismissed as peddlers of doom and gloom. Unfortunately, history shows that life on Earth is far from being 'nice'. In our competitive Darwinian world, people are doing unpleasant thing to other people all the time (for examples, see ROMANIA, BULGARIA, GERMANY, RWANDA).

8.1 The Permissive Society

Once upon a time, when someone did something antisocial or unkind, or broke the law or even a promise, he or she was punished. The essential purpose of the punishment was education and deterrence, to ensure that the offender and observers were aware of the consequences of such behaviour, and to reduce the chance of it being repeated. Within a family, punishment of a naughty or disobedient child was part of the learning process. Other animals behave similarly, as when, for example, a lion cub annoys an adult. At the highest national level, punishment of wrong-doers was considered necessary to maintain the rule of law and to protect law-abiding and well-behaved citizens. The early US prisons were "penitentiaries", in which wrong-doers could repent.

Normally, the punishment suited the crime, Old Testament fashion; thus murder, treason or mutiny merited death, whereas lesser offences such as injury, theft or defiance of authority earned physical pain, public shame or loss of liberty (e.g. a beating, detention in the stocks, or prison, respectively), to a degree appropriate to the gravity of the misdeed. In practice, punishment often exceeded the demands of cool justice, particularly when the sentence was influenced by emotion. Fear, as when a ruler or a sea-captain felt threatened by revolt or mutiny, could lead to a ringleader being skinned or burned alive (among a wide variety of unspeakably awful fates), as an example to enforce authority. Hatred, common in racial or religious conflicts, provoked slow and excessively painful deaths, properly classified as torture, in revenge. Greed, on the other hand, if a bribe was offered, could treat an offender to a complete whitewash.

In some parts of the world, notably in developing countries, severe punishments are still in order. In CHINA, capital punishment is commonplace. Some Muslim countries dispense 'Sharia' justice, which dictates Draconian sentences such as amputation of a limb for thieves or stoning to death for female adulterers or even female rape victims. In the West, however, the 20th century saw sentimentality infiltrating punishment and traditional disciplinary behaviour.

SLOVENIA   Area 16,200 km² to 1947 then 20,250 km²

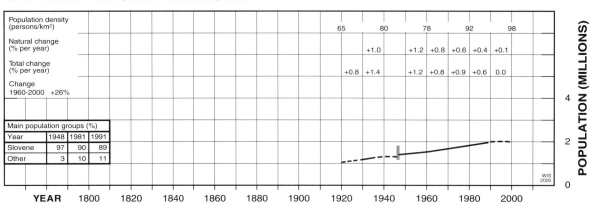

			65	80	78	92	98
Population density (persons/km²)							
Natural change (% per year)			+1.0	+1.2	+0.8	+0.6	+0.4 +0.1
Total change (% per year)			+0.8 +1.4	+1.2	+0.8	+0.9	+0.6 0.0
Change 1960-2000 +26%							

Main population groups (%)			
Year	1948	1981	1991
Slovene	97	90	89
Other	3	10	11

The Slovenes are descended from Slavs who displaced Avar tribes and settled this mountainous region in the 6th century AD. Slovenia, northernmost of the 6 republics of Tito's Yugoslavia, was never conquered by the Turks. From the 14th century until World War One it was ruled by the Habsburg dynasty of Austria, acquiring the Catholic religion and Germanic culture. Even so, the Slovenes shared with other Slavs the ideal of unification, which was achieved in 1918. In the Kingdom of Serbs, Croats and Slovenes, which became Yugoslavia (Land of the South Slavs) in 1929, Slovenia was disguised as the province of Dravska, without its Adriatic coastline. This was regained from Italy when Slovenia was reborn as a republic of communist Yugoslavia after World War Two.

Yugoslavia's population doubled during its lifetime (1929–91), sharpening ethnic tensions between the republics. Slovenia, like Croatia, resented Serb domination and the transfer of income from Slovenia's thriving economy to support the more backward southern republics. In 1991 it abandoned communism and multiculturalism and seceded peacefully from Yugoslavia. Population growth ceased, apart from an influx of 60,000 Bosnian refugees in 1992, and the economy, based on manufacturing, services and tourism, has prospered.

Strong agitation against capital punishment (death) began late in the 18th century, led by Quakers in Britain and the USA, who believed "there is that of God in every man", so even the most evil criminal could reform and become a good citizen. In the 20th century the death penalty was abolished in many Western nations and replaced by imprisonment "for life" (which in practice usually meant release after one or two decades). Also at the end of the 18th century there were calls for more humane punishment for lesser offenders. Corporal punishment, the infliction of physical pain, was outlawed as a legal punishment in Britain long ago, but it remained in the educational system as a judicial caning, or a slap on the hand with a rubber shoe, or similar, until the 1990s. Thereafter, in Britain, almost any form of corporal punishment was frowned upon, even in the home as part of family discipline. Persuasion, 'rehabilitation' (psychological pressure) and terms in prison replaced the "barbaric physical assaults on the person" that had served as punishment for millennia.

Whether the modern system achieves better deterrence is much debated. Certainly there is still plenty of crime. Britain's prisons are overcrowded, and most educators claim that school discipline has deteriorated and vandalism much increased. Critics lament that the authorities care more for criminals than for their victims. "If your house is burgled", sentimentalists say, "you are as much to blame as the burglar if you tempted him by leaving the door ajar". Whenever possible, the accused is given the "benefit of the doubt". "We are all guilty", claim apologists for law-breakers. Commonsense justice, essential to a civilised society, has been undermined.

In England, in 2002, a girl aged 11 was arrested for the 36th time in a year on charges including assault, theft, criminal damage and breach of bail, and a boy of the same age for the 151st time. He, it was said, had "stolen a car almost every day for the past two years". The courts could not punish them because of their age, and both were freed (*The Guardian*, 6.4.2002).

Draconian punishment was only one facet of Western society that relaxed in the 19th and 20th centuries, slowly at first but accelerating after World War Two. In the 1960s, what to many people were 'falling standards' ushered in the years of 'permissiveness'. Easy-going

SOMALIA Area 637,700 km²

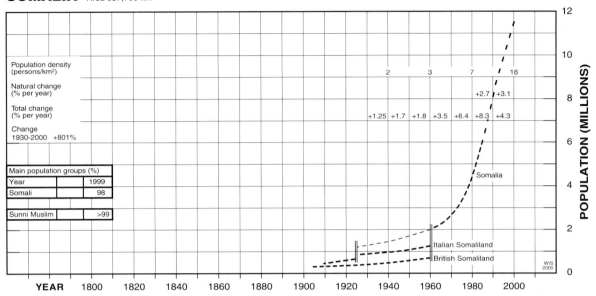

The Somali people, nomadic pastoralists of Hamitic and Bantu origin, have occupied this hot semi-arid to arid region (known to the ancient Egyptians as Punt) for at least 2000 years. Arab traders introduced Islam in the 8th century AD and established prosperous coastal cities. Somali-speakers had spread westwards into Ethiopia and Kenya when Britain and Italy declared protectorates in, respectively, north and south Somalia in the late 1880s. Italian Somaliland acquired Somali-speaking Jubaland from Kenya in 1925, and was a base from which Italy attacked and occupied Abyssinia (Ethiopia) in 1935, creating Italian East Africa which was dismantled in World War Two.

In 1960 independent Somalia was formed by uniting British and Italian Somalilands. Its history has been disastrous. Initial ambitions to unite all Somali-speakers in a 'Greater Somalia' were rejected by Kenya and Ethiopia, leading to intermittent warfare with the latter and inflicting floods of refugees on Somalia's fragile economy already beset by droughts. By the 1990s intertribal fighting was rife between individual clans within the Somali ethnic supergroup. The country

had no effective government. Between 1991 and 1993 some 50,000 Somalis were killed and about 300,000 starved to death. A UN peacekeeping force arrived in 1993, failed to achieve its objectives and withdrew in 1995. Somalia is now the playground of trigger-happy warlords and their private armies, one of whom has attempted to secede in Somaliland (the original British Somaliland). The economy depends on money sent home by a million Somalis working abroad. Some areas are unoccupied because of the danger from land mines. Many civilians are confined to internal refugee camps, suffering chronic hunger and disease. Drought and famine recurred in 1998–2001. Female circumcision is almost universal.

Population estimates since independence are, to say the least, unreliable. The graph, based on published data, requires total growth of 6% to 8% annually in the 1970s and 80s, figures that seem unlikely even with massive refugee movements. Compared to the 1999 UN prediction of 11.532 million for 2000, the Population Reference Bureau cut its estimate of 10.7 million in 1998 to 7.1 million in 1999.

behaviour that previously had been suppressed became tolerable. British examples: gambling, abortion and homosexuality made legal, sexual freedom (the pill), widespread drug use, suicide not a crime, relaxed censorship on matters of 'public decency', divorce made easy, etc.

Why the change? Was it a natural consequence of human social evolution, or is there significance in the way it coincides with the start and development of the WROG period (section 4.5) which saw the lifting of all kinds of restraints on human growth? The latter is more likely. Before the WROG period began in the mid-18th century, death in the family was a commonplace event. Populations could not increase, so the large number of births had to be equalled by deaths, usually from disease or malnutrition. Physical pain and suffering were normal in most people's everyday lives. But in the developed world, from about 1750 onwards, people found that death could be cheated and pain reduced. Populations rocketed (the DC surges). Humanitarians invented themselves to target pain and death, and when political correctness became a potent force after World War Two, painful and fatal punishments were eliminated in most Western nations.

So far, so good. But the WROG period will soon end (section 7.3) and strong restraints on growth will return. So strong, in fact, that they will force populations to shrink, for lack of resources. Human life will cheapen. The Darwinian struggle will resurface, with individuals

1.8 – 18 million

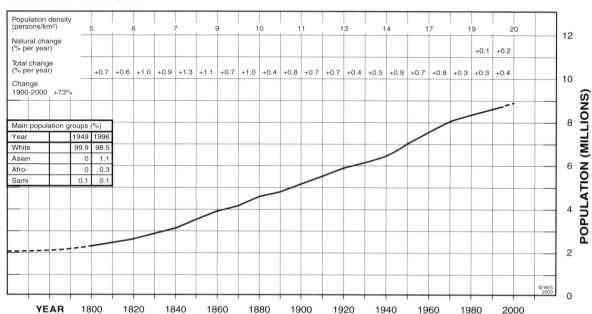

SWEDEN Area 450,000 km²

Germanic tribes lived in Sweden in Roman times. Their descendants were the Swedish Vikings who raided and settled coastal Russia in the 9th and 10th centuries AD. The country was united (except for Sami in the north and Danes in the south-west) in the 11th century. Finland was conquered in the 13th and Sweden united with Denmark and Norway in the 14th. In the 16th century Sweden regained independence and became a great military power in the 17th, with an empire including all Scandinavia and extending deep into Russia, but in the Great Northern War (1700–21) most of it was lost. Present boundaries were fixed when a union with Norway (1814–1905) was dissolved.

The slow steady rise in population is normal for prosperous developed nations, but it lacks the slowdown in recent years that is often seen. A recent increase in immigration, linked to a brief surge in the birth rate (1989–94) is at least partly responsible. In 2002 4% of the population was Muslim. Censuses: at 5 to 20 year intervals from 1800.

and groups seeking to survive at the expense of others. Each group will be tightly organised to win its battles. Anyone whose lifestyle prejudices the well-being of his group, or is of no value to it, can expect to be judged a traitor or a shirker, and be eliminated or expelled from the group. The permissive society, an artefact of the easy life available to well-off people during the WROG period, will be well and truly dead.

8.2 'The Wonderful Years Ahead'

If you believe the media, including popular science programmes on TV and articles in innumerable magazines, the future for humanity is blissful. No more household chores; the washing, cooking, cleaning and lawn mowing will be done by robots, who will even repair themselves and the other gadgets when they malfunction. In your car, you speak the destination and then relax while the computer takes you there, keeping the car a safe distance from other vehicles in the flow of traffic. Research into ageing will ensure that you remain fit and active well into your second century. You can holiday on the Moon, under a vast dome, and you may even emigrate to Mars when it is fully 'Terraformed'. Or, if you prefer, you can don virtual reality kit and experience every imaginable place and sensation without leaving home.

What tosh! Look at the graphs in this book, reflect for a moment, and you will understand why all those optimistic stories are fantasy, impossible to realise.

In the first place, such luxury doesn't come cheap. Everyone would have to be rich. Everyone? At the start of the 21st century the same media, in sombre mood, reminded us that 20% of the world's population controlled 90% of its income. The other 80%, overwhelmingly in developing countries, would have to catch up with the developed world by actually completing their development.

That is the first impossibility. Already the richer 20% of world population (1.2 billion

SWITZERLAND   Helvetia   Area 41,200 km²

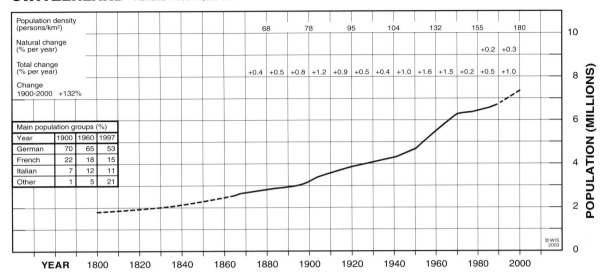

Population density (persons/km²)						68	78	95	104	132	155	180							
Natural change (% per year)												+0.2	+0.3						
Total change (% per year)				+0.4	+0.5	+0.8	+1.2	+0.9	+0.5	+0.4	+1.0	+1.6	+1.5	+0.2	+0.5	+1.0			
Change 1900-2000 +132%																			

Main population groups (%)			
Year	1900	1960	1997
German	70	65	53
French	22	18	15
Italian	7	12	11
Other	1	5	21

POPULATION (MILLIONS)

© WIS 2003

YEAR   1800   1820   1840   1860   1880   1900   1920   1940   1960   1980   2000

After the Helvetii tribes had been subdued by Julius Caesar in 58 BC they enjoyed Roman rule for 4 centuries. Later invasions and conquests by Germans, Franks, Holy Romans and Habsburgs provoked the formation of a defensive alliance, the Swiss Federation, in 1291. This was the basis of modern Switzerland, which has been neutral in all subsequent wars. The Swiss now have the highest standard of living in the world. There is little friction between the three main linguistic groups.

The graph shows the fairly slow and steady population rise typical of Western European nations. Most of the growth of recent decades was due to immigration, which in the 1990s amounted to more than two-thirds of the total. The 'other' category in the table represents the immigrants. By 1999 many Swiss opposed further immigration, wishing to preserve their national culture.

Most of the country is high mountains; the population density in relation to habitable land is 450. One third of Swiss forests is being damaged by atmospheric pollution. In 2001 there were fears that global heating would destabilise the mountain tops by melting permafrost in them. Censuses: roughly every tenth year from 1865.

people), with their profligate lifestyles, have begun to seriously heat the global atmosphere and oceans, changing the climate and threatening to displace a billion people from their low-lying homes as sea level rises. If the poorer 80% suddenly 'developed', everything that is going wrong now, globally, would go wrong at least five times as fast. That includes climate change, sea level rise, forest destruction, wildlife loss, atmospheric pollution, resource depletion (especially oil) and so on. Earth's ecosystems can't cope with 1.2 billion fairly well-off people; they would simply implode under the weight of 4.8 billion extra Westerners. The brutal fact is, developing nations are fated to stay developing.

And, to a large extent, it is their own fault. The populations of most developing countries are increasing at 2% to 4% per year, as their graphs show. Very few of them have economies that are expanding faster than that. So they are sinking deeper into poverty, as dictated by the Micawberish Rule (section 6.4). With every year of fast population growth their chances of ever completing their development are receding. (No wonder so many of their people try to escape by emigrating.) But if they slowed or reversed their population growth, there could be hope.

Another factor is the growing number of rising graphs that end at a sharp peak and trough, as in BOSNIA, RWANDA, etc. Most of these are populations that reached Violent Cutback Level (VCL, section 4.2). Their WROG period of relative prosperity and peace has ended. From now on their main concern will be the Darwinian struggle to survive. Household gadgets, or holidays on the Moon, are out.

Nor should Western nations be complacent. The Micawberish Rule applies to them also. Economic recession coupled with fast population growth increases poverty in any nation. In due course the competition for newly scarce resources may escalate into civil war and Malthusian cutback, especially if the population growth is multicultural immigration, which makes inter-group rivalry more likely. And holidays on the Moon less so.

1.8 – 18 million

SYRIA Area 185,200 km²

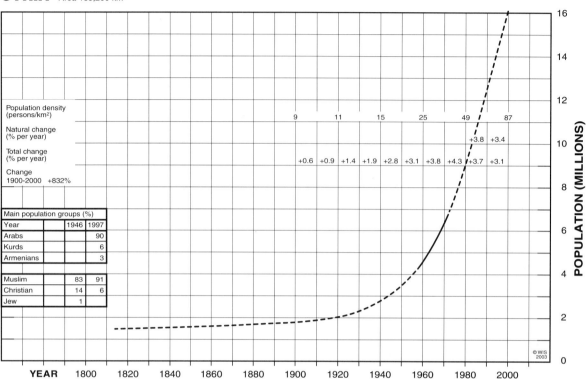

| Population density (persons/km²) | | | 9 | 11 | 15 | 25 | 49 | 87 |

| Natural change (% per year) | | | | | | | | +3.8 | +3.4 |

| Total change (% per year) | | | +0.6 | +0.9 | +1.4 | +1.9 | +2.8 | +3.1 | +3.8 | +4.3 | +3.7 | +3.1 |

| Change 1900-2000 | +832% |

Main population groups (%)

Year	1946	1997
Arabs		90
Kurds		6
Armenians		3
Muslim	83	91
Christian	14	6
Jew	1	

© WIS 2003

POPULATION (MILLIONS)

YEAR 1800 1820 1840 1860 1880 1900 1920 1940 1960 1980 2000

Syria was part of south-west Asia's 'fertile crescent' within which, as long as 20,000 years ago, humans were beginning to domesticate wild animals and harvest wild grain. It is said that the transition from hunter-gathering to farming was caused by overpopulation pressure: cereals grown as tended monocultures will yield far more grain per square metre than the same cereals gathered from the wild, but more human effort is required. By 8000 BC the region supported agricultural settlements, which developed into villages and city states.

After 2000 BC Syria formed part of, successively, Babylonian, Hittite, Assyrian, Persian, Macedonian, Roman and Byzantine empires. Natural forests, including famous species like the Cedars of Lebanon, were cut down, and farmed hillsides lost their soil. In the 7th century AD Arab armies spreading Islam conquered Syria and established Damascus as the capital of their great Muslim Empire. Syria has remained Arab and Muslim ever since, repelling

Christian Crusader incursions between the 11th and 13th centuries. Ottoman Turks added Syria to their empire in 1516 and maintained control until the end of World War One, when it became a French protectorate. Since independence in 1946 Syria has been involved in a series of ethnic/religious wars against Israel, losing the Golan Heights in 1967 but intervening powerfully in Israeli-Lebanon conflicts after 1976. A tentative merger with Egypt as the 'United Arab Republic' in 1958 broke down after 3 years. The fundamentalist Muslim Brotherhood party rebelled in 1982 but was crushed with 20,000 deaths.

Syria is not lavishly endowed with natural resources and much of the country is arid. Nearly half the arable land is irrigated from the Euphrates river, whose flows may soon be depleted by irrigation in Turkey. Tourism is increasingly valuable. The economy is fragile, relying significantly on aid from oil-rich states, and is bound to weaken further if the extremely high population growth rate is maintained.

8.3 The Life-saving Charities

"Every 60 seconds, 10 people in the developing world die from dirty water. In the time it took you to read this paragraph, someone, somewhere, has died. Yet a gift of just £2 a month from you today can help bring a family a supply of safe, clean water...". So runs an appeal on behalf of WaterAid (registered charity no. 288701) that arrived with my household water bill in March 2001. Similar appeals reach me and, I suppose, most other householders in the UK, from Christian Aid, Concern, Save the Children, Action Aid and a variety of other charities. They all want financial backing for their efforts to feed the starving and heal the sick, mainly, judging from the photographs of grateful recipients, in sub-Saharan Africa and the Indian sub-continent – regions in which populations are growing very fast.

"Treatable diseases such as tuberculosis, meningitis and pneumonia are still the leading causes of death in the developing world" says Médecins Sans Frontières, which dates from the 1960s, in its appeal leaflet of August 2001. MSF, the world's largest international medical relief charity, with a £200 million annual budget (*Encarta Encyclopedia 2000*), rushes humanitarian

TAJIKISTAN Area 143,100 km²

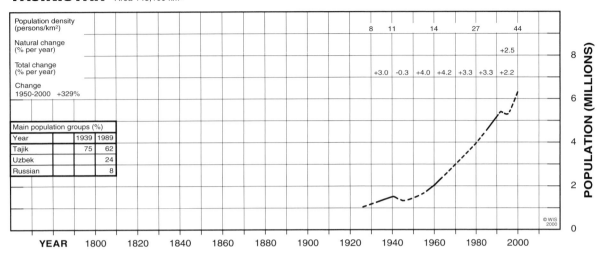

		1939	1989
Population density (persons/km²)			
Natural change (% per year)			
Total change (% per year)			
Change 1950-2000	+329%		

Main population groups (%)			
Year		1939	1989
Tajik		75	62
Uzbek			24
Russian			8

Eastern Tajikistan includes most of the sparsely populated Pamir mountains, the 'Roof of the World'. Huge glaciers nourish rivers that water fertile valleys in the west. The Tajik ethnic group, linked to Iran and Afghanistan, was in occupation when invading Arabs spread Islam in the 8th century AD. Mongol hordes overran the region in the 12th century, and their influence remained until Russia annexed it in the 19th century.

British fears that the Russians would move on into India were eased by creating a buffer zone, the Pamir Gap, in which Afghan territory met that of China. The Tajik Soviet Socialist Republic was established on an ethnic basis (see KAZAKHSTAN) in 1924.

Development, mainly intensive cotton farming, and population growth, advanced rapidly, but when the Soviet Union collapsed in 1991 and independence was achieved, civil war broke out between the communist government and Muslim rebels based in Afghanistan. More than half a million refugees fled the south-western valleys, and violence continues in spite of a truce in 1997. Droughts in 2000 and 2001 shrivelled crops and a UN emergency food aid programme is operating.

In 2001 Tajikistan refused entry to 10,000 Afghan refugees, suspecting that many of them were rebels. Most of the population is Sunni Muslim.

aid to "over 80 countries worldwide where the casualties of war, disaster and epidemic look to us to provide assistance". If MSF's dedicated specialists looked to the future and addressed the overpopulation that causes the poverty that generates warfare and disease, their work could have lasting, not just ephemeral, value.

In January 2000 the Director of Oxfam, David Bryer, wrote to me (and all the others) asking me to donate £2 per month to help prevent 840,000 Bangladeshi children dying of poverty every year. In the same month, on BBC Radio 4 (24 January) the Bangladeshi Environment Minister blamed the West for rising sea levels and warned that the West will have to accept at least 20 million environmental refugees when one fifth of Bangladesh is flooded. Actually, if Oxfam succeeds in saving all those children the refugee problem will be far worse, because they will augment Bangladesh's annual population increase to 2,800,000 (see BANGLADESH). Oxfam's appeal doesn't mention family planning. Nor, apparently, has the Minister considered employing it to *reduce* Bangladesh's population density to the point that people do not have to live in flood-prone regions.

Oxfam's work began in 1942, aimed at "relieving poverty, distress and suffering in all parts of the world and raising public awareness of the problems" (*Encarta Encyclopedia 2000*). How has Oxfam performed? World population in 1942 was 2.5 billion. Now it is 6 billion, at least half of whom lack basic sanitation and survive on less than 2 dollars a day, according to the World Bank in 2000. In purely numerical terms, in spite of the huge sums that Oxfam has raised from well-wishers (an annual budget of $130 million is quoted in *Encarta Encyclopedia*), many more people endure poverty, distress and suffering in the world now than in 1942. You can add to these sufferings another misery, hopelessness, which most of the 20 million occupants of refugee camps must now be feeling.

So where have Oxfam and its fellow charities gone wrong? Quite simply, they have ignored population growth, especially the Micawberish Rule (section 6.4). They have opted

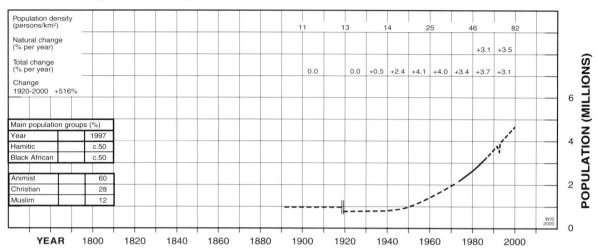

TOGO   French Togo   Area 56,800 km² (after 1919)

The long narrow state of Togo in West Africa is occupied by Hamitic tribes in the north and Black Africans, mainly Ewe-speakers, in the south. There are 37 ethnic groups. Europeans traded in slaves along the coast for centuries after Portuguese navigators arrived there in 1491. Germany established a coastal protectorate in 1884 and 'pacified' the interior by 1912, but was expelled in World War One. After the war the country was divided into French (east) and British (west) mandates; British Togoland opted to be administered by the Gold Coast (Ghana) after 1919 and merged with the Gold Coast in 1956.

French Togo achieved independence in 1960. After several coups a northern party established a dictatorial regime which suppressed opposition. In 1992 the trade unions called a general strike which paralysed the economy and generated 250,000 refugees. Dubious multiparty elections in 1993 restored the northern party to power, but disputes with Ghana and Benin continue to harm the economy, which is agricultural augmented by phosphate mineral mining.

for political correctness, the short-term satisfaction (and potent money-raiser) of *doing good* by saving lives. If, like Population Concern and Marie Stopes International, they had concentrated on family planning, and if by so doing they had stabilised world population, there would have been a wonderful opportunity for the Micawberish Rule to take effect, allowing poor economies to improve. But no, they chose the easier sentimentalist option. In consequence, as the end of the WROG period approaches, their chances of ever achieving their initial aims have shrunk to zero. And they alone are to blame.

All human history is of populations expanding when resources are available and shrinking when they are not. Charles Darwin's Malthusian friends knew it in the 1830s: "Public charities... aggravated the problem; hand-outs made paupers comfortable and encouraged them to breed. More mouths, more poor, more demands for welfare – it was a vicious circle" (Desmond & Moore, 1991). The planet will not be saved until we use our intelligence to break this natural Darwinian cycle.

8.4 The Informed Society

At the end of the 20th century most people in the developed world had access to a computer, and with it the Internet and the world wide web (www). This is an almost unimaginably huge store of information that is available to any Internet user at small cost. Before the Internet, i.e. before about 1980, information was available in a variety of forms such as books, films and tapes, but it was much more laborious to obtain than simply typing a code or clicking a mouse without leaving your chair. Also, of course, there was far less of it. Much of the www consists of websites promoting ideas, policies, business opportunities or services that previously would have been found in specialist magazines or 'junk mail'.

Potentially, the www is the most fertile and diverse source of information and ideas imaginable. Potentially, today's Internet users are the best-informed generation that has ever lived. Politicians love to boast that the people they govern, who voted for them, have been educated to use computers from primary school onwards, and are now the 'informed society'.

TUNISIA   Area 160,000 km²

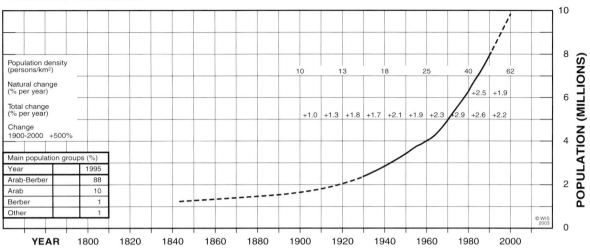

Population density (persons/km²)					10	13	18	25	40		62	
Natural change (% per year)									+2.5	+1.9		
Total change (% per year)				+1.0	+1.3	+1.8	+1.7	+2.1	+1.9	+2.3	+2.9	+2.6	+2.2
Change 1900-2000 +500%												

Main population groups (%)	
Year	1995
Arab-Berber	88
Arab	10
Berber	1
Other	1

POPULATION (MILLIONS)

YEAR   1800   1820   1840   1860   1880   1900   1920   1940   1960   1980   2000

© WIS 2003

Berbers, Phoenicians, Carthaginians, Romans, Vandals, Arabs, Turks and French have all occupied and left their mark on Tunisia. Under Turkish rule it was notorious as a base for Barbary pirates, but in 1883 it became a French protectorate and colony, achieving independence in 1956. The present population, mostly mixed Arab and Berber, is 98% Muslim. Tunisia's forests were felled in Roman times to grow grain, and after centuries of feeding the population of Rome and Italy the soil was eroded and the desert encroaching.

Agriculture is still important, though constrained by droughts, but tourism, oil and gas, and manufacturing are mainstays of the healthy economy.

The high rate of population growth after World War Two is typical of North African Muslim nations, although Tunisia with its popular tourist industry has been more developed, educated and prosperous than its neighbours. Recently its liberal version of Islam has been challenged by fundamentalist movements.

But is this really true? Pollsters quizzing samples of national populations repeatedly expose astonishing ignorance of basic subjects such as geography, history, science, current events or prominent people. In 2001, many English schoolchildren could not name common vegetables. Their verbal communication skills were poor. In 2002, 83% of young American adults could not locate Afghanistan on a blank world map, according to a National Geographic survey. Famously, even the current President of the USA seemed unsure of whether Nigeria is a continent, though he was certain that "most of our imports come from abroad". In Britain, research for the Department of Health found in 2001 that "a quarter of young people believed that the contraceptive pill could protect them from sexually transmitted infections" (*The Guardian*, 28 July 2001).

On the other hand, quiz shows on radio and television demonstrate how knowledgeable people are about sport, pop music and fashion. When the young and not-so-young spend hours 'surfing' the Internet it seems they are browsing for items that catch their passing fancy, or chatrooms where they can gossip, not for boring old stuff about how the world works, how it supports their lifestyles, and whether all is really for the best in it. Pornography websites are said to be more visited than any other.

If the Informed Society really was informed it would know all about population. Famous and influential people would know whether the population of their nation was growing or not (section 6.7). Members of think-tanks would be aware that mass immigration into overcrowded countries would generate bad trouble (section 6.7 again). Eco-warriors fighting to save the Earth (Greenpeace and Friends of the Earth, among others) would spend more time combating overpopulation than collecting waste paper for recycling, or marching on May Day to protest against poverty in the developing world. It would be generally understood that population growth is the most fearsome force driving and steering the development of human society. With population stability, the worst evils of Darwinian competition: wars, invasions, genocides, ethnic cleansings and so on could be avoided. If populations shrank, developing nations could actually develop, dangerous ethnic or religious confrontations could naturally calm down, resources would last longer, pollution would lessen

1.8 – 18 million

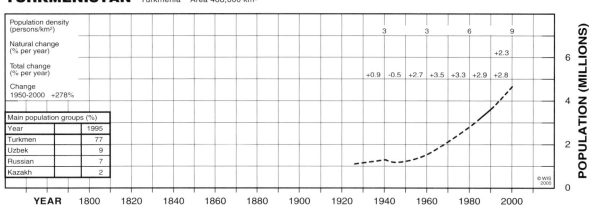

TURKMENISTAN   Turkmenia   Area 488,000 km²

		1995
Main population groups (%)		
Year		1995
Turkmen		77
Uzbek		9
Russian		7
Kazakh		2

Population density (persons/km²): 3 3 6 9
Natural change (% per year): +2.3
Total change (% per year): +0.9 -0.5 +2.7 +3.5 +3.3 +2.9 +2.8
Change 1950-2000 +278%

The Kara Kum desert, east of the Caspian Sea, occupies most of Turkmenistan. It was crossed by the Silk Route for trade between the Chinese and Roman empires, which was cut at times by invading Persians and Huns and, in the 7th century, by Arabs spreading Islam. In the 13th century Genghis Khan's hordes of Mongol horsemen swept across the Kara Kum and beheaded the inhabitants of Merv, the Silk Route city located where a river disappears under desert sands. Another Mongol, Tamerlane, perpetrated a similar slaughter at Urgench in 1372. Russia invaded in the 19th century, crushing minor khanates and creating the Turkmen Soviet Socialist Republic in 1925.

In the 1960s the Soviets built the Kara Kum Canal, 1100 kilometres long, that brought water to the desert from the Amu Darya river (the ancient Oxus) with disastrous consequences to the river's natural destination, the Aral Sea. The stagnant economy was transformed by harvests of irrigated cotton, but the Aral Sea is now only half its former size, its bed a salt desert. Turkmenistan became independent when the Soviet Union collapsed in 1991. Its large reserves of natural gas and oil still await exploitation for export. 90% of the population is Sunni Muslim.

and global ecosystems would again be able to cope with anthropogenic challenge.

Negligent is the word that best describes politicians and other decision-takers who are unaware of population growth and its consequences. The subject needs no teaching. Its existence and importance is obvious to anyone who watches, thoughtfully, developments in today's world. And, of course, a decision-taker who does not understand the train of events that led to a demographic disaster is very likely to (a) react inappropriately, as NATO did over KOSOVO, and (b) learn nothing from it to improve the quality of future decisions.

Concerned people and organisations (e.g. the Population Reference Bureau: www.prb.org) have ensured that there is no shortage of population websites on the Internet, but they cannot be heavily visited or the subject of population would be better known than it is. The reason, of course, is to do with human nature. As long as the WROG period endures (and the developed world is enjoying a particularly benign final phase of it), there is no incentive for most people to exercise their minds in serious logical thought – which is hard work. Why not enjoy the simple pleasures of gossip, sport, music or computer games when there is no apparent need to do otherwise?

Formal education today, world-wide, tends to concentrate more on turning students into productive economic units than on familiarising them with the world about them, and with the history, development and current predicament of humankind.

Thanks to the taboo on discussing population (section 6.7), to most Western people the future seems assured. Their dim vision of a lifestyle as good as or better than the present, stretching far into the 21st century, is so agreeable that they are happy to disregard warnings from 'doom and gloom merchants' like the Ehrlichs (1990). In their 'informed' minds, feelings reject reality, sentimentality rejects sense. It's human nature. When your life is easy, you tend to drop your guard.

8.5 Sentimentality Rules in the Nanny State

In an unsentimentalist world, actions have natural consequences, and causes have appropriate effects. One follows the other "as night follows day", it used to be said. When

UNITED ARAB EMIRATES   Trucial States   Area 83,700 km²

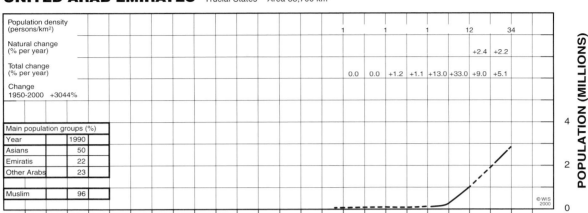

Population density (persons/km²)							1	1	1	12	34		
Natural change (% per year)										+2.4	+2.2		
Total change (% per year)						0.0	0.0	+1.2	+1.1	+13.0	+33.0	+9.0	+5.1
Change 1950-2000 +3044%													

Main population groups (%)	
Year	1990
Asians	50
Emiratis	22
Other Arabs	23
Muslim	96

YEAR 1800 1820 1840 1860 1880 1900 1920 1940 1960 1980 2000

POPULATION (MILLIONS) 4 2 0

© WIS 2000

These southern desert shores of the Arabian (Persian) Gulf have been settled since at least the 3rd millennium BC by Bedouin nomads and sailors partaking in trade between Mesopotamia, Persia and the East. They converted to Islam in the 7th century AD. Portuguese navigators arrived in 1498 and established a base dominating the Straits of Hormuz from which they controlled maritime trade into and out of the Gulf. After they departed in 1633 local sheiks continued harassing the shipping until a British naval force imposed peace on the 'Pirate Coast' in the 1820s and signed 'truces' with the sheiks.

The Trucial States became a British protectorate in 1892, the main occupations being nomadism, fishing and pearl diving. The pearl market collapsed in the 1930s, leaving the region in poverty until oilfields were discovered, first in 1958 in the largest sheikdom,

Abu Dhabi, and later in Dubai and Sharjah. Oil wealth fuelled a DC population surge in the 1960s, augmented by floods of immigrant workers attracted to the oil-based industries. In 1990 more than 90% of the work force was foreign. Britain withdrew from the Gulf in 1971, having assisted in the merging of Abu Dhabi, Dubai, Sharjah and 4 smaller sheikdoms or emirates to form the United Arab Emirates.

With the fourth largest oil reserves in OPEC, good for more than a century at present extraction rates, and immense reserves of natural gas, the UAE is rich and flourishing. Fresh water scarcity is offset by desalination of sea water, but much food has to be imported. The country is noted for its nature reserves, especially the sanctuaries for desert fauna and migrating birds. Censuses: irregular from 1968.

a criminal was caught, he or she was punished. Carelessness, like crossing the road without looking out for traffic, caused accidents. Taking a calculated risk, like bathing in a possibly polluted river, or passing beneath a cliff from which stones might fall, sometimes ended in tragedy. To the realist, that was life; some even argued that Darwinian natural selection was weeding out those least fitted to survive, maintaining thereby the fitness of the species as a whole.

Before the WROG period the death of one individual allowed another to survive, so the Darwinian argument was relevant. While WROG lasts, however, more people can survive, populations can grow, and it is less easy to argue that criminals or careless risk-takers get what they deserve. The expression "it's not fair, it wasn't his fault" is heard. Some other factor may be to blame, like the upbringing of a criminal or a risk-taker, or the tragedy may even be someone else's fault, like the car driver who wasn't prepared for a pedestrian to step unexpectedly into the road. Doubt and compassion modifies the cause/effect equation. When laws were passed to protect the health and safety of feckless or unlucky humans, the 'nanny state' was born.

Nanny states normally exist in the developed world. In them, referring to some recent British cases, a householder who tries to fight off a burglar can be judged at fault if the intruder is injured or killed, or even if the burglar is injured on the premises through his own carelessness. If a responsible person or organisation erects a notice warning the public of a hazard, and someone has an accident in spite of the notice, the erector may be blamed because the notice proves that he knew that the hazard existed yet he did not make it safe. If a doctor fails to recognise a disease, or a surgeon's scalpel slips during an operation, they can be sued for negligence or incompetence. So can any consultant whose work is anything less than perfect.

1.8 – 18 million

URUGUAY Area 176,000 km²

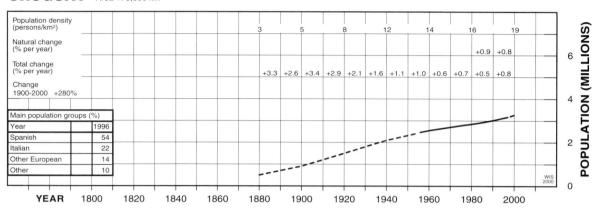

Population density (persons/km²)					3	5	8	12	14	16	19				
Natural change (% per year)										+0.9	+0.8				
Total change (% per year)					+3.3	+2.6	+3.4	+2.9	+2.1	+1.6	+1.1	+1.0	+0.6	+0.7	+0.5	+0.8
Change 1900-2000 +280%															

Main population groups (%)	
Year	1996
Spanish	54
Italian	22
Other European	14
Other	10

The first Spanish explorers to visit the grassy plains of Uruguay, in 1516, died at the hands of the local Amerindians. European settlements were not established until the late 17th century. Most of the natives were killed as Spain, Portugal, Brazil and Argentina battled for control of the country. Independence from Spain was achieved in 1828, but civil war soon broke out between factions representing Brazil and Argentina and continued sporadically until 1865 when the factions united to fight Paraguay (1865–70). From 1963 to 1973 the Tupamaros urban guerrillas fought to impose Marxism, but were defeated by the military and eventually became a political party.

The economy has always been heavily dependent on agriculture, especially stock-rearing, but tourism is becoming important. The sharply rising population graph after 1900 reflects increased European immigration, encouraged by government. In the 20th century the growth rate fell only slightly in numerical terms, but the percentage growth rate, decade on decade, decreased drastically.

VOJVODINA Area 21,500 km²

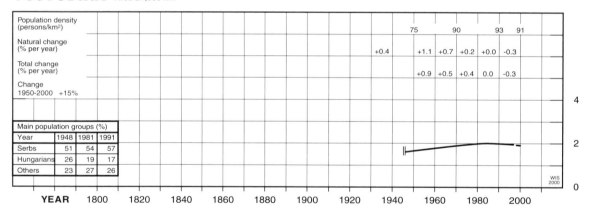

Population density (persons/km²)						75	90	93	91			
Natural change (% per year)					+0.4	+1.1	+0.7	+0.2	+0.0	-0.3		
Total change (% per year)						+0.9	+0.5	+0.4	0.0	-0.3		
Change 1950-2000 +15%												

Main population groups (%)			
Year	1948	1981	1991
Serbs	51	54	57
Hungarians	26	19	17
Others	23	27	26

Vojvodina is the northern province of Serbia, the others being Serbia Proper and Kosovo. After Roman occupation it was settled by Slavs in the 6th century and by Magyars (Hungarians) in the 10th. The Turks conquered it in 1526 and reduced the population, but they in turn were driven out by the Habsburgs in the 17th century. The country was repopulated by Hungarians and Germans from the north and by Serbs from the south and was part of Austria-Hungary until World War One. Its present boundaries were fixed after World War Two, when most of the German residents had fled and Tito's communist regime reallocated their extensive lands to farmers who had been wartime partisans.

At first the large Hungarian minority tended to assimilate into the majority Serb culture. The low rate of population growth minimised ethnic competition. In the 1990s, however, Serb resentment at the reverses they were suffering elsewhere in the former Yugoslavia intensified nationalistic pressure on the ethnic minorities.

WALES (incl. Monmouthshire)   Area 20,750 km²

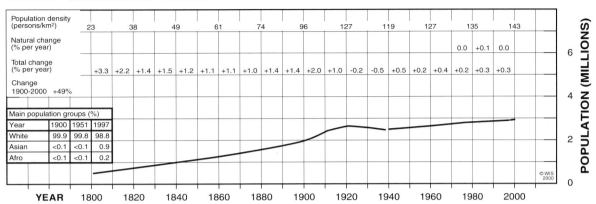

Population density (persons/km²)	23	38	49	61	74	96	127	119	127	135	143
Natural change (% per year)									0.0	+0.1	0.0
Total change (% per year)	+3.3	+2.2	+1.4	+1.5	+1.2	+1.1	+1.1	+1.0	+1.4	+1.4	+2.0	+1.0	-0.2	-0.5	+0.5	+0.2	+0.4	+0.2	+0.3	+0.3
Change 1900-2000 +49%											

Main population groups (%)			
Year	1900	1951	1997
White	99.9	99.8	98.8
Asian	<0.1	<0.1	0.9
Afro	<0.1	<0.1	0.2

© WIS 2000

POPULATION (MILLIONS): 6, 4, 2, 0

YEAR: 1800 1820 1840 1860 1880 1900 1920 1940 1960 1980 2000

Stone and Bronze Age peoples left impressive monuments in Wales, but most modern Welsh consider themselves descended from Iron Age Celts who began invading from continental Europe in the 6th century BC. The Romans found 4 powerful Celtic tribes inhabiting Wales, but subdued them by 79 AD. After the Roman departure in 409 AD England was conquered by Anglo-Saxon invaders who drove the Celtic natives westward into Wales – where they were not necessarily welcome. Welsh tribes fought off the Vikings during the later Dark Ages and, after 1066, the Normans and English. However, English armies conquered Wales in 1284 and declared it a principality, ruled by the top English prince. Welsh nationalism survived and spawned occasional uprisings such as that of Owen Glendower (1402–16), but it was not until 1999, when a Welsh Assembly with limited autonomy was elected, that Wales began to break away from England.

The population of hilly Wales was always at or near carrying capacity, growing very slowly to about 400,000, before about 1750. Then, the Industrial Revolution introduced new technologies that increased food production and hugely expanded the demand for Welsh coal, iron ore and slate. Growth was continuous until the decades after World War One when the old heavy industries declined, causing economic stagnation and social decay. Following World War Two, Wales enjoyed an expansion of light and service industries and tourism. The difference between natural and total population change shows that population growth 1970–2000 was brought about mainly by immigrants from elsewhere. Censuses: every tenth year from 1801 except 1941.

When a pupil was killed by a falling rock during a school visit to the English coast, the local council banned further visits "until the cliffs have stabilised" (i.e. for ever). In 2001, schoolteachers were advised not to take pupils on excursions after a disobedient pupil drowned and the teacher in charge was found guilty of manslaughter. This 'blame culture' is the down-side of the nanny state. Children can be deprived of educational or recreational travel, and the opportunity to learn self-reliance as regards personal safety. Those responsible for old mines or quarries, cliffs, caves, lakes, deep rivers or other challenging but potentially dangerous sites may refuse access for study or recreation. Consultants may charge exorbitant fees to cover the cost of insurance against mistakes which, being human, they sometimes make.

Health and safety legislation, the mark of the nanny state, plays its part in protecting the careless and the helpless but detracts from the enjoyment of life when it forces people to shun adventurous pursuits which involve risk-taking, for fear of legal consequences. Enterprising individuals who feel stifled may console themselves that the nanny state is an artefact of the WROG period which will not last much longer.

8.6 Chiefs, Slaves and Colonists

When Portuguese sailors made their first tentative voyages down the west coast of Africa in the 15th century, they found the black inhabitants organised into ethnic groups or tribes, whose distinct cultural differences were maintained by ongoing inter-tribal wars (Lopez & Pigafetta, 1591). Most tribes had a supreme chief or king, as the Portuguese found when they began trading goods for slaves with the kings of Congo (north Angola) in the 16th century (Esteves Felgas, 1958). Slavery and its close relative, serfdom, had been normal in African society, and world-wide, since ancient times (Ponting, 1993). A clay tablet in the Yale Babylon collection, 5000 years old, is a draft contract for the sale of slaves by a Sumerian trader, including a money-back guarantee if they proved unsatisfactory (*New Scientist*, 7 April 2001).

1.8 – 18 million

WEST BANK Area 5,900 km²

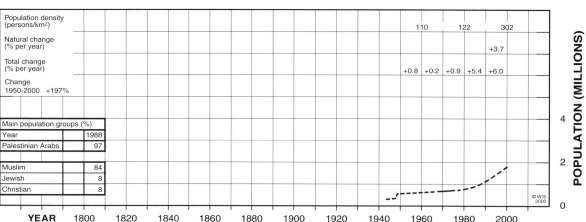

Population density (persons/km²)		110	122	302
Natural change (% per year)				+3.7
Total change (% per year)		+0.8 +0.2 +0.9 +5.4	+6.0	
Change 1950-2000 +197%				

Main population groups (%)	
Year	1988
Palestinian Arabs	97
Muslim	84
Jewish	8
Christian	8

POPULATION (MILLIONS)

YEAR 1800 1820 1840 1860 1880 1900 1920 1940 1960 1980 2000

© WIS 2000

The West Bank (see also ISRAEL) was an integral part of Palestine (to the Jews it comprised Judea plus Samaria) until 1948 when it was seized by Transjordanian (later Jordanian) forces during the First Arab-Israeli War. About 300,000 refugees fled to the West Bank from elsewhere in Palestine. Israel occupied it in the Six Day War of 1967 and has administered most of it ever since. Explosive population growth in recent years has raised Palestinian unemployment to around 25%, poverty is chronic and the economy depends on remittances from expatriate workers and on massive foreign aid. Typically, parents have many more children than they can support unaided. The Jewish population lives in settlements, many of them fortified against Palestinian attack. The number of settlers (including in East Jerusalem) has risen from 55,000 in 1977 to 376,000 in 2002. New settlements are still being established, causing intense Palestinian anger. Fresh water is already in short supply, especially for the Palestinians, because of the high population density.

The Palestinian Authority under Yasser Arafat campaigns, sometimes violently, for independence. Young zealots from Muslim fundamentalist organisations (Hamas, Hizbullah, Islamic Jihad) have attacked Jewish settlements and cities in Israel, causing many deaths. Israeli retaliation, military and economic, has worsened Palestinian unemployment and poverty. A low-key *intifada* against Israel began in 2000 and has steadily intensified, with tit-for-tat atrocities involving suicide bombers and military reprisals. Tens of thousands of children, brought up hating Israel, reach fighting age every year. Genocide and ethnic cleansing (the VCL scenario) has been prevented, so far, by Israel's overwhelming military might. The West Bank is of great strategic importance, nearly dividing Israel in half, so Israel is likely to cling tenaciously to it.

John Monteiro, a naturalist and graduate of London's Royal School of Mines, who worked and travelled extensively in Angola between about 1850 and 1875, remarked on "The great slaughter now going on in a great part of Africa… as the result of the suppression of the slave shipments from the coast… whereas formerly they were sent to the coast to be sold to the white men and exported, they are now simply murdered" (Monteiro, 1875). He contended that most African slaves were prisoners taken in battles between tribes, or alleged wrongdoers tried and found guilty by ritual ordeals involving 'fetish' (magic), for which execution was the normal penalty. Some chiefs would order as many as 15 executions in a day, but sale into slavery was preferred, because goods including guns could be bought with the proceeds. The customs Monteiro described had helped to ensure that Africa was still thinly populated 100,000 years or more after *Homo sapiens* evolved there.

Monteiro's findings confirmed those of earlier visitors to the interior of Africa, from James Bruce and Mungo Park in the 18th century to Livingstone and Stanley in the 19th (Perham & Simmons, 1942). They were unsurprised by the perpetual civil wars and battles for supremacy between tribes and local clans, but were often shocked by the casual killings and executions ordered by chiefs, whether to test the efficiency of a gun, to dispose of a potential rival, to punish an imagined insult, or simply as a demonstration of personal power.

For several centuries Europeans were content to maintain trading posts along the West African coast. Slaves were a major trading commodity for them, as well as for Arab traders in East Africa. There is powerful support for Monteiro's observations in the fact that the ending of the European-run slave trade (in the mid-1800s) did not cause sudden population increases in the countries involved. Had it done so, given the huge numbers of slaves involved, the first population estimates would have been higher, and the graphs would show an appropriately rising trend. Actually the strong increase in growth, the DC population surge, was delayed until

YEMEN   North Yemen plus South Yemen   Area 532,000 km² (North 195,000 km² , South 337,000 km²)

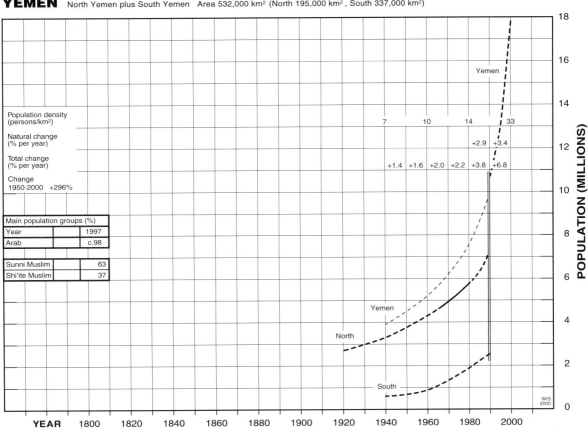

The well-watered highlands of western Yemen supported very early civilisations. Kingdoms arose and were coveted and conquered with monotonous regularity. Bilqis was the Queen of Saba (Sheba) kingdom who visited King Solomon in Jerusalem, according to tradition, around 1000 BC. The Himyarite kingdom succeeded Saba and repelled Roman and Ethiopian invaders in the early centuries AD; it became Christian but converted to Islam in the 7th century when under Persian rule. Muslim dynasties followed each other including the Rasulids (13th to 15th centuries) when Yemen excelled in the arts and sciences. However, millennia of deforestation and irrigation of crops had subjected the fertile lands to erosion and salinization.

In the 16th century Yemen fell to the Ottoman Turks, who made it coffee grower to the world. Local tribes recovered the country in the 17th century, but were weakened by repeated revolts and invasions from Saudi Arabia. The coffee market collapsed as Brazil entered it. Britain declared protectorates over South Yemen in the 19th century and the Turks fought their way back into the north, occupying it in spite of guerrilla opposition until forced to withdraw in World War One. North Yemen then became an independent kingdom, but South Yemen was controlled by the British until, capitulating to guerrilla warfare, they pulled out in 1967 and the south adopted communism. Meanwhile North Yemen had lost a territorial war with Saudi Arabia, had briefly united with Egypt and Syria, had endured persistent border conflicts with South Yemen, and had fought a bitter civil war (1962–70), after which it became a republic.

The two Yemens fought each other over border claims in 1972 and 1978–79. Then, slowly, the West-oriented north and the Marxist south (which had fought Oman and Saudi Arabia in the 1970s), became reconciled. A fierce inter-tribal war in the south (1986), the drying-up of Soviet aid, and the 1972 discovery of an oilfield straddling the border, were factors influencing the governments to merge as one Yemen in 1990. In the same year a million Yemeni workers in Saudi Arabia were sent home because of the Gulf War. 1994 saw a violent civil war in which the north defeated the south, but apart from tribal fighting and banditry the years since then have been peaceful.

Yemen's economy, based on agriculture, oil and gas, requires much foreign aid. In view of the nation's warlike traditions and the pressures of very fast population growth (the TFR was 7.6 in 1999), the future is full of hazards.

the early to middle 20th century. European imperial powers had 'scrambled for Africa' late in the 19th century, and it was their (sometimes brutal) imposition of peace on the warlike tribes, and their provision of agricultural know-how and medical care, that triggered the surges. This was not pure philanthropy, as it ensured a large healthy work force for colonial projects.

Unsurprisingly, agitation to stop the immemorial practice of slavery had begun a few decades into the WROG period, when tight Darwinian controls on the structure of European society were beginning to slacken, and compassionate action on behalf of underdogs became possible. DC surges continued after the colonists relinquished power and went home,

1.8 – 18 million

ZAMBIA Northern Rhodesia   Area 752,600 km²

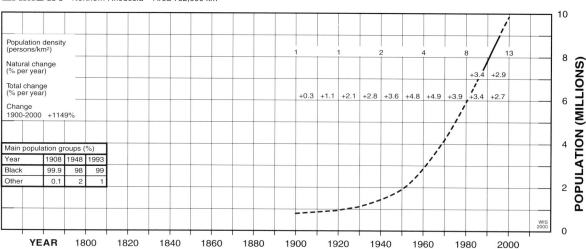

Year	1908	1948	1993
Black	99.9	98	99
Other	0.1	2	1

Main population groups (%)

Population density (persons/km²)

Natural change (% per year)

Total change (% per year)

Change 1900-2000   +1149%

Bantu people entered Zambia from the north and west long before Arab slave traders penetrated the region from the east coast in the 1500s. Slaving on a large scale continued well into the 19th century. In the 1820s the knock-on tribal invasions and expulsions initiated by the Zulu *mfecane* reached Zambia, as the Makololo peoples, driven northward out of Zimbabwe, captured land from the Tonga and Lozi tribes.

Apart from early Portuguese explorers, and David Livingstone after 1851, the first Europeans to take an interest in Zambia were British missionaries and adventurers in the late 19th century. Cecil Rhodes' British South Africa Company laid claim to the country in 1890. In 1924, soon after the discovery of the mineral-rich Copper Belt, it was designated Northern Rhodesia, a British protectorate.

European settlers in the wider region pushed to establish the

Federation of Rhodesia and Nyasaland in 1953, but in face of African nationalism it disintegrated and independent Zambia emerged, led by Kenneth Kaunda, in 1964. Subsequent development has been mainly peaceful, but the economy has suffered gravely from corruption, the cost of supporting liberation movements in adjacent countries, and the fall in the price of copper. In 2000 the country was effectively bankrupt, with a huge foreign debt.

Life expectancy in Zambia has fallen from 50 years in 1990 to 43 in 1998 due to chronic poverty and the prevalence of HIV/AIDS, which affects 20% of the population and is predicted to worsen. The graph shows the DC (death control) surge, beginning around 1930 with improved agriculture and health care, leading to a phase of very rapid population growth which may now slow down or even reverse.

between the 1950s and the 1970s. They have only ended in those countries whose populations have reached VCL, such as RWANDA, or have a major HIV/AIDS problem (e.g. SWAZILAND). They will all cease with the ending of the WROG period in the early decades of the 21st century (Chapter 9).

The WROG period has been a one-off opportunity during which, (with the benefit of hindsight) developing countries could have adopted Western practices and achieved a better standard of living. They could have developed. What actually happened was that, *after* the colonial phase was over, bodies such as the World Bank made huge loans to newly independent nations whose economies were being wrecked by DC population surges. By ending their poverty, according to the demographic transition theory (section 5.2), their DC surges could be controlled. But the lenders, and most charities, fought shy of birth control and family planning, preferring the easier sentimentalist option of *doing good* by feeding the hungry and healing the sick. In consequence, population growth has outpaced economic growth, the Micawberish Rule has operated, and the number of poverty-stricken people has doubled, or trebled, or worse.

Nor has slavery ended. In the 21st century it is still rife on the coffee, cocoa and rubber plantations of Gabon and the Côte d'Ivoire, in central and west Africa, where 200,000 children are sold into slavery every year, according to UNICEF. And even in Europe, young women who are forced into prostitution, paying a proportion of their earnings to a pimp in return for 'protection', are slaves in all but name.

8.7 Miscellaneous Sentimentalist Assumptions

Judging by media reports, one of the most serious threats to humankind in the 3rd millennium AD is 'asteroid strike'. For several years now, astronomers have been working

ZIMBABWE Southern Rhodesia Area 390,800 km²

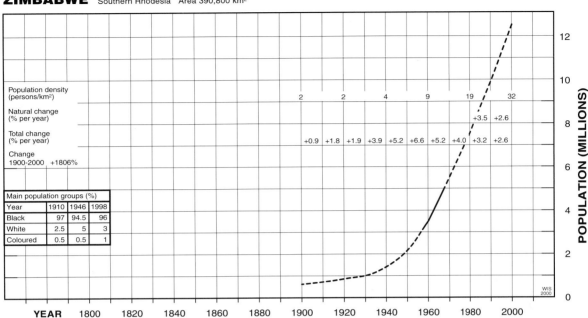

Population density (persons/km²): 2 2 4 9 19 32

Natural change (% per year): +3.5 +2.6

Total change (% per year): +0.9 +1.8 +1.9 +3.9 +5.2 +6.6 +5.2 +4.0 +3.2 +2.6

Change 1900-2000: +1806%

Main population groups (%)			
Year	1910	1946	1998
Black	97	94.5	96
White	2.5	5	3
Coloured	0.5	0.5	1

WIS 2000

POPULATION (MILLIONS)

YEAR 1800 1820 1840 1860 1880 1900 1920 1940 1960 1980 2000

Bantu peoples began migrating into Zimbabwe about 2000 years ago, displacing the long-term inhabitants, the San. The stone-built fortresses and shrines of Great Zimbabwe, southern Africa's most impressive prehistoric structures, were constructed by Bantu pastoralists, probably the Shona, between the 11th and 15th centuries. Their city-state disintegrated in the 1500s, probably due to the agricultural and social consequences of overpopulation in a dry country subject to droughts.

In the early 19th century Ndebele tribes fleeing the Zulu *mfecane* arrived from the south, subdued the resident Shona, drove out the Makololo, and established Matabeleland. A few decades later Cecil Rhodes advanced his dream of continuous British territory from the Cape to Cairo by trickily gaining control over much of the country from the Ndebele king Lobengula. By 1895 the region was white-ruled and referred to as Southern Rhodesia. After a period as a British protectorate it became a self-governing British colony in 1923. White settlement of the best farmland accelerated and 1934 saw the introduction of labour laws similar to the South African *apartheid*.

Black nationalism never subsided and posed a growing threat,

impelled by a particularly steep DC surge, that caused the white government to illegally declare independence in 1965. Britain and the UN imposed economic sanctions, but it was mainly the escalating black-white civil war that forced much white emigration and a return to British rule in 1979. Zimbabwe emerged as a prosperous independent black-ruled state in 1980. Violent tribal rivalries between the Shona majority and the Ndebele minority have persisted. In 2000 Robert Mugabe's Shona government was struggling to retain power by settling black peasant farmers on expropriated profitable white-owned plantations, and allowing poachers to destroy the wildlife of famous game parks, irrational actions in a nation that by now was almost bankrupt. By the end of 2001 the majority of white farmers had been forced off their land, by a black population ten times greater than when the land was settled.

Now, like neighbouring Botswana and Zambia, Zimbabwe is afflicted by an HIV/AIDS epidemic so serious that life expectancy fell from 56 to 49 years in the 1990s. It is predicted to fall to 31 years by 2008. In 2000 25% of Zimbabweans aged 15 to 49 were HIV-positive.

very seriously to spot large chunks of rocky or icy matter whose orbits round the Sun intersect with Earth's orbit and could, therefore, crash into our planet with disastrous results. How to avert such a threat, if it materialised, is the subject of titillating speculation, including several block-busting films. But apart from asteroids, surely it is a fair bet that clever old *Homo sapiens* can look forward to centuries of comfortable living, 'business as usual', in fact. President Bush of the USA was so confident, in 2001, that global heating and climate change are unproved fantasies that he felt able to reverse an election pledge, made 3 months earlier, to reduce American emissions of carbon dioxide. **HUMAN INVULNERABILITY** is assured, it would seem.

The geological record shows that asteroid strikes severe enough to wipe out a large nation have indeed occurred, but only once every ten million years or thereabouts. Other threats to humanity are more certain and, probably, much more imminent. They include any or all of the consequences of populations increasing at the rates shown by the graphs in this book, on a planet whose natural resources are finite and whose natural ecosystems are already unable to cope with man-made pollution. The analogy of a truck with failed brakes and an

1.8 – 18 million

AMERICAN SAMOA   Area 197 km²

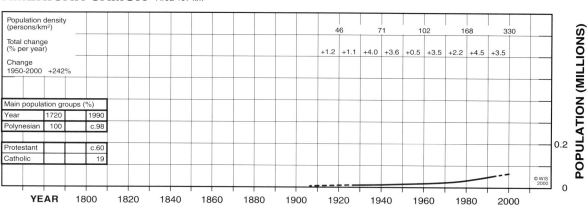

Population density (persons/km²)											46	71	102	168	330	
Total change (% per year)											+1.2	+1.1	+4.0	+3.6	+0.5	+3.5	+2.2	+4.5	+3.5
Change 1950-2000 +242%																

Main population groups (%)		
Year	1720	1990
Polynesian	100	c.98
Protestant		c.60
Catholic		19

POPULATION (MILLIONS) — 0.2 — 0

© WIS 2000

YEAR   1800   1820   1840   1860   1880   1900   1920   1940   1960   1980   2000

American Samoa, the eastern part of the Samoan Islands archipelago, 1000 kilometres east of Fiji in the south-western Pacific, is a group of 7 volcanic and coral islands. A band of seafaring Polynesians settled them about 1000 BC; a second wave some 1000 years later expelled the first arrivals. Dutch explorers in 1722 were the first European visitors. Trading posts were set up by Germany, Britain and America, who in 1899 took advantage of inter-island warfare to gain control of the archipelago. America claimed the eastern islands and has administered them ever since. They are now an unincorporated territory of the USA.

Agriculture is the main occupation and the weak economy depends on US aid. The high birth rate, unemployment rate and population density encourage emigration; in 1990 more American Samoans lived in continental USA than in the islands. Censuses: every 10 years from 1930.

idiot driver accelerating down a steep hill is appropriate.

When experts predict that a completely new London airport will be needed by 2015 to cope with the growth in air traffic, or that road traffic in Britain will double by 2025, or that non-white minorities will comprise 25% of UK population by 2040, or that world population will reach 10 billion by 2080, or that atmospheric concentrations of ozone-depleting nitrous oxide will rise 45% by 2100, they are working to a *BUSINESS AS USUAL* scenario. In other words, they extrapolate recent trends far into the future.

Thomas Malthus was influenced by the Business As Usual concept when, in 1798, he predicted imminent famines and wars as population growth outpaced food production. Half a century after the start of Britain's WROG period he had not realised the extent to which restraints on population growth had weakened. Now in the early 2000s the converse is true. In the developing world especially, strong restraints on growth are on the way back. They have already shambolicised the populations of at least 8 countries that have overshot their VCLs (e.g. RWANDA). End the international peacekeeping and food aid that these and many other countries 'enjoy' and their populations would decline.

In 2002 the restraints on population growth in Israel/Palestine were strengthening before our eyes, as the toll of violent deaths mounted. In the territories involved (ISRAEL, WEST BANK, GAZA) explosive population growth is Business as Usual. With population densities exceeding 300 on both sides, the conflict has become a war of attrition between ethnically and religiously polarised populations. The Israeli population (including a million Arabs) is growing at about 360 per day, and the Palestinian at about 320 per day. Two years into the *intifada*, the death toll has occasionally reached 50 per day (both sides together). There is still far to go, hopefully, before deaths regularly exceed 680 per day (showing that VCL has been reached). Business as Usual in the interim will be grim.

SELF-SUFFICIENCY, often linked to *ORGANIC FARMING*, is a concept that became popular in the later 20th century among idealistic Westerners. Rejecting the 'rat race' of urban life, they were seeking a slower, 'greener', lifestyle in harmony with Nature, growing their own food in the healthy old-fashioned way without the help of herbicides, pesticides or artificial fertilisers. Ideally, they would have a smallholding with a pig, a few goats, and one or two milking cows with names like Buttercup or Daisy. Their main crops would be vegetables, especially potatoes – which can keep people alive almost unaided, as in western Ireland before 1845.

ANDORRA Area 470 km²

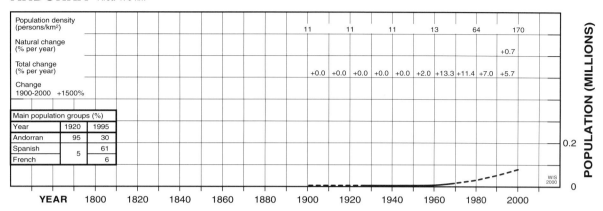

Population density (persons/km²)							11	11	11	13	64	170
Natural change (% per year)												+0.7
Total change (% per year)							+0.0	+0.0	+0.0	+0.0	+0.0	+2.0	+13.3	+11.4	+7.0	+5.7
Change 1900-2000 +1500%												

Main population groups (%)		
Year	1920	1995
Andorran	95	30
Spanish	5	61
French		6

The tiny mountainous principality of Andorra, between France and Spain, established in 1278, is ruled nominally by two 'princes', a Spanish bishop and the President of France. The centuries passed placidly in Andorra until, about 1960, the onset of large-scale tourism caused a population explosion. The very high total growth rates exceeding 13% per year derived from the arrival of immigrants, mainly Spanish, to work in the ski and duty-free shopping businesses.

Although the percentage rates of total change fell between 1960 and 2000, the actual number of new residents increased in every decade. Thus in the 1960s the average numerical increase of 800 people per year was 13.3% of the population base, whereas in the 1990s the annual increase of 2900 people was only 5.7% of the greatly increased base. Many people wrongly assume that if the percentage population growth is declining, the numerical growth must be declining too.

The truly self-sufficient life is the life of a peasant: hard monotonous physical work digging, planting, weeding and dealing with pests such as slugs, caterpillars and aphids. It is the life of most people in the developing nations, supporting a low standard of living, at best. In Britain it was the life which people abandoned, in the later 18th century, as they flocked to the industrialising towns. There was specialisation: smiths, millers, weavers, etc., whose skilled crafts and trades developed into the Industrial Revolution by mechanisation on a large scale. Self-sufficiency as practised in Britain today is less of a true lifestyle than a sentimentalist hobby, financed by personal savings or by welfare handouts.

Organic farming, on the other hand, can be profitable, when sentimentalists are prepared to pay premium prices for 'natural' food which, they feel, must be inherently more wholesome than food grown intensively. This may be true, although the average city-dweller who eats (sparingly) ordinary food seems to survive pretty well. What is certain is that organic farming without agrochemicals is relatively unproductive. In England before the Industrial Revolution, when farming was necessarily organic, the food produced was enough to keep alive a population of no more than 5 million (see ENGLAND LONG TERM), at near-starvation levels when the harvests were average.

Now, of course, if English farmers were all organic they could produce food for several times 5 million people, using their improved crop varieties and their tractors and computers instead of horses, oxen and human labourers. But they could not feed England's current population of 50 million. In 2000, England's overwhelmingly intensive farmers produced only about two-thirds of the food consumed by England's population. Britain's 'trade gap' in food, the cost of food imports minus the value of food exports, was £9 billion in 2000 (*Farmers Weekly*, 8 June 2001).

A century ago, when China's population was around 350 million and there were no artificial fertilizers, its food was organic and "depended on the use not only of animal manure, but also that of human excreta, 'night soil', which caused much recycling of human intestinal parasites" (J P Duguid, *in litt.*).

Nations like the USA, with a population density of only 30 (compared to England's 382) might go wholly organic and still, perhaps, export some surplus food. But most crowded Western nations could not do so, even now, when the oil that powers intensive and organic farms alike is still plentiful and cheap. One or two decades from now, when the WROG

ANGUILLA Area 155 km²

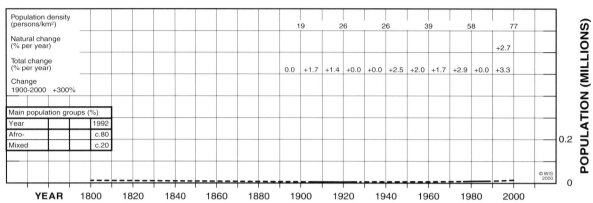

					19	26	26	39	58	77
Population density (persons/km²)					19	26	26	39	58	77
Natural change (% per year)										+2.7
Total change (% per year)	0.0	+1.7	+1.4	+0.0	+0.0	+2.5	+2.0	+1.7	+2.9	+0.0	+3.3
Change 1900–2000 +300%										

Main population groups (%)		
Year		1992
Afro-		c.80
Mixed		c.20

This small eastern Caribbean island was settled by Arawak Indians some 3500 years ago. They had been replaced by warlike Carib Indians when Europeans first visited it around 1500 AD. Britain colonised it in 1650 but it was unsuited to plantation agriculture and remained largely undeveloped. The population reached 10,500 around 1800 and then slowly declined until 1900. In 1967 Anguilla resisted an attempt to link it administratively with the nearby islands of St. Kitts and Nevis, fearing their dominance. It has remained a British dependency with a growing income from tourism. The population is almost wholly Christian.

ANTIGUA and BARBUDA Area 440 km²

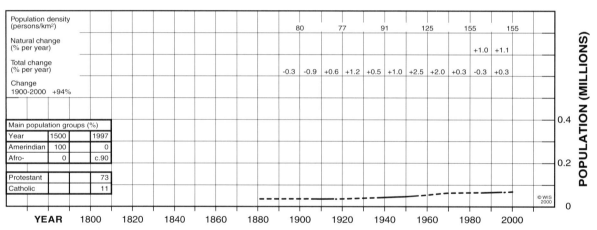

					80	77	91	125	155	155
Population density (persons/km²)					80	77	91	125	155	155
Natural change (% per year)									+1.0	+1.1
Total change (% per year)	-0.3	-0.9	+0.6	+1.2	+0.5	+1.0	+2.5	+2.0	+0.3	-0.3	+0.3
Change 1900–2000 +94%										

Main population groups (%)		
Year	1500	1997
Amerindian	100	0
Afro-	0	c.90
Protestant		73
Catholic		11

The original Arawak Indian dwellers on Antigua, in the eastern Caribbean, were forced out by invading Carib Indians around 1200 AD. Columbus sighted and named the island in 1493. Spanish and French attempts to settle did not succeed. Britain established a permanent settlement in 1632 and Antigua became a British colony in 1667. The native forests were replaced by sugar plantations worked by African slaves, which brought great prosperity until slavery was abolished in 1834. Meanwhile the smaller island of Barbuda, leased by the Codrington family, grew food for the slaves.

Tourism burgeoned after World War Two and now dominates the economy. The islands joined the West Indies Federation in 1958, seceded in 1962, and became fully independent in 1981. Since 1970 the population has been almost stable, thanks to the high rate of emigration. A succession of hurricanes severely damaged the islands in the 1990s. Fresh water is in short supply.

period is ending, the cost of energy and agrochemicals will be spiralling upward. So, especially in crowded countries, will the price of food. If voluntary organic farming is remembered then, it will be as a sentimentalist affectation of a lost golden age.

8.8 Green Revolutions: "Feeding the World"

The Industrial Revolution of the mid-18th century was the first of several technological breakthroughs which, among other achievements, vastly increased the ability of farmers to produce food. Malnutrition as a principal cause of death was undermined, and world

AZORES   Area 2,300 km²

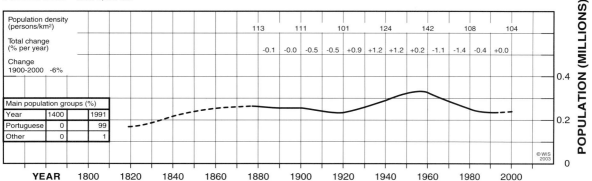

Population density (persons/km²)						113	111	101	124	142	108	104	
Total change (% per year)						-0.1	-0.0	-0.5	-0.5	+0.9	+1.2	+1.2	+0.2	-1.1	-1.4	-0.4	+0.0
Change 1900-2000 -6%													

Main population groups (%)		
Year	1400	1991
Portuguese	0	99
Other	0	1

POPULATION (MILLIONS)

0.4

0.2

© WIS 2003

0

YEAR   1800   1820   1840   1860   1880   1900   1920   1940   1960   1980   2000

Portuguese sailors reached the Azores archipelago, in the Atlantic Ocean 1500 kilometres west of Portugal, in 1427 (though the islands are said to be represented on a map of 1351) and found it uninhabited. Portuguese settlement began in the 1430s. Except for the Spanish occupation, 1580–1640 (when Spain ruled Portugal), the Azores have always been Portuguese, and they are now an autonomous Portuguese region. Tourism is beginning to boost an otherwise mainly agricultural economy. The population has been regulated by waves of emigration, mostly to North America where there are more Azoreans than in the islands. Censuses: roughly every 10 years after 1878.

population, which had been nearly 600 million in 1750, began to surge upward. By the mid-20th century it was 3 billion, and in southern Asia where population growth was greatest it was outpacing the production of grain. Starvation again threatened, but was averted by the 'Green Revolution' which introduced new varieties of rice and wheat which doubled the yield of grain (at high cost in extra fertilisers and pesticides, and in certain trace-element deficiency diseases: *New Scientist*, 30 March 1996). World population growth could continue, and after only about 40 years it was nearing 6 billion and had largely absorbed the extra output of the Green Revolution.

The ability of technical improvements to increase food production tends to be underestimated, as it was by Malthus in 1798. The Green Revolution ensured that dreadful famines predicted by Paul Ehrlich (1968) for the 1970s and 80s were relatively mild, their worst effects offset by foreign aid.

In 2000 a new revolution was promised, based on genetic engineering of farm livestock (including fish) to grow bigger and faster, and of crop plants to increase yields, create resistance to diseases, pests, and specific agrochemicals, and reduce fertiliser need by self-fixation of nitrogen. Some doomsters foresaw a nightmare world of featherless chickens, mice growing human organs on their backs, herbicide-resistant weeds, wild species out-competed by unnatural hybrids, farming controlled by monolithic biotech corporations, and plant and animal chimeras generated by mixing genes from unrelated organisms that Nature neither would nor could bring together. But no matter, more people could be fed and mass starvation held at bay for a few more decades while populations catch up again. Then what?

In 2002, however, the world had other problems. Climate change, linked to global heating, had been accepted by most scientists as a grave threat that was actually happening and rapidly developing, and a few governments were taking the first timid steps (the Kyoto protocol) along the road of damage limitation. The Genetic Revolution might well increase agricultural yields, but if in consequence world population continued to grow at 800 million per decade, all hopes of stabilising greenhouse gas emissions would vanish.

What have the green revolutions achieved? In 1750 most of the world's peoples lived on the edge of starvation (Ponting, 1993, pages 251–4). Now, 250 years later, five times as many survive in similar conditions, and most of the other 3 billion are not well off. The extra food has not ensured that people are better fed; on the contrary, it has promoted a vast increase in the numbers of hungry people. There is no reason to suppose that the Genetic Revolution, if it comes, will do otherwise.

So why bother with more revolutions? To sentimentalists they are good because they

0 – 1.8 million

BAHAMAS Area 13,900 km²

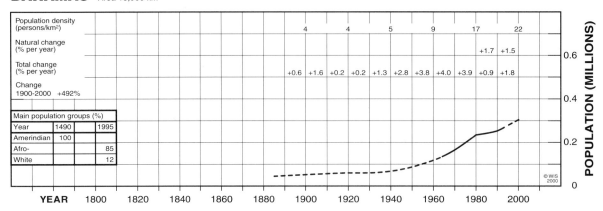

Population density (persons/km²)											4	4	5	9	17	22	
Natural change (% per year)															+1.7	+1.5	
Total change (% per year)										+0.6	+1.6	+0.2	+0.2	+1.3	+2.8	+3.8	+4.0	+3.9	+0.9	+1.8
Change 1900-2000 +492%				

Main population groups (%)		
Year	1490	1995
Amerindian	100	
Afro-		85
White		12

The shallow sea (baja mar) southeast of Florida encloses the Bahamian archipelago of more than 2000 limestone islands and cays. They were occupied by Arawak Indians when Columbus made landfall, his first in the New World, on San Salvador island in 1492. Spain deported many Indians as slave labour but the first European settlers were British, in the 17th century. Some of them were notorious pirates and wreckers, which led to conflict with, and brief occupation by, France and Spain. British cotton planters,

and American settlers in the 18th century, imported black slaves, who were freed in 1838. The scarcity of fertile soil and fresh water retarded agricultural development, but after World War Two tourism became big business. The islands are now an international financial centre. Following independence in 1973 increasing numbers of economic migrants have arrived from elsewhere in the Caribbean, especially Haiti. Censuses: every tenth year after 1970.

BAHRAIN Area 707 km²

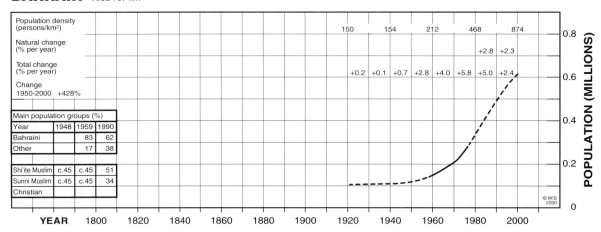

Population density (persons/km²)								150	154	212	468	874	
Natural change (% per year)											+2.8	+2.3	
Total change (% per year)							+0.2	+0.1	+0.7	+2.8	+4.0	+5.8	+5.0	+2.4
Change 1950-2000 +428%				

Main population groups (%)			
Year	1948	1959	1990
Bahraini		83	62
Other		17	38
Shi'ite Muslim	c.45	c.45	51
Sunni Muslim	c.45	c.45	34
Christian			

Bahrain Island is the largest of an archipelago of 35 small islands in the Arabian (Persian) Gulf, in a strategic position on ancient trade routes between Mesopotamia, India and the Far East. The Dilmun trading empire was based on Bahrain in the 3rd and 2nd millennia BC, giving way to Babylonian and other empires including the Portuguese, 1521–1602. The islands adopted Islam in the 7th century. When the Al-Khalifa family, which now rules Bahrain, gained control in 1783 the archipelago had a valuable pearl fishery, whose collapse coincided with the discovery of oil in 1932. Bahrain was a British protectorate (against threats from Persia and the Ottomans) from 1861 until it declared independence in 1971.

Prosperity arising from oil production and refining initiated a DC population surge from about 1950, accentuated by a huge influx of foreign workers who now comprise more than one third of the population. Native Bahrainis (who are mainly Arab with a minor Iranian component) protested violently in the 1990s against loss of jobs to immigrants and in favour of more democratic government. Crude oil resources are nearly exhausted but large natural gas reserves remain. Fresh water, nearly as scarce as oil, will be provided in future by desalination of sea water. Bahrain is preparing for the loss of oil income by diversifying into industry, commerce and tourism.

BARBADOS   Area 430 km²

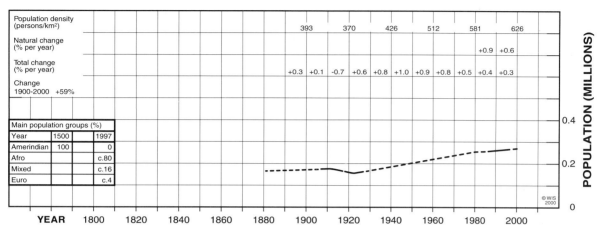

Population density (persons/km²)						393	370	426	512	581	626	
Natural change (% per year)										+0.9	+0.6	
Total change (% per year)						+0.3	+0.1	-0.7	+0.6	+0.8	+1.0	+0.9	+0.8	+0.5	+0.4	+0.3
Change 1900-2000 +59%												

Main population groups (%)		
Year	1500	1997
Amerindian	100	0
Afro		c.80
Mixed		c.16
Euro		c.4

© WIS 2000

POPULATION (MILLIONS)

0.4
0.2
0

YEAR   1800   1820   1840   1860   1880   1900   1920   1940   1960   1980   2000

The first occupants of this coral island in the eastern Caribbean were Arawak Indians who were displaced by warlike Carib Indians around 1200 AD. Spanish sailors are thought to have frightened away the Caribs, but it was a Portuguese sea captain who named the island "bearded" after the native fig trees with hanging bunches of aerial rootlets. Britain claimed the uninhabited island in 1625 and soon replaced its forests with highly profitable sugar plantations worked by African slaves. Revolts by the slaves were ruthlessly put down, and when slavery was abolished and they were freed in 1834 most were re-employed by the planters, in poor conditions. Independence came in 1966 after 3 years membership of the West Indies Federation. Sugar was overtaken by tourism in the 1970s as the island's main source of income. Censuses: irregular from 1911. The dip in population, 1911–1920, is indicated by 2 censuses. Barbados is very crowded and emigration is important.

alleviate hunger. To realists they allow an expansion of the hungry population. Starvation is postponed, and will kill more people when it finally strikes. The least painful and most reliable way to alleviate hunger is effective birth control. Remember, Europe's peasants enjoyed full bellies after the Black Death reduced the population, because those who survived had more good farmland.

8.9   Knowing When to Stop

In the 1970s the British Parliament passed laws to protect badgers. Lobbyists had claimed that people were cruelly persecuting the badger (Britain's largest wild predator, weighing 10–15 kg), and the protection they achieved was savage (up to 6 months in prison and/or a huge fine, for illegally harming a badger). Badger numbers had always been kept low by farmers to minimise damage to crops and livestock, and the new laws caused population growth at the rate of 77% in 10 years (Wilson *et al*, 1997). Farmers linked the badger population explosion to a tuberculosis epidemic in their cattle. Cereal crops, domestic gardens and small wild animals such as hedgehogs, bumble-bees, snakes and lizards suffered from excessive badger predation (Stanton, 1999). Badger lobbyists rejected all such allegations, claiming "They are not *proved*". Official attempts to obtain proof, one way or the other, by culling badgers in specified areas, were sabotaged by activists.

In parts of the USA, sentimentalists have prevented the hunting of wild deer, causing a population explosion and epidemic Lyme Disease (spread by deer ticks). Sympathy for a perceived underdog (Paxman, 1998), or for a Disney cartoon hero, drives the zealots to promote unlimited expansion of a single species, regardless of consequences. They are disrupting an ancient and satisfactory man/wildlife balance and they are not interested in stopping, or even considering it.

Failing to stop in time causes cumulative trouble. The outbreak of foot and mouth disease in Britain in 2001 revealed, to the surprise of many, the huge financial value of tourism to Britain. Visitors particularly enjoy touring through the English countryside, searching for remnants of the pastoral scenery immortalised by John Constable two centuries ago. The integrity of large parts of it was legally protected after World War Two by designations such

0 – 1.8 million

BELIZE British Honduras   Area 22,950 km²

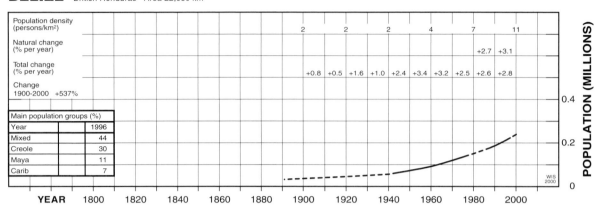

Population density (persons/km²)	2 2 2 4 7 11		
Natural change (% per year)	+2.7 +3.1		
Total change (% per year)	+0.8 +0.5 +1.6 +1.0 +2.4 +3.4 +3.2 +2.5 +2.6 +2.8		
Change 1900-2000 +537%			

Main population groups (%)		
Year		1996
Mixed		44
Creole		30
Maya		11
Carib		7

Amerindian tribes in Belize, on the eastern coast of Central America, took part in the invention of Mesoamerican farming some 8000 years ago. Abundant food stimulated the growth of advanced cultures including the Olmecs and then the Maya, who maintained great cities and vast religious monuments at the height of their development, 700–800 AD. But Mayan intensive agriculture required deforestation, which caused soil erosion, vulnerability to drought, food scarcity, population decline and cultural collapse (Ponting, 1993). Spanish colonists found Belize unattractive and the first European settlers were British buccaneers and loggers (in the re-grown forests that covered the whole country) in the 1600s. British Honduras was declared a colony in 1862. Boundary disputes with Guatemala were not resolved until the 1990s, a decade after Belize became independent in 1981.

Population began to rise sharply in the 1940s (the DC surge) with the arrival of death control in the form of modern medicines and agrochemicals. Belize is still the least developed Central American nation, with extensive unspoiled rainforests and coral barrier reefs.

BERMUDA   Area 53 km²

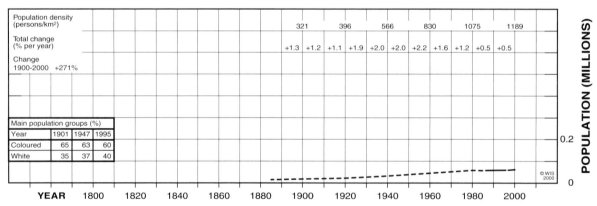

Population density (persons/km²)	321 396 566 830 1075 1189		
Total change (% per year)	+1.3 +1.2 +1.1 +1.9 +2.0 +2.0 +2.2 +1.6 +1.2 +0.5 +0.5		
Change 1900-2000 +271%			

Main population groups (%)			
Year	1901	1947	1995
Coloured	65	63	60
White	35	37	40

The tight cluster of more than 100 small islands and islets that is Bermuda, in the western North Atlantic, was uninhabited when the Spanish mariner Juan Bermudez discovered it in 1515 and remained so until 1609 when British colonists bound for Virginia were shipwrecked there and began a settlement. It became the first ever British colony in 1684. Slaves were imported to work plantations but during the 18th century the islands were best known as havens for pirates. Tourism became increasingly important after World War One. Favourable tax laws have made Hamilton, the capital, an international centre of finance and commerce. Golf courses occupy nearly 10% of the land area and the islanders enjoy the third highest income per person in the world. They rejected independence in 1995 and remain an autonomous British colony.

BHUTAN  Area 46,500 km²

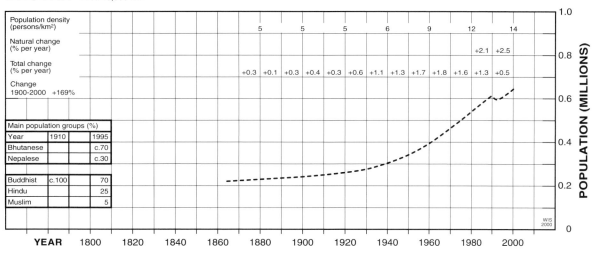

Bhutan is a small region of mountain peaks and deep valleys on the southeast slopes of the Himalayas, invaded and ruled by India, Tibet and China at various times in the second millennium AD. For centuries there was intrigue and strife between the lords of individual valleys, the most powerful of whom could be made ruler of the nation. When fighting spilled over into British India the British reacted in 1865 by expropriating the low-lying southern quarter of the country. A hereditary king was chosen in 1907, largely ending the internal conflicts. Bhutan's traditional isolationist policy was terminated in the 1960s but the present intention is to preserve the favourable ratio of natural resources to population by encouraging birth control. Some 80,000 illegal immigrants, mostly Nepali Hindus, were expelled between 1988 and 1993.

Nearly one third of the country is protected as wildlife sanctuaries (including one for yetis), natural forest reserves and national parks. They attract low-volume, high-price tourism, boosting the economy. Current population estimates vary widely, from the UN calculation of 2 million to the official government estimate of just over 600,000.

as *National Park* and *Area of Outstanding Natural Beauty*, but planning authorities have been unable to prevent creeping deterioration of the landscape ("my little bit of new development won't make any difference") by new housing, quarries, intensive farming, roads, even tourist facilities – all the consequences of growing population pressure. Unless a halt is called, the landscape will lose its appeal and the visitors will not come.

Scientific research and discovery now advance headlong, as myriads of research teams all over the developed world compete in Darwinian mode to patent their innovations and profit from them. In consequence, the effects of releasing new products into the environment are not always fully understood. Drugs such as thalidomide can have unexpected side effects. Genetically engineered organisms have the potential to disastrously contaminate and change the natural world. If human society was less crowded, moving at a slower pace, fuller consideration could be given to such risks. It would be easier, when in doubt, to stop.

If you know when to stop, and can act accordingly, you have the best possible chance of avoiding problems or disasters like HIV/AIDS, unwanted pregnancy, obesity, drunkenness, addiction, poverty, imprisonment, etc. The trouble is, you need to have puritan, or even superhuman, genes.

Most relevant to the theme of this book is knowing when (and how) to stop immigration into countries that are already overcrowded. In 2002 the West was trying to cope with fugitives from overpopulated developing nations who attempted by fair means and foul to enter and gain citizenship. Many thousands paid gangsters to smuggle them in, while equal or greater numbers claimed to be asylum seekers fleeing from persecution. Shiploads of up to a thousand refugees at a time sailed into Mediterranean ports or grounded themselves on beaches. In the USA, evading frontier guards on the Mexican border was an exciting challenge, often dangerous and sometimes, where baking deserts had to be crossed, fatal. Australia, in 2001, refused to allow boats carrying illegal immigrants to make landfall, turning them back. Sometimes the overcrowded boats capsized or sank, when hundreds drowned.

0 – 1.8 million

BOTSWANA   Bechuanaland   Area 582,000 km²

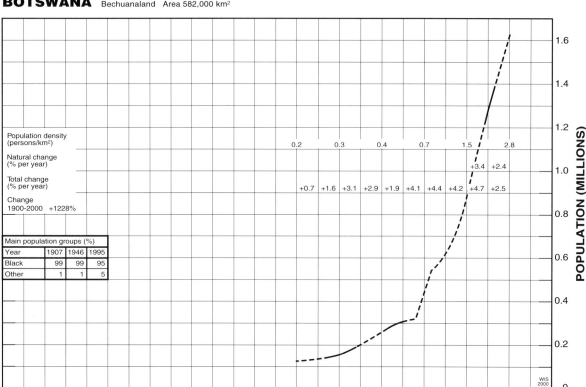

Population density (persons/km²)			0.2	0.3	0.4	0.7	1.5	2.8
Natural change (% per year)							+3.4	+2.4
Total change (% per year)			+0.7	+1.6	+3.1	+2.9	+1.9	+4.1	+4.4	+4.2	+4.7	+2.5
Change 1900-2000 +1228%								

Main population groups (%)

Year	1907	1946	1995
Black	99	99	95
Other	1	1	5

Bantu tribes moved south through eastern Botswana nearly 2000 years ago, mingling with the indigenous Bushmen and Hottentots and eventually, as settlement became denser, displacing them westward into the arid Kalahari. In the 1820s the genocidal rampage of Zulu tribes under Shaka, far to the southeast, dispossessed the Ndebele peoples who fled north and drove the local Bantu, the Tswana, towards the Kalahari. At the same time Boer farmers trekking north from the Cape began to harass other Tswana who asked the British to intervene. A British protectorate, Bechuanaland, was established in 1885.

After the Second Boer War, 1899–1902, the protectorate developed peacefully but agricultural setbacks caused by droughts and cattle diseases forced many of the male population to work in the South African mines. Nationalist agitation began in the 1940s, and 1966 saw a peaceful transition to independent Botswana. The economy prospered with the expansion of diamond mining.

Population growth in Botswana has been exceptionally fast, though somewhat irregular due to the migratory element. In the 1990s, however, HIV/AIDS took a hold on the population and by 2002 life expectancy for men and women had fallen to about 40 years. 39% of the adult population was HIV-positive.

BRITISH VIRGIN ISLANDS   Area 150 km²

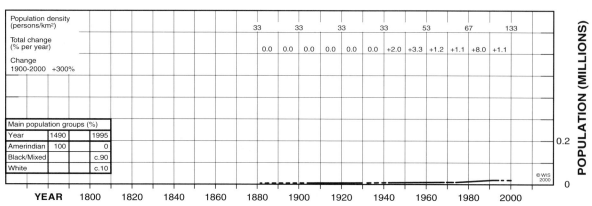

Population density (persons/km²)			33	33	33	33	53	67	133
Total change (% per year)			0.0	0.0	0.0	0.0	0.0	0.0	+2.0	+3.3	+1.2	+1.1	+8.0	+1.1
Change 1900-2000 +300%								

Main population groups (%)

Year	1490		1995
Amerindian	100		0
Black/Mixed			c.90
White			c.10

The 70 small islands and islets of the British Virgin group, a UK Overseas Territory just east of Puerto Rico, were inhabited by Arawak Indians when Columbus visited them in 1493. Spain controlled them until 1648 when Dutch settlement began, but Britain occupied and annexed them in 1672. Pirates infested them for a while. The settlers grew sugar cane on plantations worked by imported black slaves, but sugar eventually became unprofitable. In the 20th century the islands became an offshore financial service centre and are so popular with tourists that the government is acting to prevent over-development.

BRUNEI   Area 5,765 km²

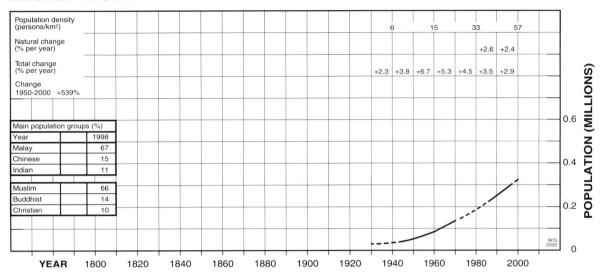

Population density (persons/km²)									6	15	33	57	
Natural change (% per year)											+2.6	+2.4	
Total change (% per year)								+2.3	+3.8	+6.7	+5.3	+4.5	+3.5	+2.9
Change 1950-2000 +539%													

Main population groups (%)		
Year		1998
Malay		67
Chinese		15
Indian		11
Muslim		66
Buddhist		14
Christian		10

POPULATION (MILLIONS)

0.6

0.4

0.2

0

WIS 2000

YEAR   1800   1820   1840   1860   1880   1900   1920   1940   1960   1980   2000

Brunei was an Islamic sultanate in the 15th century, controlling much of Borneo and adjacent islands. As its power waned in the 19th century, a dynasty of British assistants to the Sultan, the Brookes Rajas, developed the country. It became a British protectorate in 1888. Oil and gas, discovered in 1929, transformed the economy. The people now pay no income tax and enjoy free pensions, education and health care. 80% of food is imported. Brunei refused to merge with Malaysia in 1962 and became independent in 1984. The onshore oilfields are running down but big offshore fields have been found.

The percentage population growth rate (total change) has declined every decade since the 1950s, but the graph shows steady numerical growth.

Most of the target nations had laws strictly limiting the numbers of entrants, but immigration authorities were under politically correct pressure to be compassionate. Refugee support groups felt good as they welcomed anyone who was poor and unhappy, ethnic minorities saw Darwinian advantage in augmenting their numbers, and businesses used the opportunity to employ cheap labour, sometimes legally, sometimes illegally with no questions asked.

Governments were faced with difficult choices. To appear anti-immigrant invited the charge of racism, but the indigenous public in general resented the diversion of scarce public funds into accommodating and maintaining "bogus asylum seekers" (section 6.4). They doubted the sense of politically correct administrators who claimed that an unending influx of aliens, poor or not, is unquestionably good for the host country. In Britain they were angered by Foreign Secretary Cook's announcement that the British national dish was no longer roast beef, but chicken tikka massalla.

No-one, it seemed, was prepared to point out that a demand often made by refugee lobbyists: that nations should accept immigrant quotas in proportion to the size of their populations, meant that the most densely populated nations would become even more disproportionately congested. Excessive overcrowding alone is a logical reason to cry "our country is *full*" in the small target nations of Western Europe with population densities between 232 (Germany) and 468 (Netherlands). Their governments should know that the rising tide of economic migrants, spawned by enormous fast-growing populations (e.g. INDIA, PAKISTAN, BANGLADESH, which, according to the Micawberish Rule (section 6.4), are destined to grow ever poorer and more desperate), will never halt spontaneously, and very soon will swamp them unless "stop!" is cried. Unfortunately, logic is unimportant to sentimentalists, but the longer they manage to postpone the cry, the worse the ultimate consequences will be.

0 – 1.8 million

CANARY ISLANDS   Islas Canárias   Area 7,270 km²

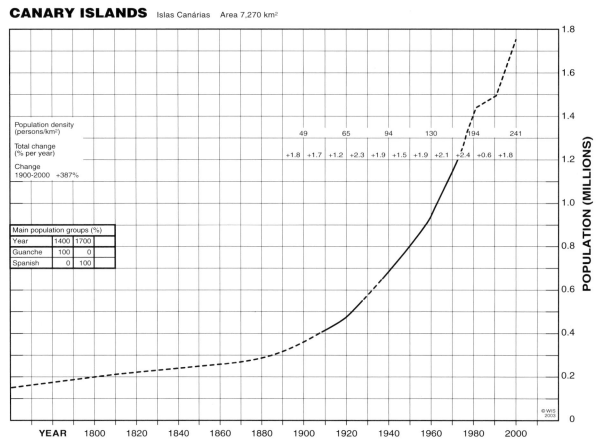

This archipelago of 13 volcanic islands off the northwest African coast was known to the Phoenicians and the Romans, who named it after the wild dogs (canes) that shared it with about 80,000 Guanches, a Berber tribe. Spain conquered the islands between 1400 and 1500, clearing the forests and enslaving the Guanches to work on sugar cane plantations. Conditions were so bad that no Guanche survived beyond 1600. Sugar cane is still grown, among other irrigated crops, especially bananas, but tourism has become the major industry. Censuses: about every ten years from 1910.

CAPE VERDE   Cabo Verde   Area 4,030 km²

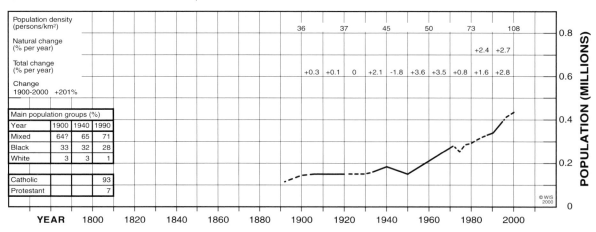

The Cape Verde archipelago of ten large and several small islands 600 kilometres off the West African coast opposite Senegal had no permanent residents when Portuguese navigators reached and claimed it in 1456. Portuguese settlers in the 15th and 16th centuries imported African slaves to work sugar and cotton plantations. The semi-arid climate required large-scale irrigation. Recurrent severe droughts together with a high birth rate forced much emigration.

Separatist movements became active in the 1950s and independence from Portugal was achieved in 1975. Many refugees from Angola arrived in the later 1970s. The birth rate is still high (total fertility rate 3.7 children per woman) and emigration reaches 10,000 in some years. There are few natural resources and the nation relies heavily on foreign aid to support its largely agricultural population.

Chapter 9

The Future

No-one who has looked at the 235 population graphs in this book, and has appreciated that the majority of them record many decades of rapid growth which today shows few signs of slowing down, can seriously believe that the graphs will go on rising indefinitely. Earth's resources are finite, and its ecosystems are failing to cope with even the present-day human population, most notably as regards atmospheric pollution (global heating).

But precedents have been set, since the 1970s, indicating the various ways in which a nation's population can stop growing. Slav nations of eastern Europe began to shrink when the collapse of the Soviet Union in 1991 relaxed the iron grip of communist domination. Prosperous nations of western Europe would now be experiencing population stability or decline, in accordance with the demographic transition theory, were it not for immigration. Fast-growing nations plagued by racial or religious divisions plunged into genocidal conflict and ethnic cleansing when their populations had risen to Violent Cutback Level (VCL, section 4.2). In southern Africa it may be (although the graphs do not yet show it) that HIV/AIDS has ended population growth in some nations. Developing nations afflicted by poverty, war, famine and disease, because their populations have outgrown their economies, would be experiencing population decline if they were denied foreign aid.

This chapter examines some of the scenarios that may curb human population growth on the planetary scale. They are, of course, speculative, but the speculation is better informed than usual because it includes concepts identified for the first time in this book, especially VCL and the WROG period. In 2002, as the world showed signs of polarising into opposing factions: Western secular conservatism against aggressively expanding Islam; the possibility of imminent VCL cutback on the global scale seemed uncomfortably real. That scenario is not examined here, but section 9.3 (b) touches on some of the possibilities.

9.1 Energy Shortage

"*As you commute along the motorway, keeping your place among lanes crowded with cars and trucks, a big helicopter lumbers overhead. High above it, the vapour trails of passenger jets criss-cross the blue sky. A high-speed train glides over a bridge, and on the far-off horizon a huge tanker and other ships seem hardly to move. The fields on one side of the motorway are dark green with lush grass leys; on the other side a tractor is spraying pesticide on the young wheat. After your day's stint in the plastics division of the giant chemicals firm you stop off at the supermarket, fill up with petrol and buy exotic fruit and vegetables from all over the world...*". An ordinary day, but what is the key factor that makes it all happen, without which you could revert to a life of poverty and hardship? Oil!

The Industrial Revolution began with the development of the coal-fired steam engine, in Britain in the mid-18th century. Coal was the supreme fossil fuel for some 150 years, but after the first oil well was drilled in 1859 the more versatile liquid fuel moved steadily to take the lead. In

CAYMAN ISLANDS   Area 260 km²

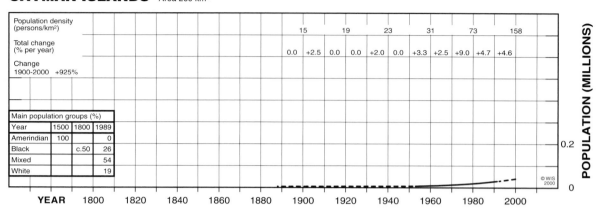

Population density (persons/km²)						15	19	23	31	73	158	
Total change (% per year)					0.0	+2.5	0.0	0.0	+2.0	0.0	+3.3	+2.5	+9.0	+4.7	+4.6
Change 1900-2000 +925%												

Main population groups (%)			
Year	1500	1800	1989
Amerindian	100		0
Black		c.50	26
Mixed			54
White			19

© WIS 2000

Columbus landed on the Cayman Islands in the Caribbean Sea south of Cuba in 1503 and saw large numbers of turtles and marine alligators ("caiman" to the Arawak Indian inhabitants). The islands became bases for pirates and privateers, many of them British, and British settlement began in the 18th century. By 1800 negro slaves comprised half the population. The weak economy based on farming and fishing provoked constant emigration until after World War Two, when tourism began to boom and the Caymans became popular tax havens. They are still a British Crown Colony. The turtles, which had been almost exterminated, are now profitably bred and farmed.

CHANNEL ISLANDS   Area 194 km²

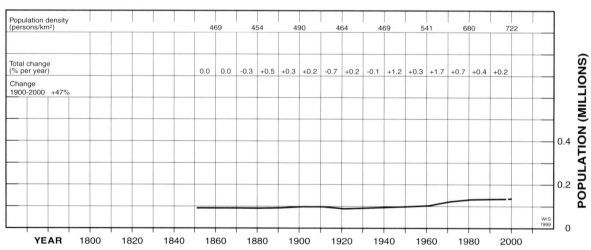

Population density (persons/km²)					469	454	490	464	469	541	680	722	
Total change (% per year)			0.0	0.0	-0.3	+0.5	+0.3	+0.2	-0.7	+0.2	-0.1	+1.2	+0.3	+1.7	+0.7	+0.4	+0.2
Change 1900-2000 +47%													

WIS 1999

The Channel Islands, close to the north coast of France, came to the British Crown with William the Conqueror, as part of his Duchy of Normandy, in 1066. Their benign climate, largely self-governing status and low taxes have proved so attractive to settlers that their population density has long been high. After World War Two the growth of tourism and the islands' tax haven value drew so many would-be settlers that financial and legal measures were taken to deter over-population. Censuses: every tenth year after 1851, except 1941.

COMOROS and MAYOTTE Areas 1,860 km² and 370 km²

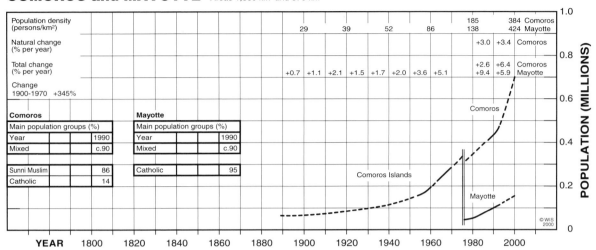

Population density (persons/km²)						29	39	52	86	185 138	384 Comoros 424 Mayotte	
Natural change (% per year)										+3.0	+3.4 Comoros	
Total change (% per year)						+0.7 +1.1	+2.1 +1.5	+1.7 +2.0	+3.6 +5.1	+2.6 +9.4	+6.4 Comoros +5.9 Mayotte	
Change 1900-1970 +345%												

Comoros			Mayotte		
Main population groups (%)			Main population groups (%)		
Year		1990	Year		1990
Mixed		c.90	Mixed		c.90
Sunni Muslim		86	Catholic		95
Catholic		14			

The four rugged volcanic islands of the Comoros group, between Madagascar and the African continent, were settled by seafarers from Indonesia early in the first millennium AD. Arab invaders converted the populations to Islam in the 7th century, and the islands were ruled by slave-owning Arab sultans when the first European sailors made landfall on them in the late 1500s. France annexed Mayotte in 1843 and declared protectorates over the other islands in the 1880s. They became a French colony, administered from Madagascar, in 1912. Their status changed to French Overseas Territory in 1947.

In a referendum the 3 Muslim islands voted for independence in 1975, whereas Catholic Mayotte opted to remain French. Since 1978 Comoros has been a Federal Islamic Republic, plagued by coups and secessionist attempts. Population growth in both territories has been astonishingly fast (if the post-1975 censuses and estimates upon which the graph is based can be believed). In consequence, the Comoros republic, with few natural resources and an economy based on subsistence agriculture, is one of the world's poorest nations. Mayotte, with French support, prospers by comparison. Both countries export large amounts of vanilla and ylang-ylang (an essential oil).

Comoros claims sovereignty over Mayotte. The islands' mixed populations are descended from African, Arab, Indonesian and Malagasy ancestors. The coelacanth, a 'living fossil' fish, is found in deep water off the Comoros coast.

1999 oil provided 32% of world energy, compared to coal's 21%. At the same time natural gas provided 22%, and nuclear fuels 6%. Biomass, mainly wood in the developing world, provided 14%, and the remaining 5% came from other 'renewable' sources (*National Geographic*, March 2001).

In section 7.3 I referred to the calculations which lead some oil experts to reliably predict a shortage of oil, beginning in the present decade, which will soon become devastating. By 2030 the potential demand will be two to three times the supply, and by 2050, four to six times. The search for alternative energy sources, which has proceeded in a desultory fashion while oil has been plentiful, will become urgent.

There is every reason to suppose that no plentiful and cheap oil substitute will be found. Reserves of natural gas are comparable, in energy value, to oil reserves (Campbell, 1997, 2003), but liquefying or compressing gas for use in vehicles is expensive, which is why its most popular use at present is generating electricity in power stations fed by gas pipelines. Coal reserves are much greater, and coal can be converted to gas by underground gasification, or to oil by hydrogenation, as in South Africa and in Germany during World War Two, but once again extra expense is involved, as it is in distilling oil from shale and tar sands. Production of liquid hydrocarbons by these methods is tedious and slow compared to the flow of petroleum under natural pressure from a well directly into a pipeline.

Oil of any origin is a feedstock for organic chemicals: plastics, paints, man-made fibres, agrochemicals, explosives, solvents, synthetic rubber, etc, but the other major non-renewable fuel, uranium, cannot so be used. Nor can the renewable systems that convert wind, wave, hydro, tidal and solar power into electricity. Only biomass can be transformed into oil, alcohol or plastics, but producing it as a crop is energy-expensive and reduces the area available for growing food (section 6.5). However, the mountains of organic waste, of domestic, agricultural or industrial origin, that developed nations produce, could theoretically be transformed into oil and

0 – 1.8 million

CYPRUS Area 9,250 km² (Greek 5,915 km² / Turkish 3,335 km²)

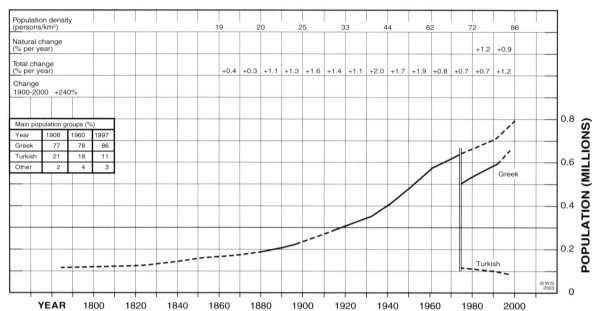

Population density (persons/km²)						19	20	25	33	44	62	72	86
Natural change (% per year)												+1.2	+0.9
Total change (% per year)						+0.4	+0.3 +1.1 +1.3 +1.6 +1.4 +1.1 +2.0 +1.7 +1.9 +0.8 +0.7					+0.7	+1.2
Change 1900-2000 +240%													

Main population groups (%)

Year	1908	1960	1997
Greek	77	78	86
Turkish	21	18	11
Other	2	4	3

The east Mediterranean island of Cyprus ('copper' in Greek) has endured a long succession of colonists including Myceneans, Phoenicians, Assyrians, Greeks, Egyptians, Romans, Byzantines, Crusaders, Templars, Venetians, Turks and British. Turkey ceded control of Cyprus to Britain in 1878. Competition between Greek and Turkish residents intensified in step with the rising population. In 1955 the Greek community, which had wanted 'enosis' (union with Greece) for more than a century, began guerrilla harassment of the British authorities. In 1960 Britain withdrew and Cyprus became independent. By 1964 the fighting between Greeks and Turks was so vicious that a UN peacekeeping force was sent in. The subsequent armed stalemate ended in 1974 when Turkey invaded to forestall seemingly imminent 'enosis'. Since then the island has been split between an agricultural Turkish north and a Greek south prospering through booming tourism and light industry. Mistrust and hatred divide the communities.

Partition, the traditional solution to interethnic civil strife, has ended the killing but has not brought reconciliation. Cyprus provides a case history of an ethnically divided, intrinsically competitive, multicultural population rising to near its Violent Cutback Level (VCL). Decisive civil war, which would naturally end with population reduction or ethnic cleansing, has been averted in the short term by partition and physical separation of the two human tribes, Christian Greeks and Muslim Turks. Censuses: at irregular intervals of 10 years or longer between 1881 and partition. Estimates of population in the Turkish north, which has seen much immigration since partition, differ widely. If Turkish soldiers and immigrants are included it may exceed 200,000.

other chemicals by pyrolysis (heating in the absence of oxygen).

Hydrogen is claimed by some to be the "non-polluting fuel of the future" (because it combusts to water), but several decades of research by hydrogen enthusiasts have so far failed to solve basic problems of future production costs, storage and transport, and of the fuel cells in which it would be most effectively used (Hoffmann, 2001). Liquefaction processes "consume about a third of hydrogen's energy content". At present the hydrogen is obtained by separation from natural gas. In the future the pure hydrogen required by fuel cells could be obtained by solar-powered electrolysis of water, but in the Los Angeles Basin about 2010, Hoffmann reports, 250 square miles of land covered by photovoltaic cells would be needed to provide hydrogen fuel for the Basin's 14 million vehicles.

As a convenient energy source, oil is uniquely cheap and effective. Westerners 'hop into the car' at a moment's notice to fetch the kids from school, visit the supermarket, or call on friends. Kenneth Deffeyes (2001) describes his awakening to the realities of energy generation without fossil fuels, pedalling on "a bicycle frame hitched to an electric generator wired to a light bulb". With a 100-watt bulb, "it took a sustained serious effort to keep the bulb glowing. I couldn't light up a 200-watt bulb".

When the price of oil threatens to skyrocket, people will react with bewilderment and anger. The possibility of imminent oil shortage has not been a consideration in forward planning. New airports are proposed, the aircraft industry is investing heavily in new models,

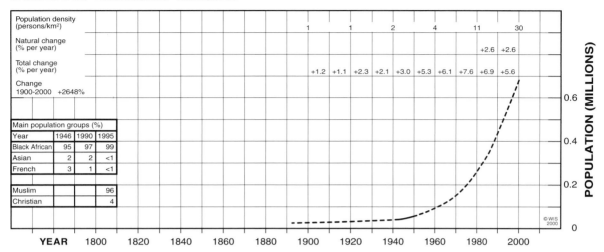

DJIBOUTI French Somaliland   Area 23,200 km²

Population density (persons/km²)						1	1	2	4	11	30			
Natural change (% per year)										+2.6	+2.6			
Total change (% per year)						+1.2	+1.1	+2.3	+2.1	+3.0	+5.3	+6.1	+7.6	+6.9 +5.6
Change 1900-2000 +2648%														

Main population groups (%)			
Year	1946	1990	1995
Black African	95	97	99
Asian	2	2	<1
French	3	1	<1
Muslim			96
Christian			4

POPULATION (MILLIONS): 0.6, 0.4, 0.2, 0

YEAR   1800 1820 1840 1860 1880 1900 1920 1940 1960 1980 2000

© WIS 2000

Djibouti, at the south end of the Red Sea, was a semi-desert region sparsely populated by nomadic pastoralists in 1888 when France annexed it as part of French Somaliland. The population had been converted to Islam by Arab missionaries and traders in the 9th century. France founded Djibouti port and promoted construction of the railway to Addis Abbaba, completed c. 1910, that became Ethiopia's vital link to the sea. After World War Two the Issa tribe campaigned to join the country with Somalia but the Afar (Danakil) tribe, ethnically Ethiopian, resisted. France renamed the country "Territory of the Afars and Issas" in 1967 and conceded independence, as Djibouti, in 1977. In 1992 the Afars fought a civil war against the Issa-dominated government which severely damaged the economy, already weakened by droughts in the 1980s and dependent on aid from France.

The high rates of total population growth from the 1950s are linked to periodic influxes of refugees from wars in Somalia and Ethiopia and of migrants seeking work in Djibouti port.

NASA has grandiose plans for space exploitation, car ownership is one of humanity's most coveted goals, and trade in goods of all kinds is heavily dependent on long-distance road transport. Forecasts of agricultural production all assume intensive farming like today's, with plentiful supplies of hydrocarbon-derived fertilisers, agrochemicals, and fuel for tractors and other farm machinery.

Initially, perhaps, people will respond rationally to oil shortage. There will be a rush to develop the alternative methods of electricity generation that have been neglected for so long: nuclear power and the renewables. Electricity will be substituted for hydrocarbons wherever possible. Campbell (1997) envisages a relatively benign initial period in the developed world: "Energy saving… is enthusiastically embraced by the populace… Everyone rides a bicycle… The airline business crashes,… but few people needed to travel anyway: electronic communications having become highly efficient… Popular overseas travel ends". People will save, not to buy airline tickets, but to buy enough petrol for a local holiday in the car. The phrase 'highway robbery' will take on a new meaning: petrol theft.

Almost immediately, the world's poorest nations will be unable to afford oil, except for the ruling elites, their backers, and the military. Rural populations will continue, or return to, subsistence farming powered by human or animal muscle. Lacking artificial fertilisers and chemicals, such surpluses as they may produce will not be enough to feed large urban populations. Unless the city-dwellers too can find land to farm, they will starve. Alternatively, they will take food from country people by force, as happened in Europe in 1316: "The poor were dying in large numbers or turned to robbery in an attempt to get food; huge bands of starving peasants swarmed across the countryside" (Ponting, 1991). Population reduction will begin as soon as foreign aid dries up.

At a rate depending on the depletion of oil supplies, and on the success or otherwise of attempts to replace them by alternative energy sources, richer nations will find themselves in similar trouble. Carrying capacity will become critical as crowded countries, lacking the wherewithal to feed all their people, collapse into chaos.

As the restraints on growth of all kinds strengthen, the global WROG period will draw

0 – 1.8 million

DOMINICA Area 748 km²

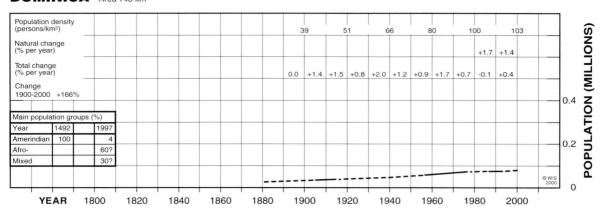

Population density (persons/km²)										39		51		66		80		100		103	
Natural change (% per year)																		+1.7		+1.4	
Total change (% per year)										0.0	+1.4	+1.5	+0.8	+2.0	+1.2	+0.9	+1.7	+0.7	-0.1	+0.4	
Change 1900-2000 +166%																					

Main population groups (%)		
Year	1492	1997
Amerindian	100	4
Afro-		60?
Mixed		30?

Mountainous Dominica Island clothed in tropical forest, in the eastern Caribbean, was sighted by Columbus on a Sunday (Domenica) in 1493. The indigenous Carib Indians, who had earlier expelled the less warlike Arawaks, fiercely opposed European attempts to settle. France and Britain fought over possession of the island, which became undisputedly British in 1805. Sugar plantations were established where the steep terrain permitted, only to be replaced by more profitable bananas. Dominica was federated with the Leeward Islands to 1939, with the Windward Islands to 1958, and with the West Indies until 1962. It became fully independent in 1978. Much of the original forest remains in the rugged volcanic interior. The population has been almost stable since 1980 in spite of the excess of births over deaths (natural change), indicating significant emigration.

EAST TIMOR Area 14,900 km²

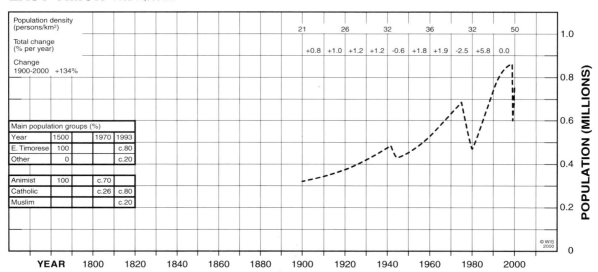

Population density (persons/km²)							21		26		32		36		32		50	
Total change (% per year)							+0.8	+1.0	+1.2	+1.2	-0.6	+1.8	+1.9	-2.5	+5.8	0.0	
Change 1900-2000 +134%																		

Main population groups (%)			
Year	1500	1970	1993
E. Timorese	100		c.80
Other	0		c.20
Animist	100	c.70	
Catholic		c.26	c.80
Muslim			c.20

When the first Europeans, Portuguese traders, reached Timor in 1512 they found an island (500 kilometres north of Australia) famed for its mountain forests of fragrant sandalwood, which for many centuries had been exported to China, India and Persia. Tradition held that the original Melanesian inhabitants had been expelled by Malay invaders from Indonesia. Portuguese and Dutch merchants competed for trade and in 1749, after more than 2 centuries of sporadic fighting between Portuguese, Dutch and Timorese, Portugal claimed the eastern half of Timor. Native resistance continued until the early 20th century, but the Portuguese language and Catholic religion took strong hold.

After the Japanese invasion and occupation (1941–45), during which some 60,000 Timorese were killed, Portugal reclaimed East Timor and began to prepare it for limited autonomy until, unexpectedly, the Portuguese withdrew from all their colonies in 1975. Later in the same year, supposedly fearing a communist takeover, Indonesia occupied the country by force, beginning a reign of terror that caused the deaths of 100,000 to 200,000 people by 1981. Guerrilla warfare and brutal repression continued and some 150,000 settlers were imported from the overcrowded islands of central Indonesia, to 'Indonesianise' and 'Islamise' the territory. Administrative and commercial jobs went to the newcomers.

Political turmoil at home weakened Indonesian control in the 1990s. In 1999, in a UN-sponsored referendum, East Timor voted for independence. Open war broke out between militias, towns were looted and burned, several thousand East Timorese were killed, and there was a huge exodus of natives and newcomers. In 2000 the country was in UN-run limbo and about 130,000 East Timorese refugees were still held in militia-controlled camps in West Timor (reduced to about 85,000 in 2001). When independence came in 2002, East Timor depended on foreign aid, with 70% unemployment, but offshore oilfields were being developed.

EQUATORIAL GUINEA   Spanish Guinea, Rio Muni, Fernando Pó   Area 28,050 km²

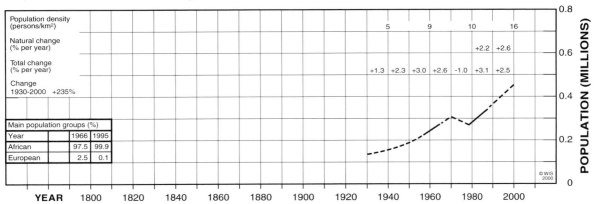

Equatorial Guinea on the western coast of Central Africa consists of an island group (Fernando Pó, now Bioka) and a mainland area (Rio Muni). Portuguese navigators in 1470 found them inhabited by Bantu tribes, Bubi on the islands and Fang on the mainland. Portugal ceded the territory to Spain in 1778. Spanish Guinea exported cocoa from huge plantations on the islands and timber from Rio Muni, using imported workers from West Africa under near-slavery conditions well into the 20th century.

Mounting Fang nationalism forced Spain to grant independence in 1968. The first president was a Fang, Macias Nguema, who began a reign of terror like that of Idi Amin in Uganda. The Spanish residents fled. Political and tribal rivals were executed. Between 1970 and 1979 more than 100,000 people were killed or fled the country. Nguema fell to a military coup and was executed in 1979, by which time he had destroyed the economy. Recovery is under way but poverty is widespread as population outgrows resources.

to a close. Overcrowded urbanised multicultural nations will be particularly vulnerable to demographic catastrophe (section 7.3). The likely course of post-WROG events is explored in section 9.3.

9.2 The Malthusian Alternative

This is the 'business as usual' scenario. Hubbert and Campbell are wrong. Oil, oil substitutes and renewable energy remain plentiful and cheap indefinitely. The WROG period endures for a generation or more, but it does end, progressively, as the rising populations of nation after nation reach VCL (section 4.2) and collapse into genocide and/or ethnic cleansing.

In the West in 2002, certain prevalent attitudes and developments were hastening the arrival of VCL breakdowns. It was politically correct to deplore cultural and/or ethnic homogeneity (as when the director of the BBC complained, in 2000, that his organisation was "hideously white") and to advocate multiculturalism (section 7.2).

Multiculturalist zealots ignore, or have not noticed, the fact that most of the recent and ongoing wars afflicting the planet have been civil conflicts within multicultural nations, in which the combatants were rival religious, political or racial groups (for examples, see SUDAN, WEST BANK, IRELAND, IRAQ) seeking to dominate or take possession of a valuable resource. There may be particular targets, such as oilfields or diamond mines, but in most cases the resource is land: the country itself, or part of it, in which the combatants live. As was demonstrated in BOSNIA, KOSOVO and RWANDA, among others, each side has two aims: to win control of the territory, and to get rid of the opposition (by genocide and/or ethnic cleansing). Usually neither side succeeds, because the UN, NATO or some other organisation intervenes. But complete elimination of the opposition, resulting in a homogeneous society, is vastly advantageous to the victor (see CANARY ISLANDS, TASMANIA, HAITI).

The rapid introduction of significantly large numbers of cultural aliens into previously homogeneous populations, as is now happening over much of the developed world, creates arenas in which Darwinian competition begins and intensifies (section 4.2). Examples are given in section 5.8. The combination of rising population with religious or racial antagonism

0 – 1.8 million

ESTONIA   Area 45,200 km² (after 1917)

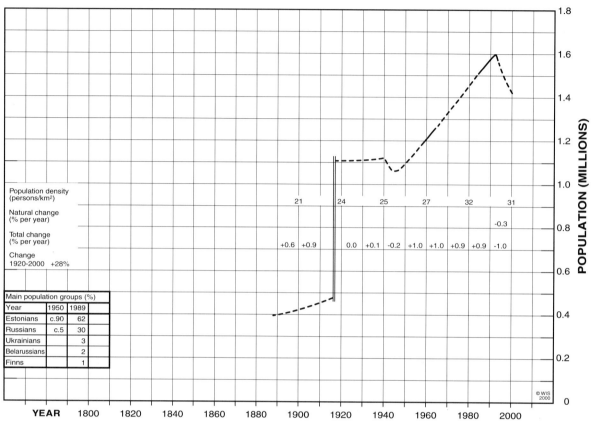

Population density (persons/km²)	21	24	25	27	32	31
Natural change (% per year)					-0.3	
Total change (% per year)	+0.6 +0.9	0.0 +0.1 -0.2 +1.0 +1.0 +0.9 +0.9 -1.0	
Change 1920-2000 +28%			

Main population groups (%)		
Year	1950	1989
Estonians	c.90	62
Russians	c.5	30
Ukrainians		3
Belarussians		2
Finns		1

For most of its history Estonia was closely linked to Scandinavia. In the first millennium AD the Finn-Slav inhabitants formed the first independent state, which was conquered by Danish Vikings in the 9th century and prospered from the 13th century within the Hanseatic League of German maritime traders. Sweden occupied Estonia in the 16th and 17th centuries, losing it to Tsarist Russia in 1721. It remained part of the Russian Empire until, following the Bolshevik takeover in 1917, an enlarged independent state was established.

Soviet Russia recovered Estonia in 1940 (and deported or murdered 100,000 Estonians), lost it to Germany in 1941, and re-conquered it in 1944. The Soviets forced collectivisation of agriculture and large-scale Russian immigration. When the Soviet Union collapsed in 1991 Estonia became independent. Population began to decline.

With its low population density the country has plentiful agricultural and forest resources, oil shale and peat, but the economy has yet to recover from the communist era.

FALKLAND ISLANDS   Las Malvinas   Area 12,170 km²

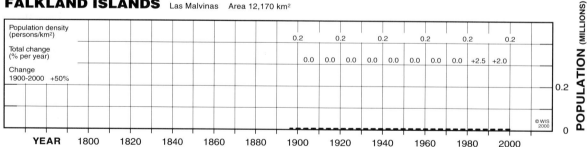

Population density (persons/km²)	0.2	0.2	0.2	0.2	0.2	0.2
Total change (% per year)	0.0	0.0	0.0	0.0	0.0	0.0	0.0	0.0	+2.5	+2.0
Change 1900-2000 +50%			

British and Dutch sailors visited the uninhabited Falklands archipelago, of some 200 bleak windswept islands and islets in the South Atlantic ocean 600 kilometres north-east of Cape Horn, between 1592 and 1690. The first French and British settlers gave way to Spain in the 1770s. Argentina claimed the islands when it became independent of Spain in 1816 but Britain annexed them as a Crown Colony in 1833. They are now an autonomous British dependency with a population of British descent. Argentina never

gave up its claim to 'Las Malvinas' and occupied them militarily in 1982, only to be expelled 2 months later by a British naval force. The military garrison outnumbers the civilian population.

In the past the economy benefited from whaling and sheep farming, but most income now derives from the sale of licences to fish in Falkland waters. Exploration for offshore oil is planned. The population is so small that large percentage changes represent only a few hundred people.

FAROE ISLANDS Sheep Islands Area 1400 km²

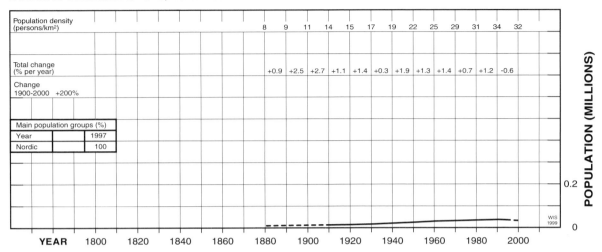

| Population density (persons/km²) | | | | | | 8 | 9 | 11 | 14 | 15 | 17 | 19 | 22 | 25 | 29 | 31 | 34 | 32 |

| Total change (% per year) | | | | | | +0.9 | +2.5 | +2.7 | +1.1 | +1.4 | +0.3 | +1.9 | +1.3 | +1.4 | +0.7 | +1.2 | -0.6 |

Change 1900-2000 +200%

Main population groups (%)		
Year		1997
Nordic		100

POPULATION (MILLIONS)

0.2

WIS 1999

0

YEAR 1800 1820 1840 1860 1880 1900 1920 1940 1960 1980 2000

The Faroe Islands between Iceland and Norway were settled in the 9th century by pastoralists from Norway, anxious to escape the 'pirates and tyrants who were ravaging the mainland'. First they reared sheep, then in the 18th century as the population grew they concentrated on fishing their territorial waters, 200 miles wide. A Danish possession from 1380, the islands became semi-autonomous in 1948. Generous financial support from Denmark promoted a standard of living said to be the highest in the world in the late 1980s. In the early 1990s depletion of the fish stocks caused high unemployment and much emigration, but the birth rate remained relatively high for Europe.

precipitates arrival at VCL and the end of WROG. The West assumes, apparently, that it is immune from such a scenario of terminal catastrophe.

In Muslim countries it is politically correct to strive for homogeneity, an Islamic world, at the expense of other religions and societies which are treated as enemies by fundamentalists. To promote multiculturalism would be seen as aberrant, even subversive. Muslim expansionism is unashamedly Darwinian, and includes aggressive breeding as a weapon in the cause (section 4.4). Muslim militants and fundamentalists feel required to impose their beliefs on other people and other nations by force – a potent recipe for violent Malthusian breakdown, as in AFGHANISTAN. Given also the fast population growth in almost all Muslim nations (Tables 3.1 and 3.2), in spite of their chronic shortage of vital resources such as water and farmland, it is not surprising that Muslims are emigrating in rapidly increasing numbers to Western nations, testing their tolerance and threatening their stability (sections 5.7, 5.8).

In nations of sub-Saharan Africa their multicultural (colonial) past is blamed for their present woes, as what appears to be a return to tribalism (homogeneity) gathers pace. The WROG period has already ended in countries like RWANDA and BURUNDI, whose populations are at or near VCL, and would have ended in many others, impoverished by corruption and senseless conflict, were it not for foreign aid. Some nations are experiencing strong restraints on growth caused by the ravages of HIV/AIDS. All in all, those parts of sub-Saharan Africa that are racked by tribal strife, poverty and disease are likely to experience the end of WROG in the present decade.

Population densities in South American nations are so low that they should be able to avoid reaching VCL for several generations, if they can escape the kind of internal strife, driven by Micawberish poverty, that affects COLOMBIA. Central America and the Caribbean have higher population densities, and already in some nations, impoverished by fast population growth and anthropogenic disasters (see HONDURAS) only foreign aid and emigration prevents Malthusian collapse.

INDIA and several of its non-Muslim neighbours in South-east Asia have fast-growing multicultural populations and are already, in some cases, densely populated. Poverty in them is extreme, widespread and worsening, following the Micawberish Rule. In East Asia, JAPAN,

0 – 1.8 million

FIJI ISLANDS Area 18,300 km²

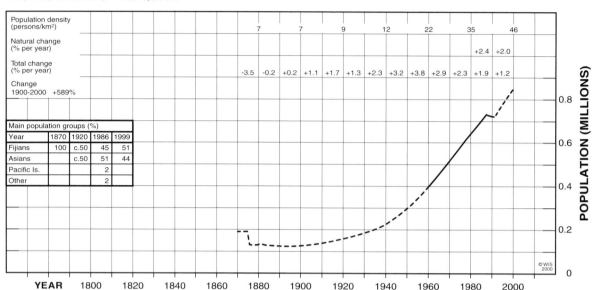

Population density (persons/km²)									7		7		9	12	22	35	46
Natural change (% per year)																	+2.4	+2.0
Total change (% per year)							-3.5	-0.2	+0.2	+1.1	+1.7	+1.3	+2.3	+3.2	+3.8	+2.9	+2.3	+1.9	+1.2
Change 1900-2000 +589%																		

Main population groups (%)				
Year	1870	1920	1986	1999
Fijians	100	c.50	45	51
Asians		c.50	51	44
Pacific Is.			2	
Other			2	

© WIS 2000

Polynesian and Melanesian seafarers discovered and settled the 800 islands of Fiji, 2000 kilometres north of New Zealand, nearly 3000 years ago, and soon hunted the islands' rare megafauna (giant pigeons, frogs and iguanas) to extinction. Population grew to around 200,000, a maximum (the VCL) dictated by incessant intertribal warfare involving ritual human sacrifice, cannibalism and casual torture. European sailors made brief visits in 1643 (Tasman), 1774 (Cook) and 1789 (Bligh), and commerce began in 1804 when, in a decade, the scented sandalwood groves were cut down and exported.

British missionaries reduced tribal fighting by conversion to Christianity, and the islands became a British colony in 1874. Old World diseases decimated the native population, as in 1875–76 when a measles epidemic killed one third of all Fijians. European settlers began developing sugar plantations in the later 19th century, importing 60,000 Indian indentured labourers between 1879 and 1916. By 1910 there were nearly as many Indians as native Fijians.

Following independence in 1970 racial tension developed between the Indians, who controlled the sugar-based economy, and the Fijians who owned most of the land but feared Indian domination. After two coups by native Fijians in 1987 nearly 70,000 Indians emigrated. A new constitution was adopted that favours a native Fijian government majority, but the ethnic tension continues and has damaged the important tourist industry. Rapid population growth is leading to problems of unemployment. In 2000 an attempted coup by native Fijians temporarily deposed the ethnic Indian Prime Minister and his government and further intimidated the Indian population. Censuses: roughly every tenth year after 1966.

FRENCH GUIANA Area 85,500 km²

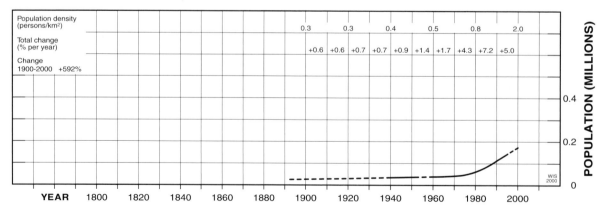

Population density (persons/km²)							0.3		0.3		0.4		0.5		0.8		2.0	
Total change (% per year)							+0.6	+0.6	+0.7	+0.7	+0.9	+1.4	+1.7	+4.3	+7.2	+5.0			
Change 1900-2000 +592%																			

WIS 2000

Europeans began settling in densely forested French Guiana, on the north coast of South America, among indigenous Arawak and Carib Indians, in 1604. The country became a French possession in 1817. African slaves were imported to work on sugar plantations, and the present population, about 90% Creole with some European and Amerindian, is largely descended from them. French convicts were sent to the notorious Devil's Island penal colony for nearly a century. The Eurospace rocket launching base began operating in 1979, coinciding with the national DC surge to cause population growth so rapid that the number of television sets per thousand inhabitants actually decreased from 1980 to 1995 – uniquely in the world. Still a French Overseas Department, it is the only non-independent territory in South America.

FRENCH POLYNESIA French Oceania, Tahiti, etc. Area 4,000 km²

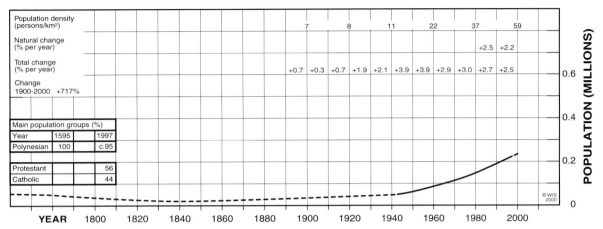

						7	8	11	22	37	59	
Population density (persons/km²)						7	8	11	22	37	59	
Natural change (% per year)										+2.5	+2.2	
Total change (% per year)			+0.7	+0.3	+0.7	+1.9	+2.1	+3.9	+3.9	+2.9	+3.0	+2.7 +2.5
Change 1900-2000 +717%												

Main population groups (%)

Year	1595	1997
Polynesian	100	c.95
Protestant		56
Catholic		44

French Polynesia consists of 5 separate archipelagos straddling a vast area of the South Pacific Ocean. The largest island is Tahiti in the Windward Islands (Iles du Vent). Polynesian seafarers settled the islands at some time before the 14th century AD. The first European visits were around 1600 and France claimed Tahiti in 1768. The islands were much visited by whaling ships and by 1800 the population was declining drastically as alcoholism and European diseases took hold. When France annexed more islands in 1840 the population of Tahiti had fallen from 40,000 in 1770 to only 9,000. The French imposed order and a DC population surge which began about 1940 continues today. Population density is still fairly low, but development and introduced species such as goats and rats have destroyed many unique native plants and animals. The economy depends on agriculture, but tourism is increasingly important. French Polynesia became a French Overseas Territory in 1958. Censuses: irregular from 1946.

SOUTH KOREA and TAIWAN are approaching population stability and are rich and culturally homogeneous but densely populated – an interesting balance. CHINA is homogeneous and industrialising, but is still basically poor, growing fast, and subject to anthropogenic disasters in the densely populated regions – another interesting balance, which suggests that Malthusian breakdown is more imminent in South-east Asia.

Ex-communist Europe is anomalous in many respects, with the advantages of stable or shrinking populations and low (for the developed world) population densities, but the communist legacy of inept administration, corruption, pollution and mistrust of the West has tended to restrict beneficial use of the rich natural resources that are, in many cases, available. Multicultural YUGOSLAVIA resolved an imminent VCL scenario in 1992 by partition, but serious ethnic/religious disputes are ongoing in GEORGIA, ARMENIA and AZERBAIJAN. In mainly Slav RUSSIA, the vicious conflict in Muslim Chechnya erupted when that state's population exceeded its VCL. The presence of UN and NATO peacekeeping forces is all that prevents renewed Slav-Muslim ethnic cleansing in BOSNIA and KOSOVO, whose populations reached VCL in 1992 and 1999 respectively. Potentially, some of the thinly populated Slav nations are in a good position to benefit from rational development of their resources, but aggressively expanding Muslim and/or Roma (Gypsy) populations in the Balkans and Caucasus will work towards Malthusian disaster in those regions.

In considering possible Malthusian futures worldwide a new and momentous factor must be included – the effects of climate change (section 6.1). They bid fair to be the ultimate anthropogenic disaster. Already some nations are experiencing unprecedented floods, heat waves, and prolonged wet or dry weather that damages agricultural productivity, but worst of all is likely to be sea level rise. Within a few decades, according to recent predictions (section 6.1) many millions of people will feel insecure in their low-lying coastal homes in BANGLADESH, EGYPT, the USA, and even Britain and the NETHERLANDS.

Refugees from developing countries will claim the right to asylum in the West, the main source, they will say, of greenhouse gas emissions that caused the oceans to rise and swamp them. Westerners will disagree, pointing out that if there is no room for the refugees in their own countries, it is because of their unrestrained fertility. They will argue that, were it not for

0 – 1.8 million

GABON
Gaboon   Area 267,700 km²

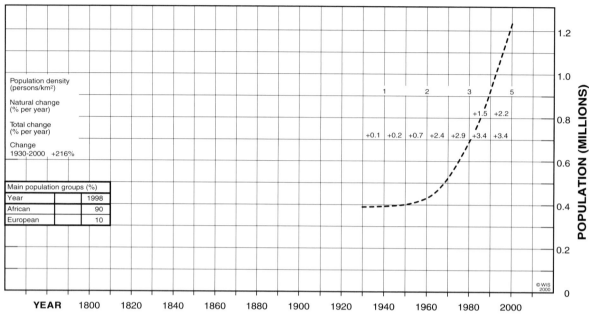

Pygmies were living in the Gabon rain forests when Bantu farmers began moving in from the north a thousand years ago. Among the last Bantu arrivals were the famously fierce Fang people. They obtained ivory and slaves for white traders who followed the first Europeans, Portuguese navigators in 1472, to explore the coastline. Slaves were procured from as far inland as the Central African Republic.

In the 'Scramble for Africa' in the late 19th century Gabon was claimed by France. It became part of French Congo in 1889 and part of French Equatorial Africa in 1910. Development was slow, with logging the main activity, but after independence in 1960 the mineral resources including oil, uranium and manganese were exploited on a large scale and boosted the economy. Population estimates show wide variation; even so-called censuses in 1970 and 1981 seem to have been very unreliable. In common with other equatorial French colonies rapid population growth, the DC surge, was delayed by the slow march of economic development.

GALÁPAGOS ISLANDS   Area 8,000 km²

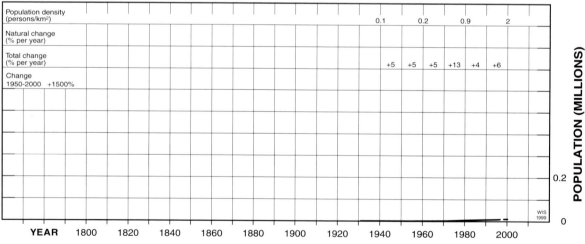

The Galápagos archipelago of 12 main and several hundred minor islands straddling the Equator in the eastern Pacific Ocean, was uninhabited, except by transient European whalers and pirates, before Ecuador annexed it in 1832. Charles Darwin in 1835 observed how certain of the indigenous animals, especially tortoises and finches, had developed physical variations specific to each main island, deducing thereby the principles of natural selection and evolution. The archipelago was declared a national park and nature reserve in 1934, and later a World Heritage Site.

In recent decades the pressures of tourism and fishing have accelerated immigration, settlement and development. Settlers have brought in dogs, cats, goats and rats which, as in so many oceanic islands, are threatening the survival of the native species including the famous giant tortoises, penguins and marine iguanas. Fishermen have attacked tourists and park officers, and held giant tortoises hostage, in disputes over quotas. Having settled in the islands, they find they cannot make a living under the national park rules. In 2001 a tanker supplying the settlers ran aground and leaked 200,000 gallons of oil, killing two-thirds of the marine iguanas on Santa Fe island.

Human occupation and population growth are fast destroying the unique environment that developed in the absence of people.

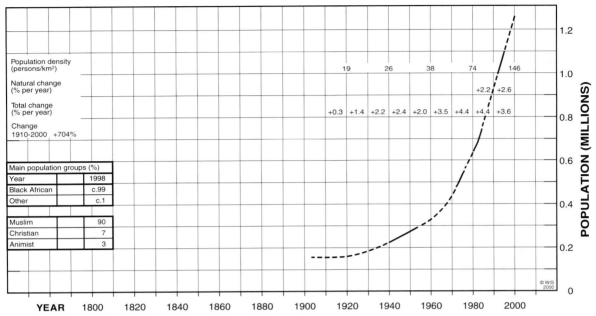

GAMBIA Area 8,600 km² (excluding water)

Population density (persons/km²)	19	26	38	74	146
Natural change (% per year)				+2.2	+2.6
Total change (% per year)	+0.3 +1.4 +2.2 +2.4 +2.0 +3.5 +4.4 +4.4 +3.6				
Change 1910-2000 +704%					

Main population groups (%)	
Year	1998
Black African	c.99
Other	c.1
Muslim	90
Christian	7
Animist	3

© WIS 2000

YEAR   1800   1820   1840   1860   1880   1900   1920   1940   1960   1980   2000

POPULATION (MILLIONS)

The first Europeans to set eyes on the Gambia river, important because it is navigable for 500 km from its mouth, were Portuguese sailors in 1455. Gambia was then a small part of the great Muslim empire of Mali. German, British and French trading posts were established on the river and were departure points for America-bound slaves. Britain declared the territory a colony in 1843 but the boundaries were not delimited until 1890. Subsistence farming supported the population, the main export being groundnuts. Independence came in 1965 and subsequent development has been mainly peaceful in spite of worsening poverty as the population expands without adequate resources.

There are 5 main tribal ethnic groups. An attempt to federate with Senegal, 1982–89, failed. Censuses: roughly every tenth year from 1944.

refugees and economic migrants, Western populations and their greenhouse gas emissions would be diminishing. They will calculate the contribution to greenhouse gas emissions made by immigrants who have settled and adopted the Western standard of living. Malthusian conflict and cutback, on a vast scale, will ensue.

9.3   Post-WROG Scenarios of Population Decline

Humankind lived through tens of millennia when populations increased very slowly, because it was normal for deaths to keep pace with births. Then came 250 years of the WROG period (section 4.5), when populations exploded because births vastly exceeded deaths. When the WROG period ends, global carrying capacity (section 4.1) will be so reduced that Malthusian cutback, the excess of deaths over births, will be on a huge scale. Traditionally, births have always been welcome, a 'blessing', whereas deaths have been unwelcome. For the first time ever, the human race will face the necessity of massive shrinkage. How will it cope with such a strange reversal of ancient custom?

9.3(a) The Oil Depletion Scenario

It is assumed here that Campbell's predictions of oil shortage beginning in the present decade (section 7.3) are valid. In a few years' time, he calculates, the demand for oil will exceed supply, causing large ongoing price rises. Every process and product that uses or depends on cheap oil will become much more expensive. Economic retrenchment will spread around the globe.

As inflation and recession begin to bite, developed nations that actually produce oil, including the USA, Canada, Russia, Norway and the UK, are likely to cut back on production to prolong the remaining life of their own reserves (like the prudent Dutch when in 1974 they

0 – 1.8 million

GAZA STRIP Area 363 km²

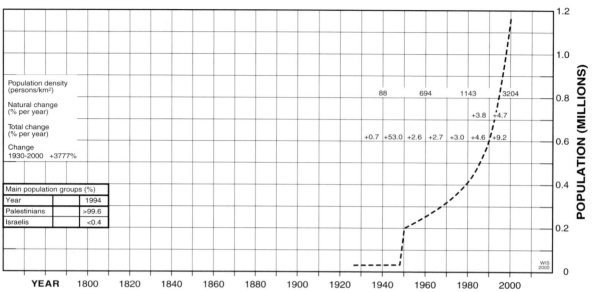

Population density (persons/km²)				88	694	1143	3204
Natural change (% per year)						+3.8	+4.7
Total change (% per year)		+0.7	+53.0	+2.6	+2.7	+3.0	+4.6	+9.2
Change 1930-2000 +3777%							

Main population groups (%)	
Year	1994
Palestinians	>99.6
Israelis	<0.4

The Gaza Strip, a small south-east Mediterranean coastal region once the domain of Philistines, then ruled by Romans, then Ottoman Turks, was part of the British mandated territory of Palestine from 1922 until 1948. During the Arab-Israeli War of 1948–49 Palestinian refugees increased the population six-fold. Egypt administered the Strip, with the help of a UN emergency force, until the Six Day War of 1967 when it was occupied by Israel. In 1994 a Palestinian administration began.

The Gaza Strip is effectively a huge refugee camp, with very high unemployment. There are several protected Israeli settlements.

Population growth is phenomenally rapid, even by Mediterranean Muslim standards (see the population density row). In 1999 70% of married couples had more than 8 children, explaining "Our weapon is our children", who will constitute the irresistible force necessary to establish a free Palestinian state. Israelis say the Gazans are breeding stone-throwers. The water in boreholes is saline or polluted by sewage. Even in the short term, conditions in Gaza are unsustainable and explosive. An *intifada* against Israel began in 2000 and is escalating as the Arab 'weapons' reach fighting age.

GIBRALTAR Area 6 km²

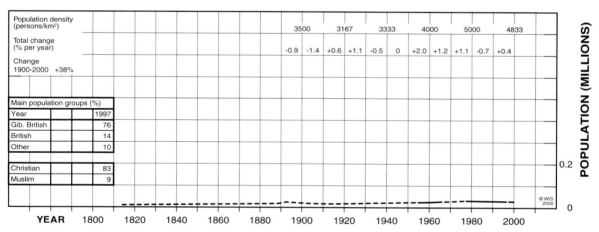

Population density (persons/km²)				3500	3167	3333	4000	5000	4833
Total change (% per year)	-0.9	-1.4	+0.6	+1.1	-0.5	0	+2.0	+1.2	+1.1	-0.7	+0.4
Change 1900-2000 +38%									

Main population groups (%)	
Year	1997
Gib. British	76
British	14
Other	10
Christian	83
Muslim	9

The Rock of Gibraltar was the northern of the two "Pillars of Hercules", as known to the ancient Greeks, which dominated the western exit from the Mediterranean Sea. Armies of Muslim Moors swept past it into Spain in 711 AD but the Spanish recaptured it in 1462. The town was sacked by Barbary pirates in 1540. Spain ceded it to Britain in 1713 and failed to recover it in a long siege, 1779–83. It became a British Crown Colony in 1830.

Spain still claims sovereignty over the territory, but in a 1967 referendum the vote was to remain British.

Gibraltar is a naval base and financial services centre. Tourism is important to the economy. The population is mainly descended from workers imported by the British after 1713, especially Italian, Maltese, Portuguese and Spanish, as well as British.

GREENLAND Area 2,175,000 km²

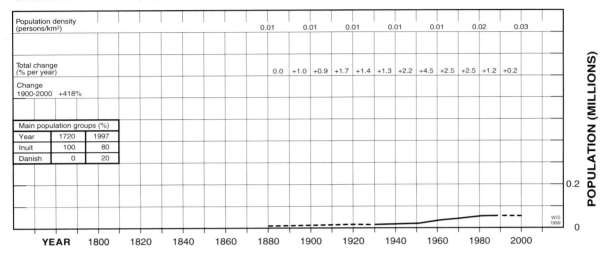

Population density (persons/km²)						0.01	0.01	0.01	0.01	0.01	0.02	0.03		
Total change (% per year)						0.0	+1.0	+0.9	+1.7	+1.4	+1.3	+2.2	+4.5	+2.5	+2.5	+1.2	+0.2	

Change 1900-2000 +418%

Main population groups (%)		
Year	1720	1997
Inuit	100	80
Danish	0	20

POPULATION (MILLIONS)

0.2

WIS 1999

0

YEAR   1800   1820   1840   1860   1880   1900   1920   1940   1960   1980   2000

Greenland, the world's biggest non-continental island, has an enormous central icecap up to 3 kilometres thick surrounded by a narrow discontinuously ice-free coastal strip of rocky or boggy tundra. Inuit peoples arrived about 5000 years ago. The first Europeans (Norse) settled along the south-west coast in the 10th century AD. A population of up to 5000 survived until the climatic deterioration of the 15th century, the 'Little Ice Age', when it died out. New Danish settlements were established from 1721. The island was a Danish colony until 1953, then part of Denmark, achieving autonomy in 1979.

The economy, and thus the population's future, depends heavily on fishing, especially for shrimp, the stocks of which are protected by a 200 mile zone of territorial waters. The effects of global heating, which could shrink the icecap and improve the climate, will be particularly interesting to Greenlanders.

reserved their Groningen gas field "for a rainy day"). There will be strong pressure on the lavishly-endowed main producers in OPEC not to follow suit. The early signs of unilateral 'might is right' action may already be appearing in 2002 with Iraqi oil the subject of serious Darwinian competition.

Rich and/or powerful nations will begin blatantly to ignore the 'rights' of poor or weak nations to an oil supply. There was a glimpse of things to come in 2001 when the USA rejected the 1997 Kyoto protocol (requiring it to cut carbon dioxide emissions) because "the American economy would suffer". The fact that consideration for others, for 'human rights', was an artefact of the WROG period (when easy living enabled political correctness to sideline Darwinian struggle) will become increasingly obvious. Victims of famine, disease, civil war and anthropogenic disasters in the poor nations, including refugees, will be neglected as foreign aid, which subsidised their DC population surges, dries up. Slogans like "Charity begins at home" will be used to justify the change of policy.

When powerful nations begin to act in ways that are brutally selfish, and benevolent international involvement in civil disputes (e.g. UN or NATO peacekeeping) ceases, potentially vicious wars that currently are held in check will flare up and run their natural courses, usually ending with elimination of the weaker side. BOSNIA, KOSOVO, RWANDA and BURUNDI are obvious candidates at present, because their populations are at or near VCL. In cases where the combatants have been restrained, so far, by fear of losing foreign aid, hostilities will be fought to a bloody finish.

The first one or two decades of oil shortage will see vigorous efforts to develop alternative energy sources. Wind, wave, solar and tidal power will generate significant amounts of electricity, at high cost (capital outlay and maintenance) and with little regard for the environmental niceties, such as unspoiled countryside, that were valued during the WROG period. As always, the physical requirements of *people* (voters and shareholders) will be paramount. If windmills in National Parks will reduce the frequency of power cuts, they will be built. The risk to health and safety posed by nuclear power stations will cease to be a consideration, and nuclear energy will come back into fashion. So will coal, the most

0 – 1.8 million

GRENADA   Area 344 km²

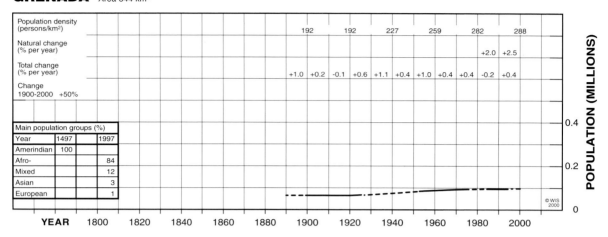

Population density (persons/km²)							192	192	227	259	282	288
Natural change (% per year)											+2.0	+2.5
Total change (% per year)						+1.0	+0.2	−0.1	+0.6	+1.1	+0.4	+1.0	+0.4	+0.4	−0.2	+0.4
Change 1900–2000 +50%												

Main population groups (%)		
Year	1497	1997
Amerindian	100	
Afro-		84
Mixed		12
Asian		3
European		1

YEAR 1800 1820 1840 1860 1880 1900 1920 1940 1960 1980 2000

POPULATION (MILLIONS) — 0.4, 0.2, 0

© WIS 2000

Grenada, in the eastern Caribbean, was the home of Carib Indians when Columbus sighted the island in 1498 and named it from a fancied resemblance to the surroundings of Granada in southern Spain. A party of British settlers was ejected by the Caribs in 1610. French settlers in 1650 drove the Caribs to commit mass suicide by jumping off sea cliffs. A British naval force took the island and its plantations worked by black slaves from the French in 1762. It became a British colony in 1783, repelled a French invasion in 1795, and enjoyed peaceful slow development as one of the Windward Islands group until independence in 1974. The notably corrupt first government was ousted in 1979 and its left-wing successors, with ties to Cuba and the Soviet Union, were overthrown by a US-led invasion in in 1983.

Censuses: irregular from 1904. Emigration has prevented significant population growth since 1980, in spite of the high birth rate. Tourism is now overtaking agriculture in importance to the economy.

GUADELOUPE   Area 1,705 km²

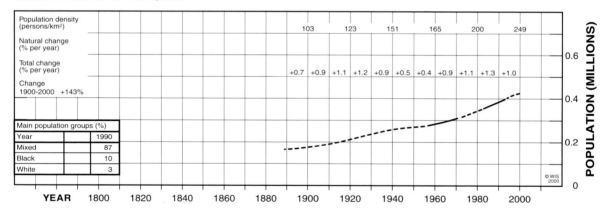

Population density (persons/km²)							103	123	151	165	200	249
Natural change (% per year)												
Total change (% per year)						+0.7	+0.9	+1.1	+1.2	+0.9	+0.5	+0.4	+0.9	+1.1	+1.3	+1.0
Change 1900–2000 +143%												

Main population groups (%)		
Year		1990
Mixed		87
Black		10
White		3

YEAR 1800 1820 1840 1860 1880 1900 1920 1940 1960 1980 2000

POPULATION (MILLIONS) — 0.6, 0.4, 0.2, 0

© WIS 2000

When Columbus visited mountainous Guadeloupe in 1493 the East Caribbean island group was inhabited by Carib Indians, who resisted Spanish attempts to settle in the early 1500s. In 1635 a French force expelled the Caribs and founded a colony with sugar plantations worked by African slaves. Britain and Sweden invaded the islands at intervals but in 1816 they became securely French. Since 1946 they have been an Overseas Department of France, with an economy based on bananas, sugar and, increasingly, tourism. Rainforests in the volcanic national park have been well protected.

GUAM Area 540 km²

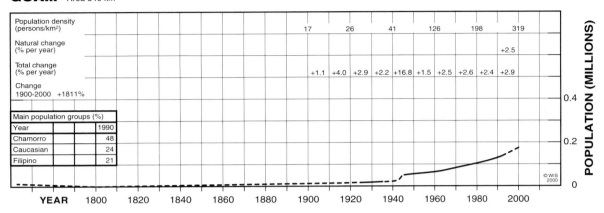

Population density (persons/km²)						17	26	41	126	198	319
Natural change (% per year)											+2.5
Total change (% per year)						+1.1	+4.0	+2.9	+2.2	+16.8	+1.5	+2.5	+2.6	+2.4	+2.9
Change 1900-2000 +1811%											

Main population groups (%)		
Year		1990
Chamorro		48
Caucasian		24
Filipino		21

POPULATION (MILLIONS) — 0.4, 0.2, 0

© WIS 2000

YEAR 1800 1820 1840 1860 1880 1900 1920 1940 1960 1980 2000

Guam is the southernmost and largest island of the Mariana Islands archipelago, in the western Pacific Ocean 2000 kilometres east of the Philippines. The indigenous Micronesian people arrived from Indonesia or the Philippines some 3,500 years ago. Magellan visited the coral/volcanic island of Guam in 1521 and claimed it for Spain. Ill-treatment and European diseases reduced the native Chamorro population from c. 80,000 in 1668, when Spanish settlement began, to only 1,500 in 1783. The USA acquired Guam in 1898 following the Spanish-American War. After the Japanese occupation (1941–44) in World War Two the USA established its main military base in the western Pacific on the island, more than doubling its population.

Guam is an unincorporated territory of the USA. Thanks to the US military establishment and its associated industries, and to tourism, the economy is flourishing and well able to support the island's high population density. The DC population surge began almost immediately after the American takeover. Censuses: every 10 years (except 1940) from 1930.

---

abundant fossil fuel, in spite of its major contribution to greenhouse gas emissions. Many coal-to-oil plants will be built, especially in countries with huge coal reserves, such as CHINA, the USA and RUSSIA. It has been suggested that arrays of solar cells covering much of the Sahara and other deserts, could supply all the world's electricity needs, but the practical and political problems of such a vast enterprise would likely be insuperable.

What will *not* happen is the burning of purpose-grown biomass crops in power stations in crowded developed countries. All their land will be devoted to growing food, because many nations, deprived of cheap oil, will no longer produce food surpluses for export. Air freight, even by airship, will become prohibitively expensive. One may suppose that marine transport, driven by ingenious hybrid power sources, will preserve a slower flow of goods between continents.

By about 2020 it will be clear whether or not sea levels are rising enough to threaten coastal lands and cities. If they are, belated serious attempts will be made to reduce the proportion of greenhouse gases in the atmosphere. Tree planting will be impractical until world population is much reduced, because the land will be needed for food crops. However, the oil industry has proved that carbon dioxide can be 'sequestrated' (removed from the atmosphere) by pumping it underground into porous rock strata. Large geological reservoirs from which oil and/or gas have been extracted are often ideal. Other proposals can only be desperate last resorts, such as seeding oceanic regions with iron to promote plankton blooms, or creating shallow 'scrubber' lagoons of lime-water which absorb carbon dioxide from the air as the wind blows over them. Calcium carbonate is precipitated, collected, and heated to drive off the carbon dioxide which is then fixed chemically in suitable rocks. A huge drawback is the energy needed to operate the lagoons which, the proposers calculate, would need to cover an area about the size of Britain! (*New Scientist*, 31 March 2001: not an 'April Fool', according to an editor). A more practical procedure would be to bake tree trunks in giant airtight ovens, producing useful alcohols and methane, and compressing the residual charcoal into briquettes which could be permanently sequestrated in suitable underwater environments.

In the West, as people become accustomed to the ending of cheap air travel to distant holiday resorts, their cherished personal mobility based on the internal combustion engine

0 – 1.8 million

GUINEA-BISSAU   Portuguese Guinea   Area 36,100 km²

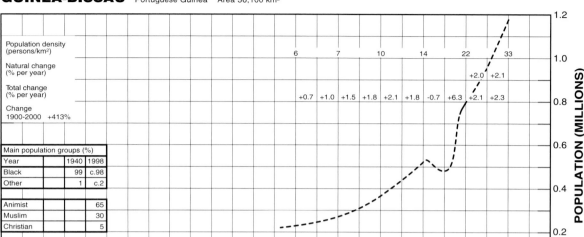

Guinea-Bissau, on the west African coast south of Senegal, was part of the Muslim kingdom of Mali when Portuguese sailors reached it in 1446. It became a centre for European slave-trading but the interior remained unsubjugated long after Portugal claimed it as Portuguese Guinea in 1879. Colonial development was slow but peaceful until separatists began guerrilla warfare in 1961. Thirteen years of draining civil war ended with independence from Portugal in 1974, when more than 100,000 refugees began to return. Subsequently, several left-wing governments have failed to lift the economy out of basic subsistence agriculture. In 1998–99 there was renewed civil war.

There are numerous ethnic tribal groups.

GUYANA   British Guiana   Area 215,000 km²

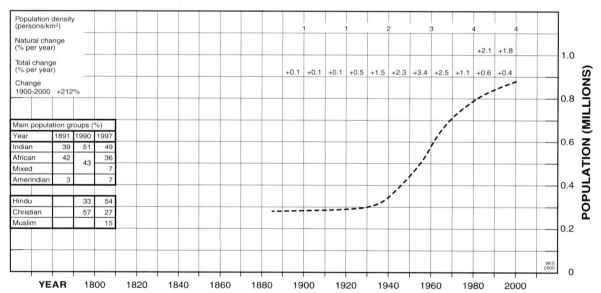

Columbus sighted Guyana, on the north coast of South America, in 1498, but it was not claimed by Spain. The first Dutch and French settlers brought in African slaves to work their sugar plantations. The native Arawak and Carib Amerindian tribes were displaced. Britain occupied the country in the Napoleonic wars and declared it British Guiana in 1831. Slavery was abolished in 1834 so Britain imported indentured labourers from India for the plantations. The country became independent, as Guyana, in 1966. The two main political parties are basically Indian and African; racial violence has accompanied allegedly rigged elections. 80% of Guyana is undeveloped tropical rainforest. Gold mining accounts for a quarter of the fragile economy. Population estimates after 1970 vary widely; some even suggest that the population is decreasing due to emigration.

HAWAII ISLANDS   Area 16,640 km²

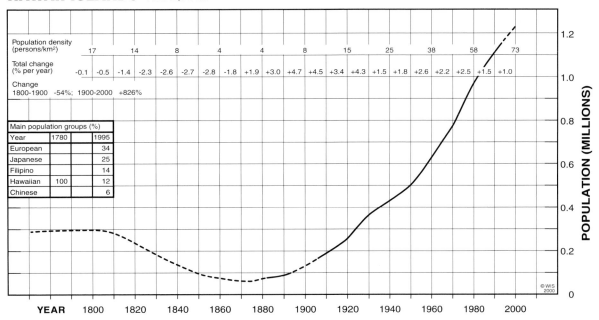

| Population density (persons/km²) | 17 | 14 | 8 | 4 | 4 | 8 | 15 | 25 | 38 | 58 | 73 |

| Total change (% per year) | -0.1 | -0.5 | -1.4 | -2.3 | -2.6 | -2.7 | -2.8 | -1.8 | +1.9 | +3.0 | +4.7 | +4.5 | +3.4 | +4.3 | +1.5 | +1.8 | +2.6 | +2.2 | +2.5 | +1.5 | +1.0 |

Change
1800-1900 -54%; 1900-2000 +826%

Main population groups (%)		
Year	1780	1995
European		34
Japanese		25
Filipino		14
Hawaiian	100	12
Chinese		6

© WIS 2000

POPULATION (MILLIONS) — 1.2, 1.0, 0.8, 0.6, 0.4, 0.2, 0

YEAR 1800 1820 1840 1860 1880 1900 1920 1940 1960 1980 2000

The Hawaiian Islands, a chain of gigantic volcanoes rising from the Pacific Ocean floor, were uninhabited by humans until about 300 AD when the first of several waves of Polynesian seafarers arrived. In 1778 the first European visitor, Captain Cook, found the islands densely populated and individual chiefs fighting to control the whole archipelago. Chief Kamehameha achieved this goal in 1795, but European settlement was already beginning, the sandalwood forests were destroyed and native workers were deported.

European diseases drastically reduced the aboriginal population.

Hawaii became a US protectorate in 1851, sugar plantations boomed and Asian labourers were brought in to work them. Population has grown fast since the islands became a USA territory in 1900 and the 50th state of the USA in 1959. Sugar, pineapples and other crops are grown but tourism is the main source of income. Much of the indigenous flora and fauna is endangered or already extinct.

will also be increasingly challenged by the scarcity and high price of fuel. Petrol rationing may be introduced to ensure that rich or powerful citizens cannot grab all the resource. Inventors will display wonderful ingenuity in harnessing a variety of power sources: batteries, fuel cells, alcohol, natural gas, vegetable oils, hydrogen, etc., to drive ever smaller and more efficient engines. Commuting long distances to work will become unfashionable, and offset to some extent by working electronically from home. The works bus, and small motor-bikes, very economical on fuel, will become popular again.

First World city dwellers, who think of food as something that comes from supermarkets, will begin to realise that it is actually produced on farms, by intensive and complex agricultural procedures for which vast amounts of oil-based products are essential (less so in the case of meat and milk, which cattle can synthesise directly from grass pasture). As food imports dry up, harsh realities will emerge. Nations will have to feed themselves. Farmers, long neglected by politicians as a minor component of society, will be told to provide all the food their countries need. They will be guaranteed first call on scarce oil, in the form of diesel fuel, petrol, agrochemicals and fertilisers.

Before long the shocking truth will become apparent: in the most densely populated developed nations, self-sufficiency in food is impossible without a drastic fall in the standard of living.

How will Westerners react when goods they had regarded as the basic necessities of life become scarce expensive luxuries? There are precedents. The populations of nations at war traditionally blamed the enemy for their deprivations and strove in unity to defeat him, anticipating that victory would bring a return of plenty, but in the post-WROG world, that opportunity is lacking. So is a tunnel with light at the end of it. People are slipping, more or less gradually but inescapably, into poverty. The closest precedents are to be found in the pre-

0 – 1.8 million

ICELAND Area 103,000 km²

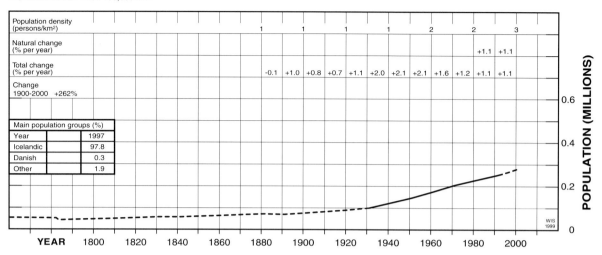

Population density (persons/km²)		1	1	1	1	2	2	3

| Natural change (% per year) | | | | | | +1.1 | +1.1 |

| Total change (% per year) | -0.1 | +1.0 | +0.8 | +0.7 | +1.1 | +2.0 | +2.1 | +2.1 | +1.6 | +1.2 | +1.1 | +1.1 |

Change 1900-2000 +262%

Main population groups (%)	
Year	1997
Icelandic	97.8
Danish	0.3
Other	1.9

Iceland, the world's 18th largest island, located on a 'hotspot' on the Mid-Atlantic Ridge, is a place of spectacular icecaps, volcanoes, hot springs and waterfalls, but is almost treeless. The first Norse colonists in the 9th century enjoyed a climate slightly warmer than the present and the population rose to nearly 80,000, but in the 'Little Ice Age' that lasted 4 centuries from 1430 survival became difficult and the population halved. In 1783 the Laki volcano belched out 30 billion tonnes of basalt lava and 90 million tonnes of sulphuric acid, enveloping the island in a poisonous haze that killed 20% of the people and 70% of the livestock. Iceland is richly endowed with 'renewable' (hydro and geothermal) energy, which generates some of the world's cheapest electrical power.

Farmland is scarce and overgrazing by livestock is causing serious soil erosion. The economy depends heavily on fishing. A huge but undeveloped resource is basalt rock. Europe's demand for crushed rock for roadstone and concrete is enormous, and it is a finite resource obtained by destroying hills and mountains. Iceland's rock is a sustainable resource, renewed with every volcanic eruption. Moreover, it could be obtainable at low cost by dredging the *sandur*, the barren coastal plains of glacial pebbles and gravel.

Iceland's natural population growth rate is the fastest in Europe outside the Muslim regions. It will hinder the national plan (2001) to eliminate fossil fuel use by replacing it with hydrogen produced using geothermal and hydroelectric energy.

ISLE OF MAN Area 571 km²

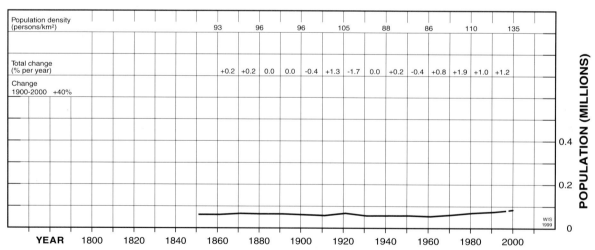

Population density (persons/km²)		93	96	96	105	88	86	110	135

| Total change (% per year) | +0.2 | +0.2 | 0.0 | 0.0 | -0.4 | +1.3 | -1.7 | 0.0 | +0.2 | -0.4 | +0.8 | +1.9 | +1.0 | +1.2 |

Change 1900-2000 +40%

A Norwegian possession from the 9th century, the Isle of Man, in the Irish Sea, became Scottish in 1266 and English in 1406. Migration, influenced by the island's semi-autonomous government and low taxes, is largely responsible for fluctuations in the essentially stable population. Crime is rare, partly because corporal punishment is legally allowed. Immigration is controlled to prevent over-population.

KIRIBATI and TUVALU Gilbert and Ellice Islands   Areas 710 km² and 26 km²

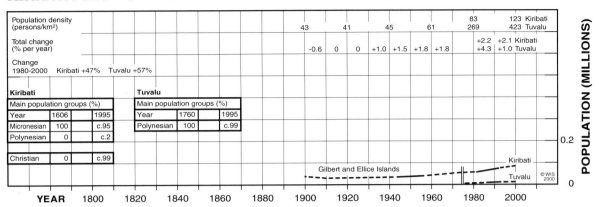

| Population density (persons/km²) | | | | | | | 43 | 41 | 45 | 61 | 83 269 | 123 Kiribati 423 Tuvalu |
| Total change (% per year) | | | | | | | -0.6 | 0 | 0 | +1.0 | +1.5 | +1.8 | +1.8 | +2.2 +4.3 | +2.1 Kiribati +1.0 Tuvalu |

Change 1980-2000   Kiribati +47%   Tuvalu +57%

Kiribati

Main population groups (%)		
Year	1606	1995
Micronesian	100	c.95
Polynesian	0	c.2
Christian	0	c.99

Tuvalu

Main population groups (%)		
Year	1760	1995
Polynesian	100	c.99

Gilbert and Ellice Islands

Kiribati

Tuvalu

© WIS 2000

POPULATION (MILLIONS)   0.2   0

YEAR   1800   1820   1840   1860   1880   1900   1920   1940   1960   1980   2000

Isolated archipelagos of mostly coral islands in the west-central Pacific, Kiribati (33 islands) and Tuvalu (9 islands) were settled by seafaring Micronesians and Polynesians respectively at least 2000 years ago. Overpopulation was inhibited by savage inter-island warfare, the warriors wearing armour made of braided coconut fibre. European sailors reached Kiribati in 1606 and Tuvalu in 1765; whaling ships frequented the islands in the 19th century. Populations were greatly reduced by European diseases and by forced emigration to work on plantations elsewhere. Britain established the Gilbert and Ellice Islands protectorate in 1892. In 1916 it became a British colony. Japanese forces occupied the Gilberts in World War Two but were driven out by Americans in 1943. Ethnic differences, Micronesians vs. Polynesians, provoked the breakup of the colony into separate independent republics in 1975–79.

Banaba (Ocean) Island in Kiribati was laid waste by phosphate miners (1900–1980) and its population of 2000 was removed to Fiji after World War Two. In 1988 and again in 1995 some 5000 people were relocated from Kiribati's overpopulated main island, South Tarawa, to other islands within the republic. Both republics, consisting largely of coral atolls, fear that sea level rise caused by global heating will render them uninhabitable in a few decades, first by seawater intrusion into their freshwater aquifers, and later by actual submergence. Tuvalu, with its high population density, is particularly vulnerable, and in 2001 the government announced evacuation plans.

WROG world of Darwinian dog-eat-dog, and in modern nations whose populations have exceeded, or are at, their VCL (see AFGHANISTAN).

At this crunch-point in the post-WROG scenario Western civilization will undergo a sea change. Political correctness will finally expire. Straightforward Darwinism will replace it, with all the selfishness, corruption, dirty tricks and cheating that distinguish the genuine struggle to survive, in which the well-being of one individual or group can only be achieved by frustrating or vanquishing another individual or group. The underdog no longer attracts sympathy; he is weak, and liable to be eliminated.

At an early stage of the sea change, when some vestiges of civilised standards survive, human rights and the sanctity of human life will be casualties. The death penalty will be introduced for any seriously antisocial act or lifestyle (such as drug addiction) because '*when a nation is painfully overburdened with people, life is a privilege, not a right*'. There will be no long expensive trials or appeals, and, as usual, a few of those executed will be innocent, but society will benefit from the severe depletion of its antisocial element. The nanny state will give way to a regime favouring quick and inexpensive corporal punishment for lesser offences. Euthanasia, abortion, and infanticide for handicapped babies will be legalised, even encouraged, and severe restrictions on family size enforced.

Central to public thinking will be the perceptions of overcrowding and resource scarcity. These are the conditions, the precedents show, in which weaker groups are harassed and ousted by stronger ones (not always minorities by majorities, e.g. RWANDA). Human nature *in extremis* being Darwinian, the most obviously alien groups, whether in appearance (e.g. skin colour) or behaviour (e.g. religion) are most likely to be targeted. When, as time goes on, scarcity becomes acute, as when vital imports are delayed, or when bad weather causes a poor harvest followed by starvation, or in many other circumstances including sea-level rise, the harassment of the weak will escalate into genocide and ethnic cleansing, the VCL scenario. Multicultural societies, born of political correctness, will be homogenised by Darwinian violence.

LIECHTENSTEIN Area 160 km²

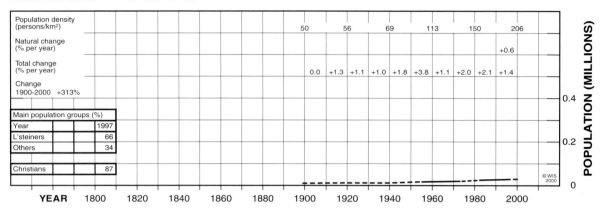

								50	56	69	113	150	206	
Population density (persons/km²)								50	56	69	113	150	206	
Natural change (% per year)													+0.6	
Total change (% per year)								0.0	+1.3	+1.1	+1.0	+1.8	+3.8	+1.1	+2.0	+2.1	+1.4
Change 1900-2000 +313%														

Main population groups (%)		
Year		1997
L'steiners		66
Others		34
Christians		87

YEAR 1800 1820 1840 1860 1880 1900 1920 1940 1960 1980 2000

POPULATION (MILLIONS) — 0.4, 0.2, 0

© WIS 2000

Germanic tribes settled this tiny region between Austria and Switzerland, on the upper river Rhine, around 500 AD. The present boundaries were established in 1434 by the merging of two counties (each ruled by a count). In 1719 they became the principality (ruled by a prince) of Liechtenstein, which stayed independent when others merged to form the German Empire in

1871. Its links with the Austrian Habsburg dynasty were severed after World War One when it forged financial and diplomatic ties with Switzerland. With few natural resources, Liechtenstein's great prosperity derives from its thriving manufacturing industry and tourism, and its status as a tax haven and financial centre. Foreign workers comprise about one third of the population.

LUXEMBOURG Area 2,586 km²

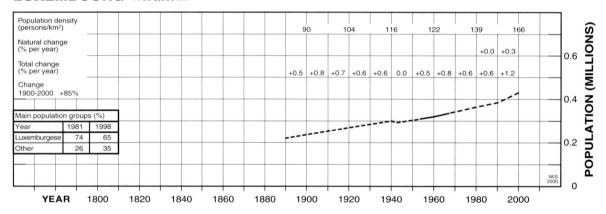

							90	104	116	122	139	166	
Population density (persons/km²)							90	104	116	122	139	166	
Natural change (% per year)											+0.0	+0.3	
Total change (% per year)							+0.5	+0.8	+0.7	+0.6	+0.6	0.0	+0.5	+0.8	+0.6	+0.6	+1.2
Change 1900-2000 +85%													

Main population groups (%)		
Year	1981	1998
Luxemburgese	74	65
Other	26	35

YEAR 1800 1820 1840 1860 1880 1900 1920 1940 1960 1980 2000

POPULATION (MILLIONS) — 0.6, 0.4, 0.2, 0

WIS 2000

In the 10th century Luxembourg was part of the Holy Roman Empire. Subsequently it was ruled by Austria, the Spanish Netherlands and France before achieving its present status, an independent Grand Duchy, in 1839.

The graph records a slow steady population increase, typical of Western Europe, only broken during the Nazi occupation in

World War Two with its deportations and executions. The sharp rise in total and natural growth rates in the 1990s results from an influx of foreign workers and their families, with relatively high fertility, mostly from southern and eastern Europe. Luxembourgese were concerned in 1999 that their traditional culture was being diluted.

MACAU Macao Area 21 km²

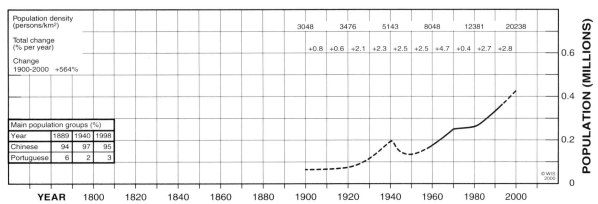

| Population density (persons/km²) | | | | | | 3048 | 3476 | 5143 | 8048 | 12381 | 20238 |
| Total change (% per year) | | | | | | +0.8 | +0.6 | +2.1 | +2.3 | +2.5 | +2.5 | +4.7 | +0.4 | +2.7 | +2.8 |
| Change 1900-2000 +564% |

Main population groups (%)

Year	1889	1940	1998
Chinese	94	97	95
Portuguese	6	2	3

© WIS 2000

POPULATION (MILLIONS)

YEAR 1800 1820 1840 1860 1880 1900 1920 1940 1960 1980 2000

Anxious to trade with rumoured 'Chin' peoples, Portuguese sailors reached Macau in 1513 and rented the tiny territory from the Chinese in 1557. Initial prosperity as the focus of trade between Europe and East Asia was followed by a long decline after Japan closed its ports to foreign trade in 1637. The 1840s and 50s saw a fundamental change when Macau became a free port and a centre for licensed gambling, and became Portuguese by treaty in 1887. Significant immigration from China began in the 1920s and continued until World War Two (when Macau stayed neutral). Much of the population growth since then has resulted from waves of immigration. In the late 1990s the booming tourist industry was hit by violent gangland rivalry for gambling revenues. Portugal ceded the territory to China in 1997.

Immigration into densely populated developed nations, especially in Western Europe, which began to escalate out of control in the late 1990s and would have been subjected to Draconian controls in the 2000s, will now be totally stopped. Borders and coastlines will be defended against illegal immigrants as if they were enemy invaders.

The undeveloped world, meanwhile, would have lost millions, even billions, of people. The poorest countries, unable to buy oil and oil-based products, quickly reverted to subsistence farming. Their cities collapsed in chaos. Lawless, starving and lacking medical facilities, populations were decimated by disease and strife between petty warlords. Famines such as were predicted by Paul Ehrlich in *The Population Bomb* (1968) ran their courses unhindered by foreign aid.

In sub-Saharan Africa the poorest national populations may fall to pre-colonial levels, except where plantations producing vegetable oils (e.g. groundnuts, oil palms) or alcohol (sugar cane, maize) are viable. Governments will disintegrate and populations reorganise themselves into tribes, ruled by warlords or chiefs. Undeveloped countries with fossil fuel reserves will maintain their rulers in great luxury while the reserves last. Developed nations will send military forces to defend their fossil fuel investments abroad, and navies will protect convoys of oil tankers and bulk carriers from piracy on the high seas.

How will the world's last superpower, the USA, respond to oil shortage? Almost uniquely in the developed world, its population density (30 in 2001) is so low that self-sufficiency in food should be possible. It is rich enough to outbid all rivals in the market, at least initially, and has huge reserves of coal and shale from which 'unconventional' oil can be distilled. It has the power, as the world becomes Darwinian, to occupy oil-producing regions 'to protect the economy'. It is well-equipped, therefore, to ride out the oil crisis as long as it maintains a low population density by excluding the inevitable floods of would-be immigrants. Russia, the other great thinly-populated nation that still has vast hydrocarbon and coal reserves, is similarly well placed in principle. It is remarkable that such an inherently intelligent race is so slow to organise an efficient economy.

By 2050, Campbell (1997, 2003) calculates, 90% of the planet's recoverable conventional oil will have been produced and consumed. Annual production will have fallen to 6 billion barrels from the peak of 25 billion barrels in the 2000s. The Gulf region will be responsible for two thirds of it. Greatly expanded production of 'unconventional' oil distilled

0 – 1.8 million

MADEIRA   Area 800 km²

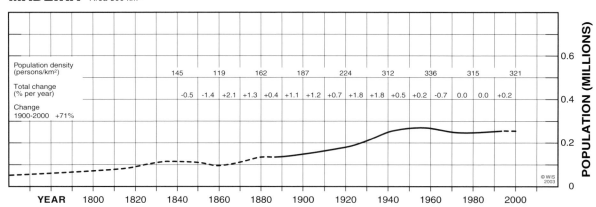

Population density (persons/km²)		145	119	162	187	224	312	336	315	321					
Total change (% per year)		-0.5	-1.4	+2.1	+1.3	+0.4	+1.1	+1.2	+0.7	+1.8	+1.8	+0.5	+0.2	-0.7	0.0	0.0	+0.2
Change 1900-2000 +71%																

Roman sailors may have visited Madeira 2000 years ago, but the archipelago of one large and 5 small rugged volcanic islands 600 kilometres off the coast of Morocco was uninhabited and totally wooded (madeira) when the Portuguese explorer João Zarco made landfall there in 1419. Portuguese settlers felled the forests and established sugar plantations worked by slaves. Vineyards followed, but tourism is now the main industry. The smaller island of Porto Santo was rendered uninhabitable in the

15th century by the escape of introduced rabbits which ate all the vegetation.

Madeira was briefly occupied by the British during the Napoleonic Wars, 1807–14, and Britain has been closely involved in the Madeira wine trade, but the islands have always been Portuguese, becoming internally self-governing in 1976. The high birth rate has forced large-scale emigration; the cumulative total of emigrants passing the million mark in the 1990s.

MALDIVES   Area 298 km²

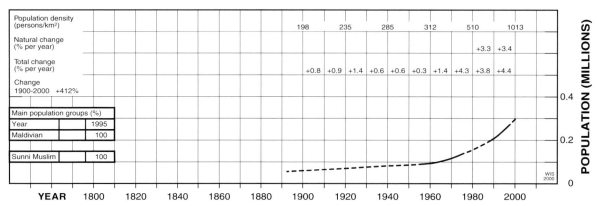

Population density (persons/km²)						198	235	285	312	510	1013				
Natural change (% per year)										+3.3	+3.4				
Total change (% per year)						+0.8	+0.9	+1.4	+0.6	+0.6	+0.3	+1.4	+4.3	+3.8	+4.4	
Change 1900-2000 +412%																

Main population groups (%)		
Year		1995
Maldivian		100
Sunni Muslim		100

The Maldives archipelago of more than 1000 tiny coral islands in the Indian Ocean has been populated for more than 2000 years. The first Buddhist immigrants from India or Sri Lanka were converted to Islam by Arab traders in the 12th century AD. Portuguese occupation in the 16th century was followed by Dutch and then British protection, the latter from 1796 when the Maldives were a dependency of Ceylon (Sri Lanka). Secessionist agitation in the 1950s led to independence in 1965. Since then, tourism has replaced fishing as the main industry.

Population density has increased drastically since 1970, causing problems with pollution, waste disposal and water supply. The government now encourages birth control, without much success. Maldivians fear the effects of global heating because, being coral atolls, the islands are all less than 3 metres above sea level. At projected rates of sea level rise they will become uninhabitable in less than a century, but the rapid population increase continues. Much coral on the reefs died in 1999, as the ocean became slightly warmer than usual. Censuses: irregular from 1920.

MALTA and Gozo Area 316 km² (after 1919)

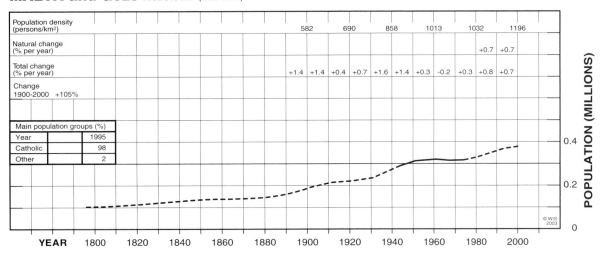

Population density (persons/km²)	582	690	858	1013	1032	1196
Natural change (% per year)					+0.7	+0.7
Total change (% per year)	+1.4 +1.4 +0.4 +0.7 +1.6 +1.4 +0.3 -0.2 +0.3 +0.8 +0.7					
Change 1900-2000 +105%						

Main population groups (%)	
Year	1995
Catholic	98
Other	2

POPULATION (MILLIONS) 0.4 0.2 0

YEAR 1800 1820 1840 1860 1880 1900 1920 1940 1960 1980 2000

© WIS 2003

The tiny densely populated island of Malta, in the Mediterranean Sea between Sicily and North Africa, has a history of occupation by all the local empire-builders from the Carthaginians to the British. The islanders withstood two famous sieges: by the Turks in 1565 and by the Axis Powers in World War Two. Malta became independent in 1964.

Soil erosion has made much of the land surface stony, dry and infertile. There were waves of emigration in the 1920s and 1960s. At other times the population was able to grow, thanks to commercial and industrial activity (including the Valetta dockyards), the installation of seawater desalination plants to double the supply of fresh water, and the development of tourism which is now the main foreign currency earner. The population is notably homogeneous, ethnically and culturally.

from oil shale, tar sands, biomass and coal will have made up part of the hydrocarbon deficit, which will worsen as natural gas reserves run low unless methods of harvesting methane hydrates can be devised. Nuclear (fission and possibly fusion) and renewable 'clean' systems will be generating electricity on a large scale, but renewable energy may be very expensive compared to 50 years earlier, as dispersed generators fail to match the economies of scale possible in conventional power stations.

Set against the hydrocarbon shortage will be the reduced demand for energy and organic chemicals (plastics, etc; section 9.1) caused by the worldwide shrinkage of population. After 50 years of climate change the planet's production of food and biomass will have been reduced by desertification, sea level rise, etc (section 6.1). A point will eventually be reached at which the supply of energy and organic chemicals from fossil fuels (dwindling) and renewables (increasing) begins to balance the waning population's demand for them.

How can we estimate world population when such a balance is reached? It could exceed the pre – Industrial Revolution maximum of 0.6 billion, because agricultural practices are so much more advanced, but growing food intensively for so many might leave insufficient oil and oil-based products to maintain a high standard of living for everyone. This would be less relevant if self-sufficient tribal societies in the new undeveloped world made up half the total population. Again, when late in the global WROG period the production of hydrocarbons peaked, the great bulk of them were consumed in the developed world, whose population was about 1.1 billion. Reduce the production by a factor of 4 or more and it could hardly support a high standard of living for more than about 0.3 billion. If, however, "quality of life" is substituted for "standard of living" (which is usually linked to conspicuously high consumption) a figure of rather more than 0.5 billion emerges.

Intuitively, a world population of 0.5 to 1 billion when the initial balance is reached seems reasonable. But it will not be a stable figure. Fossil fuel reserves will still be depleting, alternative energy supplies will be increasing, and climate may still be changing (though widespread afforestation to sequester carbon will become possible as population declines). Ideally, a stable balance will be reached, with a sustainable high quality of life possible if humanity has the wit to seize the opportunity, when world population is about 0.5 billion.

0 – 1.8 million

MARSHALL ISLANDS Area 181 km²

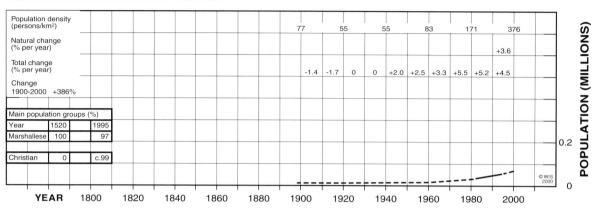

Population density (persons/km²)						77	55	55	83	171	376			
Natural change (% per year)											+3.6			
Total change (% per year)						-1.4	-1.7	0	0	+2.0	+2.5	+3.3	+5.5	+5.2 +4.5
Change 1900-2000 +386%														

Main population groups (%)		
Year	1520	1995
Marshallese	100	97
Christian	0	c.99

The Marshall Islands comprise 2 chains of coral islands and atolls, 35 small and more than 1000 very small, in the west-central Pacific Ocean about half way between New Guinea and Hawaii. Micronesian seafarers settled them about 1000 BC. The first European visitors were Spanish sailors in 1526, but the first significant European settlement was by German traders in the 19th century. Germany established a protectorate in 1885 but Japan occupied the islands in 1914. From 1920 Japan administered them, along with the Northern Marianas and Micronesia, as a League of Nations mandated territory called Nanyo. US forces captured the Marshalls in 1944 and the US administered them as part of the Pacific Islands Trust Territories until 1979. They became independent (with links to the USA) in 1986.

The US conducted nuclear tests on Bikini and Eniwetok atolls between 1946 and 1958; the population of nearby Rongelap Atoll was accidentally contaminated and had to be evacuated. Tourism and exports of copra and phosphates bolster the fragile economy which depends on US and other aid. The DC population surge began as soon as the US took control of the islands.

MARTINIQUE Area 1,130 km²

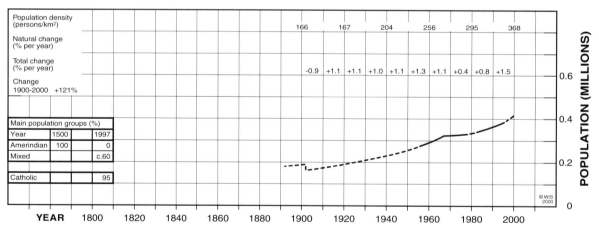

Population density (persons/km²)						166	167	204	256	295	368			
Natural change (% per year)														
Total change (% per year)						-0.9	+1.1	+1.1	+1.0	+1.1	+1.3	+1.1	+0.4	+0.8 +1.5
Change 1900-2000 +121%														

Main population groups (%)		
Year	1500	1997
Amerindian	100	0
Mixed		c.60
Catholic		95

Columbus spotted the volcanic peaks of Martinique, in the eastern Caribbean, in 1502, when the only inhabitants were warlike Carib Indians. The French established a settlement in 1635, cleared forests to plant sugar cane, and drove out the Caribs. Britain occupied the island 1794 to 1815 but returned it to France after the defeat of Napoleon. Sugar plantations became less profitable when slavery was abolished in 1848, and bananas, pineapples and especially tourism now rival sugar as the basis of the economy.

In 1902 a 'burning cloud' or pyroclastic flow from Mount Pelée raced through the city of Saint-Pierre, killing 30,000 people in a few minutes. Martinique is now an Overseas Department of France. Rapid population growth, 1980–2000, is partly due to immigration from less prosperous Caribbean islands.

MAURITIUS Area 2,040 km²

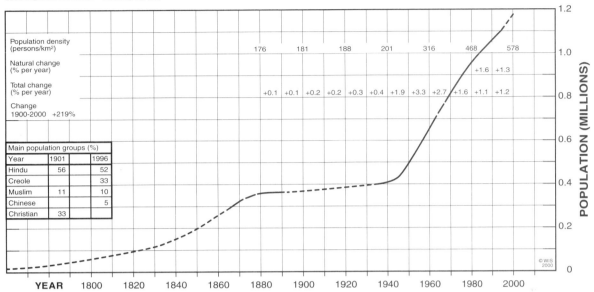

This small island 800 kms east of Madagascar was uninhabited and totally forested when Portuguese explorers visited it in the 16th century. Dutch traders settled briefly soon after and began felling the ebony trees. French settlers from 1715 established sugar plantations worked by slaves from Africa but in 1814 the island was ceded to Britain, who imported Indian indentured labourers in great numbers until 1872 when a regulatory commission was appointed. The ensuing period of slow population growth ended when a DC surge followed the elimination of malaria in the

1940s. Indians are now the largest ethnic group, dominating the government since independence in 1968. Sugar production is still the island's main industry but tourism and manufacturing are important. The high population density is recognised as a problem that is being addressed by a family planning programme. Little remains of the original forest or the unique natural flora and fauna, of which the dodo, a flightless giant pigeon that was hunted to extinction by the early settlers, is the most widely remembered.

9.3(b) The Malthusian Scenario

In this scenario the world's supply of oil or oil substitutes will last indefinitely, as many economists and others who try to predict and plan for the future seem to assume (Campbell, 1997, Chapter 10). If there are energy deficits, they will be met by 'renewable' electricity. The scenario differs in several important ways from the oil depletion scenario. The WROG period lasts longer, because no sudden lack of a vital commodity strikes all nations together. Instead, local factors such as water shortage, climate change, excessive population growth, or cultural wars, eventually bring national or regional populations separately to carrying capacity or violent cutback level (VCL). Then, in each nation or region, population growth ceases or is significantly reversed, and its WROG period usually ends.

But not always. Suppose that a culturally divided nation reaches VCL at a population of 10 million. In the civil war, 3 million are killed or expelled, and a single cultural group emerges as the victor. The nation is now homogeneous, with no internal dissensions. Population growth can resume and safely pass 10 million. Its WROG period has not ended, and its new VCL or carrying capacity may be 15 million, determined by different factors such as the availability of farmland. Overall, it is still adding to world population.

VCL conflict may be induced on a wider scale by a 'clash of cultures', as seemed to be developing after September 2001 when the USA was attacked by Muslim terrorists. Political correctness (PC) is the mindset that dictates tolerance of aliens and their beliefs and aspirations by Western populations (sections 6.7, 7.2). Conceivably, escalating terrorism could exasperate PC Western liberals, eventually driving them to Darwinism. Ethnic minorities throughout the Western world would suffer accordingly. Population growth in the developing world would decelerate or reverse as foreign aid dried up. Extreme terrorist harassment, such as large-scale biological, chemical or nuclear sabotage, could provoke massive

MICRONESIA and PALAU　Caroline Islands　Area 702 km² (Micronesia alone); 508 km² (Palau)

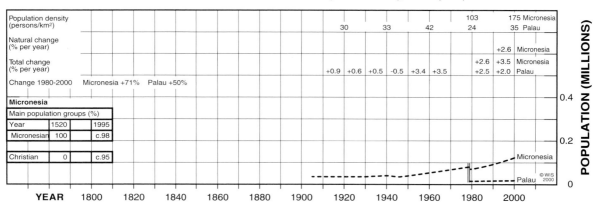

												103	175 Micronesia
Population density (persons/km²)								30	33	42		24	35 Palau
Natural change (% per year)													+2.6	Micronesia
Total change (% per year)								+0.9	+0.6	+0.5	-0.5	+3.4	+3.5	+2.6	+3.5	Micronesia
														+2.5	+2.0	Palau
Change 1980-2000　Micronesia +71%　Palau +50%													

Micronesia		
Main population groups (%)		
Year	1520	1995
Micronesian	100	c.98
Christian	0	c.95

Micronesia, an association of more than 900 widely spaced volcanic and coral islands in the western Pacific north of New Guinea, was settled by seafarers from Indonesia or the Philippines around 1000 BC. The first European visitors were Portuguese sailors in 1525. Spain claimed the archipelagos, by then known as the Carolines, in 1874, only to sell them to Germany in 1899. Japan seized them in World War One and colonised them as part of 'Nanyo' under a League of Nations mandate until World War Two when the Japanese were expelled by American forces. From 1947 to 1990 they were administered, with the Marshall and Northern Mariana Islands, by the US as the Pacific Islands Trust Territory.

The 300 islands of Palau became a semi-autonomous nation in 1979, and the remnant, the Federated States of Micronesia, achieved autonomy in 1986. Both states retain strategic links with the USA and their weak economies are heavily dependent on US aid. The beneficial US administration starting in 1947 initiated a DC population surge which still continues, to the great detriment of the largely unique native flora and fauna. Palau claims that some islands are already damaged by rising sea level.

MONACO　Area 2 km²

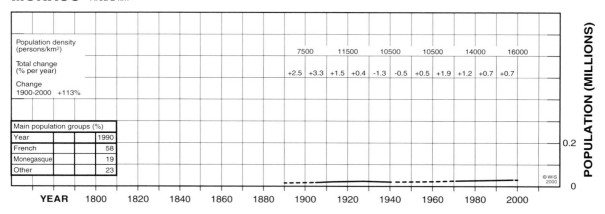

Population density (persons/km²)							7500	11500	10500	10500	14000	16000	
Total change (% per year)							+2.5	+3.3	+1.5	+0.4	-1.3	-0.5	+0.5	+1.9	+1.2	+0.7	+0.7	
Change 1900-2000　+113%													

Main population groups (%)		
Year		1990
French		58
Monegasque		19
Other		23

Now a tiny urban enclave on the French Mediterranean coast, the principality of Monaco was much larger before, in 1848, eastern lands including the town of Menton seceded to join France. Since 1297 it has been ruled by the Grimaldi family, originally Genoese, and it was at various times dominated by the French, Spanish and Sardinians. France now controls the succession and may annex Monaco if it is left without a male heir. It is now the world's most densely populated independent state, enjoying great prosperity as an upmarket tourist resort including the Monte Carlo casino, a tax haven and a centre for finance and light industry.

MONTENEGRO Area 9,400 km² to 1913, then 13,800 km²

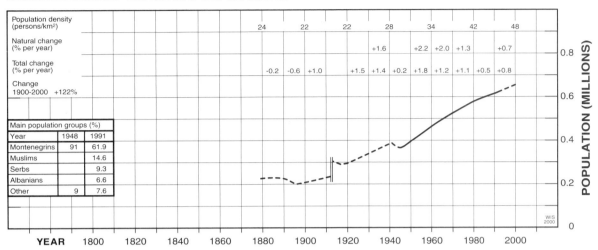

Mountainous Montenegro, the smallest semi-autonomous republic of Tito's Yugoslavia, was never fully subdued by the Turks. The Montenegrins are descended from Slavs who arrived to colonise the crumbling Roman Empire in the 5th century AD. Their little independent principality acquired parts of Kossovo and Albania after the 1913 Balkan War and then joined the Kingdom of Serbs, Croats and Slovenes which became communist Yugoslavia. It remained linked with Serbia after Yugoslavia disintegrated in the 1990s, but was agitating for independence in 2002.

Montenegro's population density is quite high in relation to the sparse supply of farmland. Emigration, the cause of actual population decrease in the 1890s, has always been important.

undiscriminating retaliation against nations suspected of harbouring or encouraging the terrorists. National populations would plummet.

As mentioned earlier, many decades may elapse before nations that currently are sparsely populated reach carrying capacity or VCL. With plenty of cheap oil, intensive agriculture can go on producing abundant food. The nations most at risk are the densely populated and culturally or ethnically divided ones. The risk is greatest when growth is fast, or exacerbates cultural diversity and thus Darwinian rivalry.

The UK provides precedents. In multicultural Northern IRELAND, neither total population nor cultural make-up has changed drastically since 1921, and the intensity of cultural (religious/ethnic) strife is not significantly worse. In ENGLAND, on the other hand, non-white ethnic minorities increased tenfold between 1951 and 1991, and religious/ethnic tension, negligible at the end of World War Two, is now a dangerous problem. One third of the total population increase of 7 million, 1951–1991, was non-white. In the general election campaign of 2001 the main political parties strove to avoid the subject of race, because the ethnic minorities, in theory, held the balance of political power (section 4.4).

Several decades into the 21st century, VCL events will be commonplace in all the inhabited continents except, perhaps, sparsely populated North and South America and AIDS-ravaged Africa. If the UN and NATO still exist they no longer intervene as they did in BOSNIA and KOSOVO. They have learned that all they achieve is a delay in the genocidal or ethnic cleansing process, which resumes as soon as their peacekeeping forces depart. Most nations at VCL remain culturally divided and experience constant guerrilla warfare or terrorism, as in RWANDA, BURUNDI and AFGHANISTAN, which maintains their chronic poverty and ensures that they cannot afford to buy the oil necessary for intensive farming. In consequence their populations cannot grow, though without effective government and services they may shrink considerably and revert to tribalism.

Other nations will have emerged from VCL chaos with reduced but homogeneous populations, able to organise effective government, repair the destruction of war, rebuild the economy and resume population growth supported by oil imports.

By 2050, on this basis, the WROG period will have ended for rather more than half of

MONTSERRAT   Area 102 km²

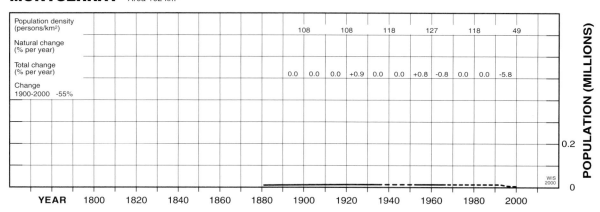

Population density (persons/km²)					108	108	118	127	118	49	
Natural change (% per year)											
Total change (% per year)				0.0	0.0	0.0	+0.9	0.0	0.0	+0.8	-0.8	0.0	0.0	-5.8
Change 1900-2000 -55%											

Columbus sighted this small mountainous island in the eastern Caribbean in 1493. Its skyline reminded him of the Montserrat monastery in Spain. Britain introduced Irish settlers in 1632 and the island was soon growing sugar cane on a scale that required thousands of African slaves. French forces briefly occupied Montserrat three times before 1783. After Britain abolished slavery in 1834 the island's economic importance waned and it was administered as part of the Leeward Islands colony from 1871 to 1958. In 1952 when offered autonomy it opted to remain a British Crown Colony.

The Soufrière volcano suddenly erupted in 1995 and has continued active, covering the south half of the island with ash and pyroclastic flows. The population was evacuated to a northern 'safe area' and some 7000 people emigrated. Those remaining depend on foreign, largely British, aid. They are mostly of mixed African-Irish stock, and Christian.

NAMIBIA   South West Africa   Area 824,000 km²

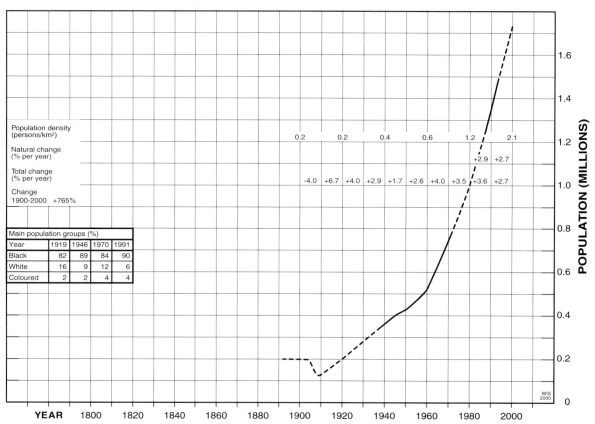

Population density (persons/km²)		0.2	0.2	0.4	0.6	1.2	2.1				
Natural change (% per year)						+2.9	+2.7				
Total change (% per year)	-4.0	+6.7	+4.0	+2.9	+1.7	+2.6	+4.0	+3.5	+3.6	+2.7	
Change 1900-2000 +765%											

Main population groups (%)

Year	1919	1946	1970	1991
Black	82	89	84	90
White	16	9	12	6
Coloured	2	2	4	4

Although the first European, Bartolomeu Dias, sailed along the Namibian coast in 1487, European powers took scant interest in this arid sparsely populated country until 1884, when Germany annexed it. German ranchers and farmers dispossessed Nama and Herero tribes from their lands and put down their rebellion in 1904 with such brutality that when it ended in 1907 nearly half the population had died. The Germans were expelled in World War One and South West Africa became a mandate of South Africa, subject to the same race laws. The apartheid policy was introduced in 1966. Armed resistance by SWAPO guerrillas began in 1968 and was complicated by South African involvement with UNITA rebels in Angola, where SWAPO had established bases.

South Africa refused to relinquish control of the country when the UN terminated the mandate in 1964, but after 2 decades of political and military confusion peace talks were held that led to an independent Namibia, with black majority rule, in 1990. Separatist guerrillas were active in the Caprivi Strip in 1999.

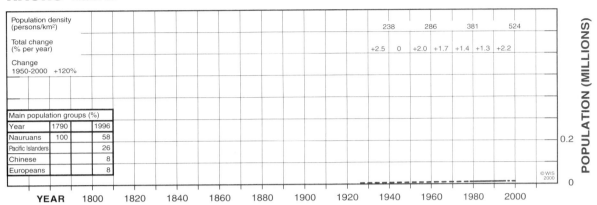

NAURU Area 21 km²

Nauru is a tiny isolated island in the southwest Pacific, between Papua New Guinea and Kiribati. Polynesians and Melanesians had settled it long before the first European set eyes on it in 1798. Germany annexed it in 1888 and soon discovered that the central plateau was entirely composed of phosphate rock, a valuable natural fertilizer, as a surface layer about 15 metres thick. A British firm began mining in 1907, using Chinese labour, and had removed 60 million tonnes by the 1990s, leaving 80% of the island as a sterile rocky wasteland. Australia had annexed Nauru in 1914 and it became independent in 1968. The islanders then became involved in the mining. When the phosphate rock is used up in the early 2000s they could all be rich but their island will be desperately overcrowded and they will have to import almost everything they need, even drinking water. In fact, their riches were mostly dissipated before 2001 by incautious government investment.

the world's nations. These are the ones that have sunk into post-WROG tribalism, with diminished poverty-stricken populations and ongoing conflict between warlords. If they include the great countries of South-central and South-east Asia, world population will have fallen significantly. Nations for which the VCL experience has meant ethnic cleansing and population reduction to achieve homogeneity number less than half of the remainder, many of them in Europe. They are economically viable and their populations may not be growing because (and this is wishful thinking) they have realised that growth leads to the unpleasantness of VCL. The remaining nations had such low population densities to start with that they have yet to reach VCL. The clever ones have introduced Draconian immigration and family planning laws to ensure that they never do reach it.

In the brave new world of the 2050s, the new developing world will be the wretched post-WROG nations. The new developed world will be those nations which are still enjoying WROG conditions and have, in some cases, prolonged them by introducing laws (oxymoronically but based on comprehension of the need) to strongly restrain growth.

World population in 2050 will be high in the Malthusian scenario, compared to the oil depletion scenario, perhaps around 4 billion. Importantly, a large proportion of it will still be consuming oil and oil substitutes as if there were no shortage (which there isn't, by definition). Probably, therefore, the greenhouse effect will be intensifying, evacuees from low-lying islands and coasts will be seeking asylum (possibly settling in post-WROG countries by force) and climate change will be determining agricultural production and driving desertification, flooding, the spread of diseases like malaria, and so on. Not much will be left of the world's virgin forests or other natural habitats, but there may be a few havens for wildlife in neglected depopulated regions of post-WROG tribal societies.

The great imponderables in this scenario include the extent to which populations are modified by factors such as HIV/AIDS, or by campaigns of conquest if powerful developed nations see advantage in recolonising parts of the new undeveloped world, or by the denial of asylum to people displaced by rising sea levels. Some developed nations may have poor access to hydrocarbons, and they may threaten others for control of these or any other vital resource. All-out war leading to mutual destruction should, with luck, be avoided.

So in the late 21st century the world could show certain resemblances to how it was two

NETHERLANDS ANTILLES incl. ARUBA   Dutch West Indies   Area 993 km² (incl. Aruba 193 km²)

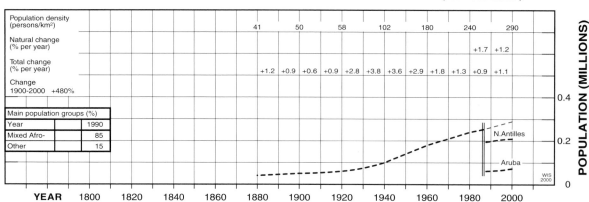

Population density (persons/km²)					41	50	58	102	180	240	290		
Natural change (% per year)										+1.7	+1.2		
Total change (% per year)					+1.2	+0.9	+0.6	+0.9	+2.8	+3.8	+3.6	+2.9	+1.8	+1.3	+0.9	+1.1
Change 1900-2000 +480%													

Main population groups (%)		
Year		1990
Mixed Afro-		85
Other		15

These two groups of small islands 800 kilometres apart in the south and east Caribbean were reached by Spanish mariners late in the 15th century. Spanish settlements and sugar plantations were taken over by the Dutch in the 17th century. The islands are now Netherlands autonomous overseas territories. The indigenous Arawak Indians were mostly exterminated on Spanish plantations and replaced by African slaves.

The DC surge after 1920 was triggered by improved agriculture, oil refining and tourism. Aruba separated in 1986 when population density had reached 250.

NEW CALEDONIA   Area 18,600 km²

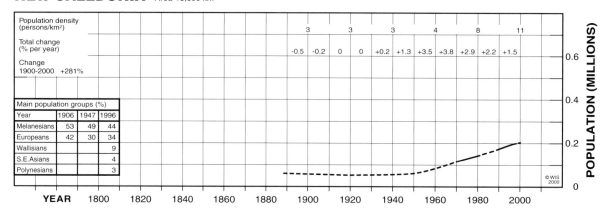

Population density (persons/km²)					3		3		3		4		8		11	
Total change (% per year)					-0.5	-0.2	0	0	+0.2	+1.3	+3.5	+3.8	+2.9	+2.2	+1.5
Change 1900-2000 +281%													

Main population groups (%)			
Year	1906	1947	1996
Melanesians	53	49	44
Europeans	42	30	34
Wallisians			9
S.E.Asians			4
Polynesians			3

When Captain Cook sighted and named New Caledonia (which reminded him of rugged misty Scottish coasts) in 1774 the island group half way between Australia and Fiji was a southern outpost of the tropical Melanesian peoples. English and French settlers arrived in 1840 and France annexed the territory in 1853. From being a French penal colony the main island, Grande Terre, developed into the world's third largest producer of nickel, and still contains the largest known deposits of that metal. The infertile soils are so rich in nickel and other minerals that a unique flora has evolved during its 80 million years as an island. Only Madagascar, Hawaii and New Zealand have, or had, more diverse island wildlife. New Caledonia became a French Overseas Territory in 1958. The growing numbers of foreign settlers, mainly French, provoked such resentment among the native Melanesian Kanaks that, while still the largest ethnic group, they began a separatist revolt in the 1980s that was put down by French troops. A referendum rejected independence. Christianity has always been the dominant religion. The DC population surge began after World War Two.

NORTHERN MARIANA ISLANDS   Area 470 km²

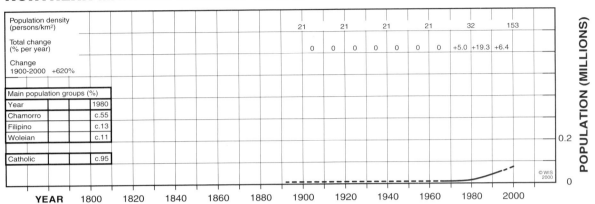

Population density (persons/km²)							21	21	21	21	32	153
Total change (% per year)							0	0	0	0	0	0	0	+5.0	+19.3	+6.4
Change 1900-2000 +620%												

Main population groups (%)		
Year		1980
Chamorro		c.55
Filipino		c.13
Woleian		c.11
Catholic		c.95

POPULATION (MILLIONS): 0.2 — 0

YEAR: 1800   1820   1840   1860   1880   1900   1920   1940   1960   1980   2000

© WIS 2000

The Mariana Islands archipelago, in the western Pacific half way between Japan and New Guinea, consists of the Northern Marianas and, at the south end, GUAM. The 16 islands are a chain of volcanoes along the edge of the Mariana Trench, the deepest (11 kilometres) part of the Pacific Ocean where one crustal plate is sliding under another. Micronesian seafarers from Indonesia or the Philippines reached the Marianas about 3500 years ago. Ferdinand Magellan, the first European visitor in 1521, claimed the islands for Spain and named them the Ladrones (thieves). The natives became Christians and the islands were renamed when Spanish occupation began in the 17th century.

Germany bought the Northern Marianas from Spain in 1899 after the Spanish-American War. Japan administered them under a League of Nations mandate from the end of World War One until 1944 when US forces captured them. After World War Two they were administered by the US as part of the Trust Territory of the Pacific Islands until 1978, when they became a commonwealth in union with the USA.

The population increase after 1980 is so rapid that it must be related to immigration. The native Chamorros are mixed Micronesian, Spanish and Filipino. The weak economy is supported by US aid and tourism. Censuses: every 10 years from 1970.

centuries earlier. It is dominated by the industrial developed world which may be exploiting the backward nations. One superpower dominates the political and military scene, unless it was mortally damaged in its VCL event, when it acquired a homogeneous population. But world population is twice the size it was in the later 19th century. Between a quarter and a half of it aspires to a high standard of living, and is consuming resources and polluting the planet faster than ever before. As the decades pass, world population will continue to decline, either involuntarily as climate change and the other anthropogenic ills render the planet less habitable and nations fight to occupy the best remaining regions, or voluntarily by planned birth control in the developed world.

Eventually, thanks to unlimited oil, the planetary environment is likely to become far less hospitable to life than under the oil depletion scenario. Human population could fall to a few tens of millions, at which level natural ecosystems, if they still function as they used to, could begin to reassert themselves (section 6.1, ***CLIMATE CHANGE***). Would there be any *wild* animals left? Probably not, apart from cockroaches and their ilk.

Perhaps fortunately, this scenario of oil and oil substitutes lasting indefinitely, though currently popular, is an unlikely one.

9.4 A lesson learned?

According to the above scenarios, the 21st century will see the world's human population peak at 7 to 8 billion and then catastrophically decline to less than one billion. The graph of human population change on Earth, by the year 2100, will closely resemble the graph of rising and falling oil production, the famous 'Hubbert's Pimple' (section 7.3). It will show world population rising oh-so-slowly to reach 0.6 billion in 1750, then shooting up to 7 billion in 2010 only to fall even more steeply to about 0.6 billion in 2100. It will be, effectively, the WROG pimple. Then what? Will the graph continue level beyond 2100, will it fall, or could another pimple follow the first?

There would be severe physical restraints on the growth of another pimple. As Sir Fred Hoyle pointed out long ago (Duncan, 1997), the present 'industrial civilisation' is using up all

QATAR Area 11,400 km²

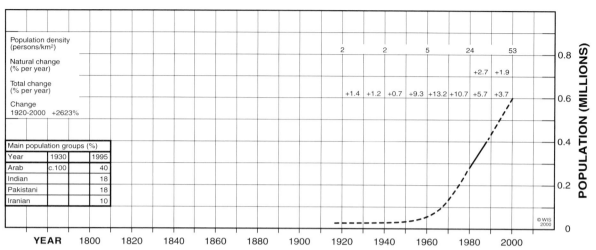

Population density (persons/km²)								2	2	5	24	53
Natural change (% per year)											+2.7	+1.9
Total change (% per year)								+1.4	+1.2	+0.7	+9.3	+13.2	+10.7	+5.7	+3.7
Change 1920-2000 +2623%												

Main population groups (%)		
Year	1930	1995
Arab	c.100	40
Indian		18
Pakistani		18
Iranian		10

YEAR　1800　1820　1840　1860　1880　1900　1920　1940　1960　1980　2000

Qatar, a desert peninsula in the southern Arabian (Persian) Gulf, was part of the Dilmun trading empire more than 4 millennia ago. It has been Muslim since the 7th century AD. Apart from coastal trade, and associated piracy, the main occupation was pearl diving until the 1930s when cheaper cultured pearls became popular. After short periods of domination by Bahrain, and then Ottoman Turkey, Qatar was a British protectorate from 1916 until independence in 1971.

Oil was discovered in 1939 and production began after World War Two. Oil-related industries boomed, attracting great numbers of foreign workers, mainly Arab and Asian. The steep rise of the graph from the 1950s is mainly due to this immigration. In 1999 less than a quarter of the population was native Qatari, whose *per capita* wealth is very great. Oil reserves are expected to run out around 2025 but natural gas reserves are vast and should last much longer.

RÉUNION Île Bourbon Area 2,510 km²

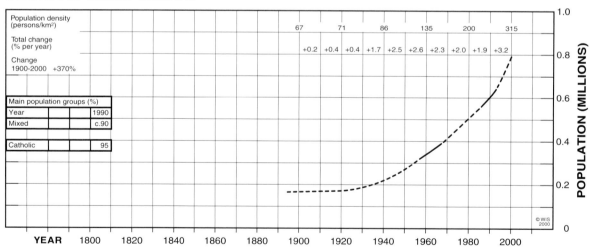

Population density (persons/km²)							67	71	86	135	200	315
Total change (% per year)							+0.2	+0.4	+0.4	+1.7	+2.5	+2.6	+2.3	+2.0	+1.9	+3.2
Change 1900-2000 +370%												

Main population groups (%)		
Year		1990
Mixed		c.90
Catholic		95

YEAR　1800　1820　1840　1860　1880　1900　1920　1940　1960　1980　2000

When Portuguese navigators first reached this mountainous volcanic island, part of the Mascarenes archipelago in the Indian Ocean 600 kilometres east of Madagascar, in 1513, it had no inhabitants. France founded a penal colony there in 1638, annexed the island in 1642 and named it Bourbon after the French monarchy. French settlers established sugar plantations worked by African slaves from 1660 onwards; after 1848 when slavery was abolished workers were brought in from India and Vietnam. The island was renamed in 1793 and became a French Overseas Department in 1946.

Sugar and sugar derivatives such as rum are still the main exports. Tourism is booming, but population growth has outpaced the available work leading to about 37% unemployment in 1997. The present population is descended from the French settlers and their workers. A strong DC surge began in the 1930s.

ST KITTS and NEVIS Area 262 km²

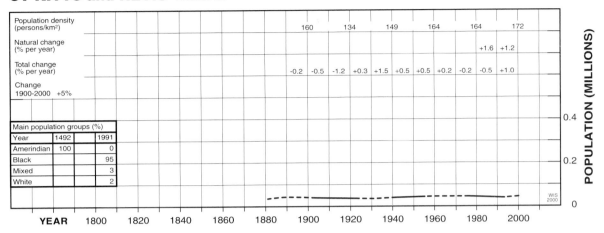

Population density (persons/km²)								160	134	149	164	164	172	
Natural change (% per year)												+1.6	+1.2	
Total change (% per year)								-0.2	-0.5	-1.2	+0.3	+1.5	+0.5	+0.5	+0.2	-0.2	-0.5	+1.0
Change 1900-2000 +5%														

Main population groups (%)		
Year	1492	1991
Amerindian	100	0
Black		95
Mixed		3
White		2

POPULATION (MILLIONS)

0.4

0.2

WIS 2000

0

YEAR 1800 1820 1840 1860 1880 1900 1920 1940 1960 1980 2000

Columbus sighted this pair of eastern Caribbean islands in 1493, naming one after his patron saint, Christopher. British and French settlers arrived from 1623, massacred the native Carib Indians, fought each other, and developed prosperous sugar plantations worked by African slaves. France dropped her claim to the islands in 1783. Slavery ended in 1834 but sugar is still very important to the economy, together with tourism. The islands were part of the West Indies Federation (1958–62) and became independent in 1983. Subsequent governments have suffered from corruption and civil unrest. Stability of the population is maintained by emigration of the natural surplus. A hurricane left more than half of the population homeless in 1998.

the planet's *easily accessible* mineral resources, especially hydrocarbons, coal and metal ores. In consequence, Hoyle argued, human civilisation is "a one-shot affair". If humankind loses its grip on technology and reverts to tribalism, the raw materials necessary to begin a second industrial revolution will not be lying around at the earth's surface, as they were for the first. Natural geological processes will take tens of millions of years to replace the oil, gas, coal and metal ores that we will have used up in about 3 centuries.

Suppose that in 2100 world population is about 0.6 billion (half China's current population), divided more or less equally into developed and undeveloped societies. Energy comes mainly in the form of electricity, from renewable (wind, photovoltaic, tidal, etc) and possibly nuclear sources. Developed societies have all the cars, tractors and trucks they need, with fuel cells depending on high-purity hydrogen produced by electrolysis of water, or else they have electric motors with efficient batteries. Pollution of land, air and water is wonderfully reduced. Plastics, textiles, paper, chemicals, fertilizers and oils are obtained from biomass, grown as specialised crops like sugar cane and canola, or as wood in the vast new forests that are purifying the atmosphere by absorbing its carbon dioxide into tree trunks. Coal and oil shales are being converted to oil and chemicals. Metal ores and other industrial minerals are available, in mines and old garbage dumps, but usually at such low grades that huge amounts of energy are expended in processing them. Wildlife is recovering in the regions from which humans have largely withdrawn because the energy needed to farm them intensively is scarce. Fish abound in the oceans. Climate is on the change again – for the better, as the greenhouse effect diminishes.

Potentially, a Utopian existence seems possible, thanks to the new balance between humans and nature enforced by the exhaustion of fossil fuels and the end of the WROG period. The question is: can human nature rise to the challenge? Humans cannot help being instinctively Darwinian, so individuals or groups will still be tempted to try to better themselves at the expense of other individuals or groups, just as they were before the WROG period, when enterprise, aggressive breeding and conquest were the keys to wealth and power.

At first, people will be very conscious that they are survivors. All around them there will be evidence of the famines and killings that eliminated nine tenths of the world's population:

ST LUCIA   Area 617 km²

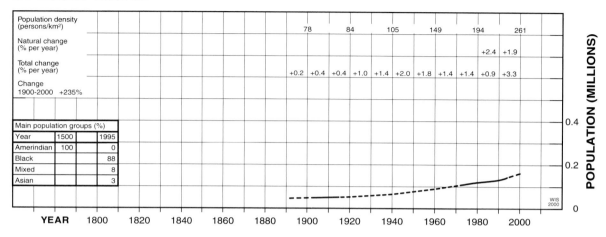

Population density (persons/km²)							78	84	105	149	194	261
Natural change (% per year)											+2.4	+1.9
Total change (% per year)							+0.2	+0.4	+0.4	+1.0	+1.4	+2.0	+1.8	+1.4	+1.4	+0.9	+3.3
Change 1900-2000 +235%												

Main population groups (%)		
Year	1500	1995
Amerindian	100	0
Black		88
Mixed		8
Asian		3

POPULATION (MILLIONS)   0.4   0.2   0

YEAR   1800   1820   1840   1860   1880   1900   1920   1940   1960   1980   2000

Spanish sailors are said to have sighted this mountainous volcanic eastern Caribbean island on Saint Lucia's day in the early 1500s, when the inhabitants were Carib Indians who had driven out peaceful Arawaks. British essays at settlement in 1605 and 1638 were thwarted, with much bloodshed, by the Caribs. French settlers arrived in the 1640s, overcame Carib resistance, imported African slaves and planted sugar cane. Britain still claimed the island and launched frequent attacks until 1814, when France pulled out. St Lucia joined the Windward Islands Federation and achieved independence in 1979.

The main export crops now are bananas, coconuts and cocoa. Tourism is booming. As a result of the island's early history the population is overwhelmingly Catholic.

ST VINCENT and the Grenadines   Area 389 km²

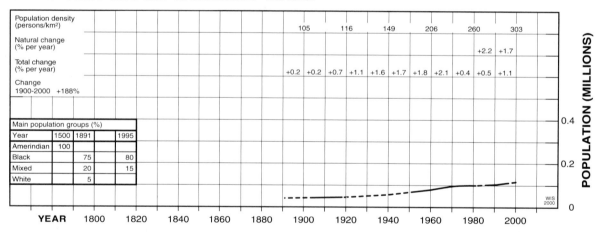

Population density (persons/km²)							105	116	149	206	260	303
Natural change (% per year)											+2.2	+1.7
Total change (% per year)							+0.2	+0.2	+0.7	+1.1	+1.6	+1.7	+1.8	+2.1	+0.4	+0.5	+1.1
Change 1900-2000 +188%												

Main population groups (%)			
Year	1500	1891	1995
Amerindian	100		
Black		75	80
Mixed		20	15
White		5	

POPULATION (MILLIONS)   0.4   0.2   0

YEAR   1800   1820   1840   1860   1880   1900   1920   1940   1960   1980   2000

Columbus reached and named the mountainous principal island of this eastern Caribbean group on St Vincent's Day, 1498. It was a stronghold of warlike Carib Indians, who had ousted the original Arawak Indian inhabitants. The Caribs prevented permanent European settlement until the early 1700s. Previously, interbreeding between Caribs and shipwrecked African slaves had created the 'Black Carib' race which became hostile to the indigenous 'Yellow Caribs'.

Rival claims to the islands by France and Britain were resolved in Britain's favour in 1783. France backed a savage revolt by both Carib groups in 1795 which led to deportation of all the Black Caribs. Coffee and cocoa plantations prospered, worked by African slaves, until 1812 when half the plantations were destroyed by a volcanic eruption that killed most of the remaining Yellow Caribs on their reservation. Further eruptions of La Soufrière in 1902 and 1979, together with hurricanes, effectively ended plantation agriculture.

Bananas, grown on small farms, are now the mainstay of the economy except on the outlying Grenadine Islands where tourism is dominant. Britain granted independence in 1979. Most of the population is Christian, mainly Protestant.

SAMOA Western Samoa   Area 2,830 km²

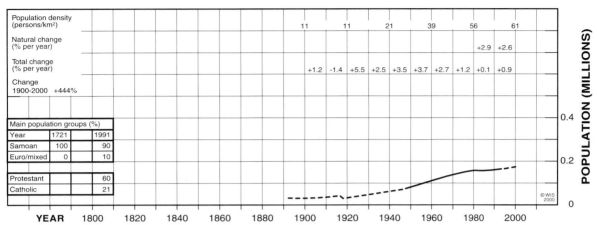

Population density (persons/km²)									11	11	21	39	56	61
Natural change (% per year)													+2.9	+2.6
Total change (% per year)									+1.2	-1.4	+5.5	+2.5	+3.5	+3.7	+2.7	+1.2	+0.1	+0.9
Change 1900-2000 +444%														

Main population groups (%)

Year	1721	1991
Samoan	100	90
Euro/mixed	0	10
Protestant		60
Catholic		21

© WIS 2000

POPULATION (MILLIONS)

0.4
0.2
0

YEAR 1800 1820 1840 1860 1880 1900 1920 1940 1960 1980 2000

Samoa (Western Samoa until 1997) consists of 2 large and 7 small volcanic and coral islands, 1000 kilometres east of Fiji in the southwest Pacific. It is the western part of the Samoan Islands archipelago; American Samoa is the eastern part. Polynesian seafarers reached and settled Samoa about 3000 years ago. Inter-island warfare and invasions from other island groups became chronic as populations grew to near carrying capacity; unification was sometimes achieved, as under the legendary Queen Salamasina in the 15th century. Dutch sailors were the first European visitors, in 1722. Subsequently, German, British and American trading posts exploited inter-island rivalries until, in 1889, Germany obtained Western Samoa as a protectorate. A New Zealand force expelled the Germans in World War One. In 1918 an influenza epidemic wiped out nearly one quarter of the population. New Zealand administered the islands as a mandate until independence in 1962.

Samoa's economy is weak, being based on agriculture. Coconut products, bananas and cocoa are the main exports. Tourism is an important source of income but the standard of living still depends on foreign aid. Since the 1970s, increased emigration has eased population growth. Censuses: every 5 to 10 years from 1951.

cities, towns and villages abandoned and crumbling, rusting vehicles littering deserted overgrown roads and, off the beaten tracks, the scattered remains of humans and domestic animals. Mass graves and piles of skulls in BOSNIA and RWANDA are precedents. In the developed world, people will concentrate in widely-spaced towns linked by electric railways and main roads. They will farm the countryside as intensively as possible, given the limited availability of agrochemicals, fertilizers and the fuel to drive farm machinery. Only the best land will be so used, producing food crops and biomass. Some of the remainder will support domestic livestock in relatively small numbers (because overwintering them requires energy-expensive hay, silage and protein), but most of it will be forest or natural grassland, managed for timber, biomass and wild livestock such as deer and buffalo.

Meanwhile, the undeveloped world will be much as it was before the arrival of industrial civilisation, its tribal societies dependent on subsistence agriculture in difficult conditions and its populations kept low by disease, warfare and the vagaries of climate. Scattered through it will be small industrialised towns, sustained by the energy from nuclear reactors or concentrations of 'renewables' and by biomass from plantations where the land is fertile.

It should not be assumed that the developed and undeveloped worlds of the 22nd century will be the same as their counterparts today. The future developed world will comprise nations that were best able to preserve competent administrations as the trappings of the WROG period collapsed around them – perhaps including a few powerful states which survive, like slave-making ants, by dominating and exploiting their prudent and industrious neighbours.

It is hard to believe that the human species will ever be able to suppress its inbuilt Darwinian competitiveness, even after such a savage lesson as the population crash of the 21st century. Maybe in the aftermath people will be sufficiently chastened to set up a World Government, as so many science fiction writers seem to consider inevitable, but the evidence of history is that the authority of such a body (e.g. the League of Nations or the United

0 – 1.8 million

SAN MARINO Area 61 km²

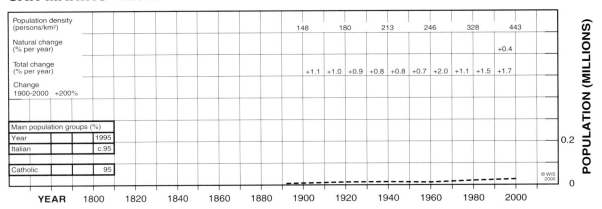

Population density (persons/km²)									148	180	213	246	328	443
Natural change (% per year)														+0.4
Total change (% per year)									+1.1	+1.0	+0.9	+0.8	+0.8	+0.7	+2.0	+1.1	+1.5	+1.7
Change 1900-2000 +200%														

Main population groups (%)			
Year			1995
Italian			c.95
Catholic			95

© WIS 2000

POPULATION (MILLIONS) — 0.2 — 0

YEAR   1800   1820   1840   1860   1880   1900   1920   1940   1960   1980   2000

Founded in the 4th century by Marinus, a Christian stonemason who sought refuge in the rugged terrain of what is now the world's oldest republic, San Marino (100 km east of Florence) is an anachronism in the modern world: the only survivor from medieval times when the Italian peninsula was occupied by many disunited states. Italy has protected it by treaty since 1862. There is agriculture and light industry, but most national income derives from tourism and the sale of postage stamps. Owing to an error in the Treaty of Westphalia (1648) San Marino was in a state of war with Sweden from then until 1996, when the mistake was rectified.

SÃO TOMÉ E PRINCIPE Area 1,001 km²

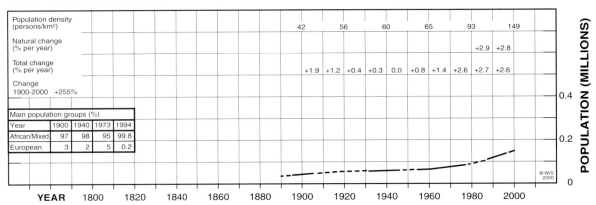

Population density (persons/km²)									42	56	60	65	93	149
Natural change (% per year)														+2.9	+2.8
Total change (% per year)									+1.9	+1.2	+0.4	+0.3	0.0	+0.8	+1.4	+2.6	+2.7	+2.6
Change 1900-2000 +255%														

Main population groups (%)	1900	1940	1973	1994
Year				
African/Mixed	97	98	95	99.8
European	3	2	5	0.2

© WIS 2000

POPULATION (MILLIONS) — 0.4 — 0.2 — 0

YEAR   1800   1820   1840   1860   1880   1900   1920   1940   1960   1980   2000

São Tomé (Saint Thomas) is the larger of this pair of volcanic islands off the coast of Gabon, equatorial Africa. They were uninhabited when Portuguese navigators landed there in 1470 and soon established sugar plantations with African slave labour. Under Portuguese rule São Tomé became a major staging post for slaves in transit from Africa to America. After the slave trade was abolished in the 19th century coffee and cocoa were grown on large plantations using contract labour.

The islands became a Portuguese Overseas Province in 1951 but nationalist movements and labour unrest intensified until independence was amicably negotiated in 1975. Most of the Portuguese residents departed. The first government had Marxist leanings and was supported by Angola, but the declining economy and the collapse of the Soviet Union led pragmatically to multiparty elections in 1991.

SEYCHELLES Area 455 km²

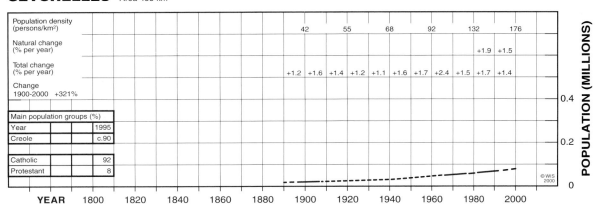

	1800	1820	1840	1860	1880	1900	1920	1940	1960	1980	2000
Population density (persons/km²)						42	55	68	92	132	176
Natural change (% per year)										+1.9	+1.5
Total change (% per year)						+1.2 +1.6 +1.4 +1.2 +1.1 +1.6 +1.7 +2.4 +1.5 +1.7 +1.4					

Change 1900-2000 +321%

Main population groups (%)	
Year	1995
Creole	c.90
Catholic	92
Protestant	8

POPULATION (MILLIONS): 0.4, 0.2, 0

© WIS 2000

The widely-dispersed Seychelles islands, in the western Indian Ocean north of Madagascar, comprise a mountainous "Granitic" group of 40 in the north and a low-lying "Coralline" group of 80 further south. Most of the population is Granitic. There were no indigenous islanders when Portuguese navigator Vasco da Gama landed there in 1502. France claimed the islands in 1756 and French settlers established plantations worked by African slaves, but Britain captured them in the conflicts sparked by the French Revolution of 1789 and annexed them in 1814. When slavery was abolished in 1835 immigrant workers came from Mauritius (the seat of government), China and India. Independence was achieved in 1976. A coup in 1977 installed the first of a series of left-wing one-party governments. Tourism has recently surpassed coconut products and fisheries as the main source of export earnings. The Creole population is descended from French settlers and African slaves.

Nations) tends to be disregarded by individual nations or cultural groups who want to go their own ways and are strong enough to do so (section 6.3).

More likely, on precedent, is that one nation will dominate the developed world by force, establishing and ruling a global empire. Such an action would be appropriately Darwinian, and the empire could be stable if the ruling power imposed cultural homogeneity and a common language, legal system, currency, etc. and maintained an effective police force, secret or otherwise, to detect and crush internal Darwinian competition. Given that some quite minor despots, such as Saddam Hussein or Robert Mugabe, have suppressed opposition for long periods, the Ruler of the World should find it easy enough. The empire would have to be run on Draconian lines, for the benefit of the ruling power. It could be partly democratic, but Darwinian competition between nations, such as aggressive breeding, would be prevented. In a necessarily imperfect world, everyone could benefit from a wise benevolent dictatorship that maintained, to ensure internal harmony, a good standard of living for all its citizens. The *Pax Romana* imposed by ancient Rome could be an example to follow and improve on.

Is there a nation that might seem destined for such a role? The USA, currently the only superpower, is an obvious candidate, but how it will fare in the global population crash, given its large ethnic minorities, remains to be seen. And in a world where genetic engineering has become commonplace, brute force may not be the only route to world conquest. Biological aggression, such as the creation and dispersal of a germ that is lethal to all humans except selected genetic strains, might be easier, more effective, and available to any small but technically smart nation. The scope for errors would be profound: if two nations tried it simultaneously the whole human race could be snuffed out in a historical instant. Perhaps that is why we have never yet detected any trace of intelligent extraterrestrial life, and why our distant descendants have never travelled back in time to contact us: Darwinian evolution anywhere in the Universe may create beings who, when they attain a certain degree of smartness (as distinct from wisdom), exterminate themselves by accident.

If the brave new world envisaged two paragraphs previously, wherein one nation rules the world for its own benefit, seems uncomfortably reminiscent of what Adolf Hitler set out to achieve in the 1930s, at least it offers the *potential* for the continued survival of humankind,

0 – 1.8 million

SOLOMON ISLANDS   Area 28,400 km²

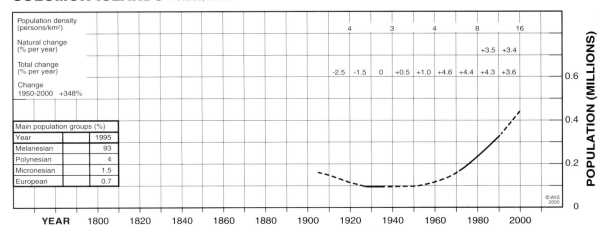

Population density (persons/km²)											4	3	4	8	16	
Natural change (% per year)														+3.5	+3.4	
Total change (% per year)						-2.5	-1.5	0	+0.5	+1.0	+4.6	+4.4	+4.3	+3.6		
Change 1950-2000 +348%																

Main population groups (%)		
Year		1995
Melanesian		93
Polynesian		4
Micronesian		1.5
European		0.7

This archipelago of 992 islands in the west-central Pacific east of Papua New Guinea has been occupied for at least 5000 years by Melanesian peoples who came originally from south-east Asia. The first European visitors, Spanish sailors from Peru in 1568, failed to colonise the islands and they remained unexploited until the late 18th century. Then, European traders reported the prevalence of inter-island warfare, head hunting and cannibalism. The arrival of Old World diseases began a drastic population decline.

Britain declared a protectorate in 1893, extended in 1899 to include islands ceded by Germany. Administrators and missionaries

between them suppressed the traditional way of life. Japan occupied the Solomons in World War Two and was expelled after bloody fighting with US forces in Guadalcanal. A 'cargo cult' (prophesying the arrival of shiploads of free goods) developed out of American generosity and high spending. Britain granted and financed independence in 1978. Rapid population growth is beginning to threaten the small family farms upon which the islanders' self-sufficient affluence is based. Inter-island civil war over land rights (Guadalcanal vs Malaita) began in 1998 and is bankrupting the fragile economy.

SURINAME   Dutch Guiana   Area 164,000 km²

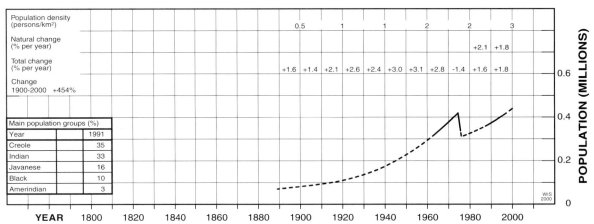

Population density (persons/km²)						0.5		1		1		2		2		3	
Natural change (% per year)															+2.1	+1.8	
Total change (% per year)						+1.6	+1.4	+2.1	+2.6	+2.4	+3.0	+3.1	+2.8	-1.4	+1.6	+1.8	
Change 1900-2000 +454%																	

Main population groups (%)		
Year		1991
Creole		35
Indian		33
Javanese		16
Black		10
Amerindian		3

Arawak, Carib and Warrau Indians lived in Suriname (named after a local Indian tribe) on the north coast of South America when the first Dutch settlers arrived in 1581. English traders followed in the early 17th century and Britain claimed possession but ceded it to the Dutch in 1667 in exchange for New Amsterdam (New York). African slaves and Asian labourers were imported to work sugar and coffee plantations. Suriname became part of the Netherlands in 1922, but

nationalist activity developed and independence was declared in 1975. Many people, who had Dutch citizenship, emigrated to the Netherlands. Those remaining have formed elected governments alternating with military dictatorships. A politico-ethnic civil war lasted from 1986 to 1992 and guerrilla activity continues.

Three quarters of Suriname's export earnings derive from the sale of bauxite, the ore of aluminium.

SWAZILAND Area 17,400 km²

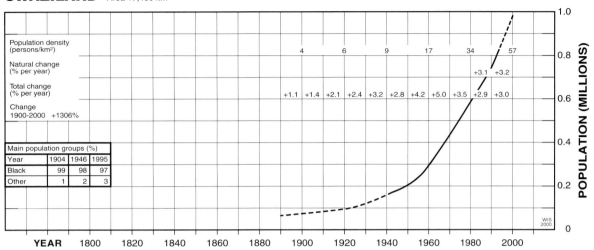

Population density (persons/km²)						4	6	9	17	34	57
Natural change (% per year)										+3.1	+3.2
Total change (% per year)						+1.1	+1.4	+2.1	+2.4	+3.2	+2.8	+4.2	+5.0	+3.5	+2.9	+3.0
Change 1900-2000 +1306%											

Main population groups (%)			
Year	1904	1946	1995
Black	99	98	97
Other	1	2	3

Bantu people had long been established in Swaziland, between South Africa and Mozambique, when Swazi tribes moved in from the east and became the dominant group, ruled by King Mswati, during the 17th century. Europeans took scant interest in the country until 1882 when gold deposits were found. After the Second Boer War (1899–1902) Swaziland became a British protectorate. It achieved independence in 1968. Subsequent development of the kingdom has been mainly peaceful, but in the 1990s HIV/AIDS became a massive problem. In 2000 at least 25% of the population were infected and the government proposed to ban girls older than 10 wearing miniskirts in school. In late 2002 it was said that the population had begun to decline, with 38% of adults infected.

in considerable comfort and in harmony with its environment. Darwinism takes no heed of political correctness, or perceived morality, or justice. The story of *Homo sapiens* is littered with vast amoral cruelties, exterminations and injustices, forgotten except by historians. And the history taught in a future world empire would be likely to approve the actions that created that empire. History is written by survivors.

In a less imperfect world, people might realise that Earth's problems could be resolved by voluntary population reduction, and would act accordingly before time runs out. That would be amazing and glorious, because humankind would have shown itself able to confront and reject its Darwinian inheritance of aggressive selfishness, its irrational addiction to blind faith, and its withdrawal from reality into political and economic correctness, by exercising its supreme gift: intelligence.

• • • • • • • • • • • • •

"We believe this book will cause... people... to ask themselves in earnest whether the momentum of present growth may not overshoot the carrying capacity of this planet – and to consider the chilling alternatives such an overshoot implies for ourselves, our children, and our grandchildren."

(The Club of Rome, 1972).

0 – 1.8 million

TASMANIA   Area 68,050 km²

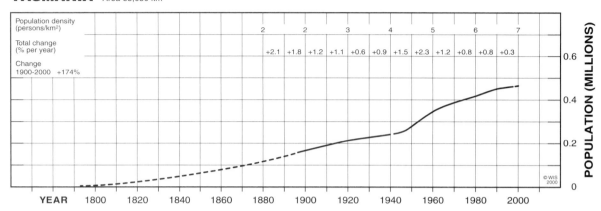

Population density (persons/km²)							2	2	3	4	5	6	7	
Total change (% per year)							+2.1	+1.8	+1.2	+1.1	+0.6	+0.9	+1.5	+2.3	+1.2	+0.8	+0.8	+0.3
Change 1900-2000 +174%													

Contrasting with most of Australia, Tasmania is green and fertile, with plentiful rain. When European settlement began in 1803 the island had some 5000 Aboriginal inhabitants, but they were persecuted so severely that the last one died in 1876. A wave of immigration to Australia after World War Two was particularly evident in Tasmania. Certain animal species unique to the island, such as the Tasmanian Tiger or Wolf, became extinct in the 20th century. Censuses: about every tenth year until 1961, then every fifth year.

TONGA   Friendly Isles   Area 750 km²

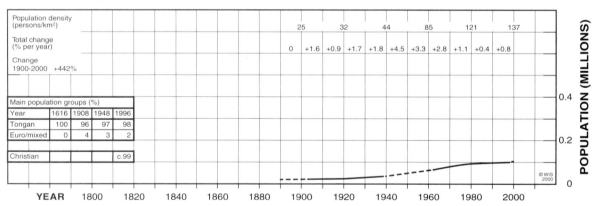

Population density (persons/km²)					25	32	44	85	121	137	
Total change (% per year)					0	+1.6	+0.9	+1.7	+1.8	+4.5	+3.3	+2.8	+1.1	+0.4	+0.8
Change 1900-2000 +442%											

Main population groups (%)

Year	1616	1908	1948	1996
Tongan	100	96	97	98
Euro/mixed	0	4	3	2
Christian				c.99

The Tonga archipelago of some 170 small coral and volcanic islands in the southwest Pacific, 2000 kilometres northeast of New Zealand, was settled by Polynesian seafarers from nearby Fiji around 3000 years ago. Tongan legends relate a history of inter-island warfare which kept the population below carrying capacity, and civil wars continued for centuries after the first Europeans, Dutch sailors in 1616, sighted the islands. Even so, Captain Cook named them Friendly Isles to commemorate their hospitality when he visited in 1773. British Methodist missionaries settled in 1826 and the islands were united under King Tupou, descendant of the legendary first Tongan king (900 AD), in 1845. Tonga became a British protectorate in 1900 and independent in 1970.

A gentle DC surge began in the 1920s and showed signs of levelling off in the 1980s, by means of emigration, when the constitutional guarantee that every Tongan male is entitled to his own plot of ground to farm was frustrated by land scarcity. The economy depends on agriculture; coconut products, yams, fruit and vegetables providing 80% of export revenue. Fishing and tourism are important. A wave power plant meets one third of the demand for electricity. Censuses: at irregular intervals from 1908.

Sources and References

Authored Books and Papers

Anderson, D. & Mullen, P. 1998. *Faking it*. The Social Affairs Unit, London.

Bahn, P. & Flenley, J. 1992. *Easter Island, Earth Island*. Thames & Hudson, London.

Bartlett, A.A. 1986. Forgotten Fundamentals of the Energy Crisis. *American Journal of Physics*, Vol. 46, 876–888.

Beebee, T.J.C. 2001. British wildlife and human numbers: the ultimate conservation issue? *British Wildlife*, October 2001, 1–8.

Berthoud, R. 2001. Teenage births to ethnic minority women. *Population Trends*, No. 104, 12–17.

Bogue, D.J. 1969. *Principles of Demography*. John Wiley, New York.

Brân, Z. 2001. *After Yugoslavia*. Lonely Planet, Melbourne.

Briggs, A. 1994. *A Social History of England*. BCA, London.

Brown, L.R. *et al.* 1997. *State of the World* 1997. Earthscan, London.

Butler, P. (ed.) 1980. *Life and Times of Te Rauparaha. By his son Tamihana*. Alister Taylor, Waiura.

Campbell, C.J. 1997. *The Coming Oil Crisis*. Multi-Science Publishing, Brentwood.

Campbell, C.J. 2003. *The Essence of Oil & Gas Depletion*. Multi-Science Publishing, Brentwood.

Carey, P. & Bentley, G.C. (eds.) 1995. *East Timor at the Crossroads*. Cassell, London.

Carson, R. 1963. *Silent Spring*. Hamish Hamilton, London.

Churchill, W.S. 1951. *Their Finest Hour*. The Reprint Society, London.

"Club of Rome" see Meadows et al, 1972.

Coleman, D. 2001. Identity Crisis. *The Spectator*, 6 January 2001. London.

Cunliffe, B. 2001. *The Extraordinary Voyage of Pytheas the Greek*. Allen Lane, The Penguin Press.

Dahlby, T. 2001. Indonesia – Living Dangerously. *National Geographic*, March 2001, 74–103.

Darwin, C. 1859. *The Origin of Species*. (1985 reprint, Penguin Books).

Darwin, C.G. 1953. *The Next Million Years*. Doubleday, New York.

Deffeyes, K.S. 2001. *Hubbert's Peak, the Impending World Oil Shortage*. Princeton University Press, USA.

Desmond, A. & Moore, J. 1991. *Darwin*. Michael Joseph, London.

Dickens, C. 1850. *David Copperfield*. (1907 reprint, Everyman, London).

Duguid, J.P. 2002. *Population, Resources, and the Quality of Life*. Population Policy Press, Wales.

Duncan, R.C. 1997. *The Olduvai Theory*. In Campbell, 1997, 106–107.

Duncan, R.C. & Youngquist, W. 1998. *Encircling the Peak of World Oil Production*. Institute on Energy and Man, Seattle.

TRINIDAD and TOBAGO   Area 5,120 km²

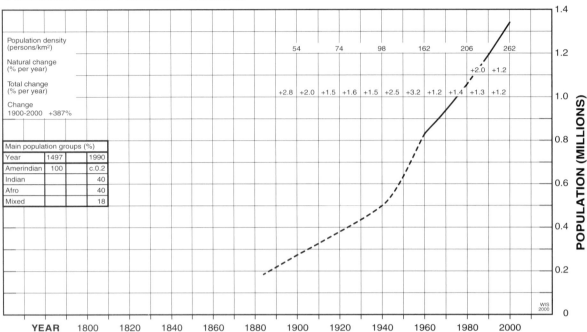

Population density (persons/km²)		54	74	98	162	206	262
Natural change (% per year)						+2.0	+1.2
Total change (% per year)		+2.8	+2.0	+1.5	+1.6	+1.5	+2.5	+3.2	+1.2	+1.4	+1.3	+1.2

Change 1900-2000   +387%

Main population groups (%)		
Year	1497	1990
Amerindian	100	c.0.2
Indian		40
Afro		40
Mixed		18

When Columbus sighted Trinidad, off the Venezuelan coast, in 1498, and named it for three prominent mountain peaks, the island's inhabitants were Arawak Indians. Nearby Tobago island was occupied by Caribs. Spain deported many Indians to slavery and established a settlement in 1592, but Trinidad was largely neglected until Britain captured it in 1797 and developed sugar plantations. Tobago on the other hand was settled from the 1630s by Dutch, British, French and others and ownership was disputed, often violently, until Britain took over in 1802. African slaves were replaced by Indian immigrant workers after 1834 when Britain banned the slave trade. Oil, asphalt and petroleum products began to overtake sugar and cocoa as the main exports after 1900. The nation became independent in 1962. Governments have been beset by conflicts between the main ethnic groups, culminating in a failed coup by Muslim fundamentalists in 1990. Censuses: irregular from 1901. The kink in the graph, 1940–60, is unexplained.

TURKS & CAICOS ISLANDS   Area 450 km²

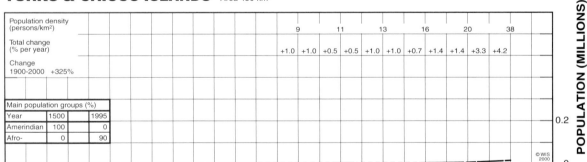

| Population density (persons/km²) | | 9 | 11 | 13 | 16 | 20 | 38 |
| Total change (% per year) | | +1.0 | +1.0 | +0.5 | +0.5 | +1.0 | +1.0 | +0.7 | +1.4 | +1.4 | +3.3 | +4.2 |

Change 1900-2000   +325%

Main population groups (%)		
Year	1500	1995
Amerindian	100	0
Afro-	0	90

The small Turks and Caicos archipelago at the southeast end of the Bahamas, consisting of about 30 mostly barren low-lying islands surrounded by coral reefs, supported a few Arawak Indians before the first Europeans, Spanish sailors, landed in 1515. The islands were of little interest to Europeans other than Bermudans who visited them for salt, obtained by evaporating sea water. Britain acquired them in 1766. Settlers from North America brought negro slaves whose descendants make up most of the population today. Fishing and tourism are the main occupations, but the islands, a British Crown Colony since 1973, have become an important financial centre.

Ehrlich, P. 1968. *The Population Bomb*. Ballantine, New York.

Ehrlich, P. & Ehrlich, A. 1990. *The Population Explosion*. Hutchinson.

Esteves Felgas, H.A. 1958. *História do Congo Português*. Carmona, Angola.

Ferguson, A. 1998. The Carrying Capacity and Ecological Footprints of Nations. *Optimum Population Trust Magazine*.

Flannery, T.F. 1996. *The Future Eaters*. Secker & Warburg, London.

Goodall, J. 2001. *Beyond Innocence*. Houghton Mifflin, Boston.

Griffiths, I. 1994. *The Atlas of African Affairs*. Routledge, London.

Hardin, G. 1993. *Living Within Limits*. Oxford University Press.

Hoffmann, P. 2001. *Tomorrow's Energy*. MIT Press, Cambridge, Mass.

Knight, I.J. 1994. *Warrior Chiefs of Southern Africa*. Firebird Books, Poole.

Leakey, R. & Lewin, R. 1996. *The Sixth Extinction*. Weidenfeld, UK.

Leggett, J. 1999. *The Carbon War*. Allen Lane, The Penguin Press.

Lopez, D. & Pigafetta, F. 1591. *Relação do Reino de Congo* (trans. 1951). Lisbon, Portugal.

Lovelock, J. 1991. *Gaia: the Practical Science of Planetary Medicine*. Gaia Books, UK.

McEvedy, C. & Jones, R. 1978. *Atlas of World Population History*. Penguin Books.

Meadows, D.H., Meadows, D.L., Randers, J. & Behrens, W. 1972. *The Limits to Growth*. Universe Books, New York.

Monteiro, J.J. 1875. *Angola and the River Congo*. Macmillan & Co., London.

Morgan, K.O. (ed.) 1984. *The Oxford Illustrated History of Britain*. Oxford University Press.

Parsons, J. 1993. Population Optimisation in British Politics. *Optimum Population Trust Magazine*, No. 2, 11–18.

Parsons, J. 1998. *Human Population Competition*. Edwin Mellen Press, New York.

Paxman, J. 1998. *The English*. Michael Joseph, London.

Percival, J. 1995. *The Great Famine: Ireland's Potato Famine 1845–51*. BCA, London.

Perham, M. & Simmons, J. 1942. *African Discovery, an Anthology*. Travel Book Club, London.

Plumb, J.H. 1950. *England in the Eighteenth Century*. Penguin Books, Harmondsworth.

Ponting, C. 1991. *A Green History of the World*. Sinclair-Stevenson, UK. (Penguin edition 1993.)

Reader, J. 1997. *Africa, a Biography of the Continent*. Hamish Hamilton, London.

Ronson, J. 2002. *Them: Adventures with Extremists*. Picador, London.

Scott, A., Pearce, D. and Goldblatt, P. 2001. The sizes and characteristics of the minority ethnic populations of Great Britain – latest estimates. *Population Trends*, No. 105, 6–15.

Shaw, C. 2000. 1998-based national population projections for the United Kingdom and constituent countries. *Population Trends*, No. 99, 4–12.

Singleton, F. 1985. *A Short History of the Yugoslav Peoples*. Cambridge University Press.

Stanton, W. 1990. Inert Landfill Sites: New Pressures and Options. *Nature in Somerset 1990*, 12–14.

Stanton, W.I. 1999. Problems caused by Badger Activity in Westbury sub Mendip and elsewhere in Somerset. *Proceedings of the Bristol Naturalists' Society*, Vol. 57, 77–97.

UNEP, 1999. *Global Environmental Outlook 2000*. Earthscan, London.

Vickers, M. 1998. *Between Serb and Albanian, a History of Kosovo*. Hurst & Co., London.

Voltaire, 1758. *Candide* (1947 trans). Penguin Books.

Webb, P. & von Braun, J. 1994. *Famine and Food Security in Ethiopia*. John Wiley, Chichester.

Willey, D. 1993. Poverty, Population and the Planet. *Optimum Population Trust Magazine*, No. 2, 20–21.

VANUATU   New Hebrides   Area 13,000 km²

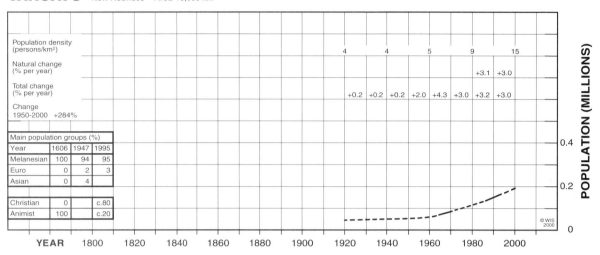

Population density (persons/km²)				4	4	5	9	15	
Natural change (% per year)							+3.1	+3.0	
Total change (% per year)				+0.2	+0.2	+0.2	+2.0	+4.3	+3.0	+3.2	+3.0
Change 1950-2000 +284%									

Main population groups (%)			
Year	1606	1947	1995
Melanesian	100	94	95
Euro	0	2	3
Asian	0	4	
Christian	0		c.80
Animist	100		c.20

The Vanuatu archipelago, about 80 volcanic islands in the south-west Pacific 500 kilometres north-east of New Caledonia, was occupied by seagoing Melanesians at least 5000 years ago. Portuguese sailors were the first European visitors in 1606 and Captain Cook named them New Hebrides, from their rugged silhouette, in 1774. Traders and missionaries brought European diseases which ravaged the native population in the 19th century. Many of the remaining people were exported to work on plantations elsewhere. In 1906 France and Britain assumed joint control of the islands through a Condominium, which led to independence in 1980.

The DC population surge began soon after World War Two and continues, with a TFR of 4.7 and 40% of the population less than 15 years old in 2001, but the population density is still low. The economy is based on agriculture, tourism and financial services. Three quarters of the land is still forested.

VIRGIN ISLANDS (U.S.)   Area 350 km²

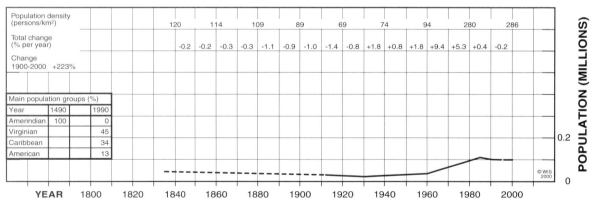

Population density (persons/km²)			120	114	109	89	69	74	94	280	286	
Total change (% per year)			-0.2	-0.2	-0.3	-0.3	-1.1	-0.9	-1.0	-1.4	-0.8	+1.8	+0.8	+1.8	+9.4	+5.3	+0.4	-0.2
Change 1900-2000 +223%												

Main population groups (%)		
Year	1490	1990
Amerindian	100	0
Virginian		45
Caribbean		34
American		13

Columbus visited this small eastern Caribbean archipelago, between Puerto Rico and the Leeward Islands, in 1493 when it was inhabited by Arawak or Carib Indians. The first European settlers were Danish and British sugar planters, but France invaded the islands in 1650. Denmark regained control in 1733, establishing the Danish West Indies in which sugar plantations, worked by African slaves, prospered. The population declined through the later 19th century as the sugar industry reacted to the abolition of slavery in 1848. In 1917 Denmark sold the islands to the USA for 25 million dollars. Since World War Two tourism has become the main industry, followed by financial services and stock rearing. Six desalination plants supplement the hard-pressed water resources. The population is mostly black and mixed. Censuses: roughly every tenth year from 1917.

Wilson, G., Harris, S. & McLaren, G. 1997. *Changes in the British Badger Population, 1988 to 1997.* People's Trust for Endangered Species, London.

Witherick, M.E. 1990. *Population Geography.* Longman Group, UK.

Zebrowski, E. 1997. *Perils of a Restless Planet.* Cambridge University Press.

Periodicals (dates embrace the periods studied)

Statesman's Yearbook, 1886–2002. McMillan, London. (annual).

Whitaker's Almanack, 1910–2000. Whitaker, London. (annual).

World Population Data Sheet, 1978–2002. Population Reference Bureau, Washington DC. (annual).

Better World, 1992–2000. Optimum Population Trust. (occasional).

Population Trends, 1990–2002. Office for National Statistics, London. (quarterly).

National Geographic, 1988–2002. National Geographic Society, Washington DC. (monthly).

New Scientist, 1980–2002. Reed Business Information, London. (weekly).

Time, 1993–2002. Time Warner, Amsterdam. (weekly).

Farmers Weekly, 1997–2001. (weekly).

Western Daily Press, 1988–2000. Bristol United Press, Bristol. (daily).

The Guardian, 2000–2002. Guardian Newspapers, London. (daily).

BBC Radio 4. (daily).

BBC World Service. (nightly).

Encyclopedias etc.

Cambridge Paperback, 1993. Cambridge University Press.

Encarta, 2000. Microsoft Corp.

Longman, 1989. Guild, London.

Hutchinson Guide to the World, 1998. Helicon, Oxford.

Travel Guidebooks (series)

Baedecker Guides

Blue Guides

Cadogan Guides

Fodor's Guides

Lonely Planet Guides

Rough Guides

0 – 1.8 million

WALLIS and FUTUNA   Area 270 km²

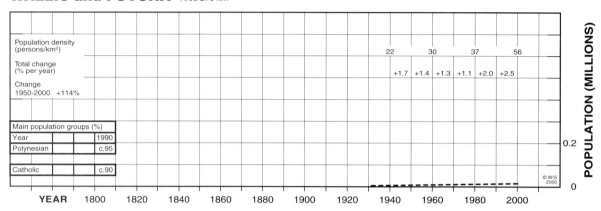

Population density (persons/km²)			22	30	37	56
Total change (% per year)			+1.7	+1.4	+1.3	+1.1	+2.0	+2.5
Change 1950-2000 +114%						

Main population groups (%)		
Year		1990
Polynesian		c.95
Catholic		c.90

Two small groups of volcanic islands in the west-central Pacific, 300 kilometres north-east of Fiji, Wallis and Futuna had been inhabited by Polynesians for several millennia when the first Europeans sighted them in 1617. French missionaries settled there in the early 19th century and France declared a protectorate in 1887. The islands became a French Overseas Territory in 1961 and the people are French citizens. Subsistence farming is the main occupation.

WESTERN SAHARA   Spanish Sahara, Rio de Oro   Area 267,000 km²

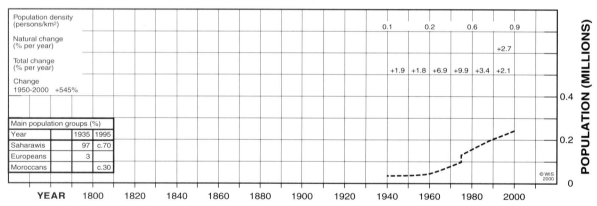

Population density (persons/km²)			0.1	0.2	0.6	0.9
Natural change (% per year)						+2.7
Total change (% per year)			+1.9	+1.8	+6.9	+9.9	+3.4	+2.1
Change 1950-2000 +545%						

Main population groups (%)		
Year	1935	1995
Saharawis	97	c.70
Europeans	3	
Moroccans		c.30

Portuguese explorers seeking the source of West African gold sailed along this desert coast in 1436 and named an inlet, optimistically, Rio de Ouro. All they found were a few Arab-Berber nomadic pastoralists. Spain annexed the region as a colony, Spanish Sahara, in 1884 but did little to develop it until the discovery of huge rock phosphate reserves in the 1960s. Morocco had laid claim to Western Sahara in 1956 and invaded unsuccessfully in 1957. In 1973 Sunni Muslim separatists of the Polisario Front declared independence as the Saharawi Democratic Republic and began guerrilla activity. When Spain decided to withdraw in 1975 Morocco and Mauritania divided the territory between them, Morocco enforcing its claim by a 'Green March' across the border by 350,000 civilians, many of whom stayed. Polisario attacks continued, backed by Algeria. Mauritania abandoned its claim and withdrew in 1979. The guerrilla conflict with Morocco is ongoing, with nearly 200,000 refugees, mostly the Polisario women and children, camped in Algeria.

Appendix 1

The 'Nation Sets' Used in Tables 3.1 and 3.2

In the tables, the world's nations (and distinct parts of nations) are grouped into sets on geographic or other grounds, necessarily somewhat arbitrarily in a few cases. In particular, Muslim nations are considered to be those in which recent censuses show the Muslim proportion of the population to be greater than any other religious group, although they may not be a majority of the total population. (Conversely, many nations defined thus as non-Muslim have large Muslim minorities.)

'Developed' World incl. ex-USSR, excl. Muslim nations (52 nations)

Andorra, Armenia, Australia, Austria, Belarus, Belgium, Bulgaria, Canada, Croatia, Cyprus, Czech Republic and Slovakia, Denmark, Estonia, Finland, France, Germany, Georgia, Gibraltar, Greece, Hungary, Iceland, Ireland, Israel, Italy, Japan, Latvia, Liechtenstein, Lithuania, Luxembourg, Macedonia, Malta, Moldova, Monaco, Netherlands, New Zealand, Norway, Poland, Portugal, Romania, Russia, Singapore, Slovenia, South Korea, Spain, Sweden, Switzerland, Taiwan, Ukraine, UK, USA, Yugoslavia (excl. Kosovo).

South and Central America (25 nations)

Argentina, Bolivia, Brazil, Chile, Colombia, Costa Rica, Cuba, Dominican Republic, Ecuador, El Salvador, French Guiana, Guatemala, Guyana, Haiti, Honduras, Jamaica, Mexico, Nicaragua, Panama, Paraguay, Peru, Puerto Rico, Suriname, Uruguay, Venezuela.

Sub-Saharan Africa excl. Muslim nations (33 nations)

Angola, Benin, Botswana, Burundi, Cameroon, Cape Verde, Central African Republic, Comoros/Mayotte, Congo Democratic Republic, Congo Republic, Côte d'Ivoire, Equatorial Guinea, Ethiopia, Gabon, Ghana, Guinea-Bissau, Kenya, Lesotho, Liberia, Madagascar, Malawi, Mozambique, Namibia, Rwanda, São Tomé, Sierra Leone, South Africa, Swaziland, Tanzania, Togo, Uganda, Zambia, Zimbabwe.

Central to South-east Asia, excl. Muslim nations (13 nations)

Bhutan, Cambodia, East Timor, India, Kazakhstan, Laos, Myanmar, Nepal, Papua New Guinea, Philippines, Sri Lanka, Thailand, Vietnam.

East Asia excl. 'developed' nations (3 nations)

China, Mongolia, North Korea.

All Muslim nations (49 nations)

Afghanistan, Albania, Algeria, Azerbaijan, Bahrain, Bangladesh, Bosnia, Brunei, Burkina Faso, Chad, Djibouti, Egypt, Eritrea, Gambia, Gaza Strip, Guinea, Indonesia, Iran, Iraq, Jordan, Kosovo, Kuwait, Kyrgyzstan, Lebanon, Libya, Maldives, Mali, Mauritania, Malaysia, Morocco, Niger, Nigeria, Oman, Pakistan, Qatar, Saudi Arabia, Senegal, Somalia, Sudan, Syria, Tajikistan, Tunisia, Turkey, Turkmenistan, United Arab Emirates, Uzbekiston, West Bank, Western Sahara, Yemen.

Ex-USSR (non-Muslim) and East European satellites (15 nations, repeats from rows above)

Armenia, Belarus, Bulgaria, Czech Republic and Slovakia, Estonia, Georgia, Hungary, Kazakhstan, Latvia, Lithuania, Moldova, Poland, Romania, Russia, Ukraine.

Index